Bio-based Wood Adhesives

Preparation, Characterization, and Testing

Bio-based Wood Adhesives
Preparation, Characterization, and Testing

Editor

Zhongqi He

Southern Regional Research Center
USDA, Agricultural Regional Service
New Orleans, Louisiana, USA

CRC Press
Taylor & Francis Group
Boca Raton London New York

CRC Press is an imprint of the
Taylor & Francis Group, an **informa** business

A SCIENCE PUBLISHERS BOOK

Cover images courtesy of Nicolas Brosse, Zhongqi He, Ningbo Li, Antonio Pizzi, Sarocha Pradyawong, Hui Wan, and Donghai Wang

CRC Press
Taylor & Francis Group
6000 Broken Sound Parkway NW, Suite 300
Boca Raton, FL 33487-2742

First issued in paperback 2020

© 2017 by Taylor & Francis Group, LLC
CRC Press is an imprint of Taylor & Francis Group, an Informa business

No claim to original U.S. Government works

ISBN-13: 978-1-4987-4074-6 (hbk)
ISBN-13: 978-0-367-78228-3 (pbk)

Library of Congress Cataloging-in-Publication Data

Names: He, Zhonggi, editor.
Title: Bio-based wood adhesives : preparation, characterization, and testing / editor, Zhonggi He, Southern Regional Research Center, USDA, Agricultural Regional Service, New Orleans, Louisiana, USA.
Other titles: Wood adhesives
Description: Boca Raton, FL : CRC Press, [2016] | "A science publishers book." | Includes bibliographical references and index.
Identifiers: LCCN 2016051817| ISBN 9781498740746 (hardback : alk. paper) | ISBN 9781498740753 (e-book : alk. paper)
Subjects: LCSH: Wood--Bonding. | Adhesives. | Biomass. | Agricultural biotechnology.
Classification: LCC TS857 .B56 2016 | DDC 668/.3--dc23
LC record available at https://lccn.loc.gov/2016051817

Visit the Taylor & Francis Web site at
http://www.taylorandfrancis.com

and the CRC Press Web site at
http://www.crcpress.com

Preface

Adhesive bonding is playing an increasingly important role in the forest product industry and is a key factor in the efficient utilization of timber and other lignocellulosic resources. As synthetic wood adhesives are mostly derived from depleting petrochemical resources and have resulted in increasing environmental concerns, natural product- and byproduct-derived adhesives have attracted much more attention in the last decades. Although adhesives made from plant and animal sources have been in existence since ancient times, increased knowledge of their chemistry and improved technical formulation of their preparations are still needed to promote their broader industrial applications. The primary goals of this book are to (1) synthesize the fundamental knowledge and latest research on bio-based adhesives from a remarkable range of natural products and byproducts, (2) identify areas of research needs and provide directions towards future bio-based adhesive research, and (3) help integrate research findings for practical adhesive applications for maximum benefits. This book should be of interest to university faculty, graduate students, research scientists, agricultural and wood engineers, international organization advocates and government regulators, who work and deal with the utilization of agricultural and forest products and byproducts.

This book covers general information on a variety of natural products and byproducts and includes the latest research on formulation, testing and improvement of the relevant adhesives. All this is covered in 15 chapters, written by an international group of accomplished contributors. Chapter contents include both raw material-oriented comprehensive reviews and adhesive-making-oriented case studies. The first seven chapters address the source, preparation, characterization, application, and modification of various agricultural products and byproducts, for the purpose of wood bonding and composite board making, with a focus on seed proteins as the major component. Chapters 8 and 9 examine and discuss two specific natural compounds, tannin and citric acid, related to their chemical nature, production, and application in wood adhesive additives and binders. The next three chapters are based on forest products. Chapters 10 and 11 review the research on the liquefaction and pyrolysis of forestry and other lignocellulosic biomass materials and the subsequent utilization of the liquefied products and pyrolysis-derived bio-oil in wood bonding, respectively. Chapter 12 reviews the general properties and quality of Amazonic white pitch produced from the tree trunk (Protium heptaphyllum) and advocates its potential use as "green" adhesives. Chapter 13 presents and analyzes the effects of rheology and viscosity of bio-based adhesives on bonding performance. Chapter 14 presents examples on the applications of nano-materials to improve qualities of wood composites. The last chapter (Chapter 15) makes concluding remarks. It briefly reviews the progress on the developments and utilization of bio-based wood adhesives including some activities and accomplishments not covered by the individual chapters. This chapter then presents outlooks on some emerging research issues that are worth future exploration by bio-based adhesive researchers.

This book project was initiated by the publisher. It is my honor to be invited to serve as editor of the book. Chapter contributions were invited by the editor. Whereas the chapter contributors and I have made efforts to avoid repeated coverage in topics and contents, some literature and data have been cited in more than one chapter, to warrant each chapter's integrity and stand-alone. All chapter manuscripts were subjected to peer reviewing and revision processes. The completion of this project involved tremendous efforts from a team of authors, reviewers and editor. I would like to express my sincere thanks to all the authors for their commitment and dedication, and timely contributions. I profusely thank the ad-hoc reviewers for their time and insight in providing many valuable comments which certainly improved the quality of this book. Special thanks go to the publisher's staff (Drs. Raju Primlani, Vijay Primlani, and Amanda Parida) for their initiative, coordination, and editorial support throughout the process of executing this book project.

Contents

About the Editor

Zhongqi He is a Research Chemist at the United States Department of Agriculture-Agricultural Research Service (USDA-ARS), Southern Regional Research Center, New Orleans, Louisiana. Prior to the USDA-ARS tenure, he was a recipient of the (US) National Research Council Postdoctoral Fellowship hosted by the United States Air Force Research Laboratory, Tyndall Air Force Base, Florida. He is the author and co-author of over 200 peer-reviewed journal articles, patents, and book chapters, and has actively pursued basic and applied research in chemistry and biochemistry of agricultural products, byproducts, and plant nutrients. His current research activities are focused on the enhanced utilization of cottonseed products as wood adhesives. Dr. He has organized and served as the sole or principal editor of eight books on sustainable agriculture, renewable resources, and the environment. He has provided peer review services to more than 90 journals and served as associate editor and board member of several scientific journals and book committees of professional societies. He received a B.S. degree (1982) in applied chemistry from Chongqing University, China, M.S. degrees (1985 and 1992) in applied chemistry from South China University of Technology, Guangzhou, and in chemistry from the University of Georgia, Athens, USA, and has a Ph.D. degree (1996) in biochemistry from the University of Georgia, Athens, USA. He is a Fellow of American Society of Agronomy.

1

Protein-based Wood Adhesives

Current Trends of Preparation and Application

Birendra B. Adhikari, Pooran Appadu, Michael Chae and
*David C. Bressler**

ABSTRACT

The current wood adhesive industry relies mostly on petrochemicals, and is dominated by formaldehyde-based adhesives. However, petrochemicals are obtained from non-renewable resources, and formaldehyde is a known carcinogen. Hence, there is a growing interest surrounding the utilization of renewable resources for the development of environmentally friendly adhesive systems for production of bonded wood products. Due to their inherent adhesive properties, proteins are gaining popularity in this sector. The purpose of this chapter is to review the potential application of protein for the prospective development of industrial wood adhesives. Herein, we summarize the historical perspectives concerning the development of proteinaceous wood adhesives, potential proteinaceous feedstock resources for large-scale production of protein, and recent developments in protein-based wood adhesive formulations.

The proteinaceous feedstocks for adhesive development can be sourced as single cell-based proteins, plant-based proteins, and animal-based proteins. Among these, the plant-based proteins, specifically soy protein, have been widely explored in recent years for development of proteinaceous wood

Department of Agricultural, Food and Nutritional Science, Faculty of Agricultural, Life & Environmental Sciences, University of Alberta, Edmonton, AB, Canada, T6G 2P5.
* Corresponding author: dbressle@ualberta.ca

adhesives. On the other hand, only a handful of reports are available on utilization of whey protein and other waste animal protein for development of protein-based adhesive formulations. Thus far, no such studies have been conducted on single cell-based proteins.

Current bio-based adhesive technologies utilize such methods as protein denaturation by chemical, enzymatic, or thermal treatment, chemical modification of end functional groups, chemical crosslinking of protein molecules, co-reacting or blending of protein with a prepolymer, or a combination of any of these methods. Formulations developed by chemical crosslinking of denatured protein fragments or by co-reacting protein, after denaturation, with a phenol-formaldehyde prepolymer, have resulted in protein-based adhesive systems performing comparable to urea-formaldehyde- and phenol-formaldehyde-based wood adhesives. Both types of formulations using soy protein isolate have been successfully used for commercial production of composite wood products.

The high cost of soy protein isolate limits the widespread application of these products and therefore, are largely unable to replace petrochemical-based resins. Realizing this limitation of soy protein as well as other edible proteins, there has been renewed interest in waste animal protein as a renewable protein feedstock for formulation of wood adhesive systems. Waste animal protein is an undervalued protein stream that has shown great potential in wood adhesive applications. As newer and better technologies are being developed for efficient valorization of by-product streams in diverse industries, production of bio-based wood adhesives from such undervalued protein streams is inevitable in the not too distant future.

Introduction

Biomass represents an abundant, biodegradable, renewable, and potentially carbon-neutral resource for the production of energy and biomaterials. It is obtained from a wide range of sources including agricultural, industrial and forestry residues, as well as municipal wastes. Agricultural residues constitute a major part of the total annual production of biomass and are important feedstock sources for industrial purposes. United Nations Environment Programme (UNEP) estimates that the amount of agricultural biomass waste generated annually amounts to 5 billion metric tons (UNEP, 2011). This large volume of biomass is a tremendous resource and could potentially serve as renewable raw materials that could substantially displace the use of fossil fuels in technical industries. Furthermore, waste agricultural biomass is commonly disposed of through incineration or landfilling, which significantly contribute to air pollution, water and soil contamination, and/or global warming through emission of greenhouse gases (GHGs). Thus, utilization of such renewable resources would not only contribute to the longevity of non-renewable resources, but would also support efforts to mitigate environmental issues associated with the disposal of waste agricultural biomass.

At the beginning of the 20th century, raw materials or feedstocks for technical industries were obtained solely from renewable resources. The development of cheaper and better-performing petrochemical-based polymers in the 1930s led to a dramatic drop in the use of biomass-based feedstocks and eroded the biopolymer market. The oil crisis in the 1970s saw a large increase in the price for a barrel of crude, which led to an enhanced interest in alternative resources such as biomass. However, industrial interest in renewable resources declined when crude oil prices dropped again in the mid 1980s (Eggersdorfer et al., 1992; Mecking, 2004). Over the last decade, the desire to reduce dependence on diminishing petrochemical resources that are associated with price volatility and growing environmental concerns has been a major driver for the development of sustainable resources. Despite the fact that petroleum derived chemical products are predominant in today's chemical industries, the global renewable chemicals market is on the rise. MarketsandMarkets, a global market research company that publishes premium market research reports annually, has estimated that the renewable chemicals market including alcohols, platform chemicals, and biopolymers was worth 49.0 billion USD in 2015, and at a compound annual growth rate of 11.47 percentage, is projected to reach 84.3 billion USD by 2020 (MarketsandMarkets, 2015b).

Even though the current renewable chemicals industry is facing considerable challenges from petroleum-based chemicals, the move from non-renewable to renewable resources would provide economic benefits to the biomass sectors and would also address the rising environmental concerns associated with petroleum products. In recent years, the scientific community and the industrial sector have witnessed a paradigm shift in research towards the utilization of renewable resources in various industrial applications. However, the recent growing interest in biomass is not simply a matter of an increased desire for renewable versus non-renewable resources. The rising interest in biomass feedstocks is also due to the incredible flexibility and adaptability of these feedstocks for industrial applications (Lipinsky, 1981; Schnepp, 2013). Biomass is composed of biopolymers including polysaccharides, lignin, polyphenols, and/or polypeptides. These biopolymers are multifunctional and their properties can be tailored by numerous mechanisms to produce chemically diverse macromolecules (Schnepp, 2013).

Among the various biopolymers, proteins are the most abundant class of macromolecules and function as the principal organic building blocks in living organisms. Over 50 percentage of the dry weight of cells is composed of proteins (Cozzone, 2010). In addition to being a food source, proteins have potential application in the development of a number of products currently supplied by the synthetic polymer industry. The biggest advantage that proteins offer compared to the petroleum-based polymers is that they can be processed in a similar manner as synthetic polymers, but are biodegradable and produce a smaller carbon footprint. Because of this, there has been an increased interest in using proteins from various sources for several novel, non-food applications such as the manufacturing of plastics, coatings, surfactants, biocomposites, and adhesives (Audic et al., 2003; Bressler and Choi, 2015; De Graaf, 2000; Kumar et al., 2002; Mekonnen et al., 2013c; Mekonnen et al., 2014; Montano-Leyva et al., 2013; van der Leeden et al., 2000).

Global production of wood-based panel products—veneer sheets, plywood, particleboards, and fibreboards—reached 388 million m³ in 2014 (FAO, 2015b). Composite wood products, such as particleboards and medium density fibreboards, are made by bonding low-value wood by-products (i.e., sawdust, wood chips) with urea-formaldehyde resins (USB, 2012). The International Agency for Research on Cancer classified formaldehyde as carcinogenic to humans in 2006 (IARC, 2006). Consequently, the recent interest in bioadhesives for woodworking has been driven by the necessity to replace the formaldehyde-based adhesives in interior wood products. Because of their unique physicochemical properties, protein-based adhesives are being explored and utilized as environmentally friendly replacements for formaldehyde-based adhesives in wood industries. For example, a formaldehyde-free soy protein-based adhesive technology has been commercialized for production of plywood and composite wood products such as particleboards, medium density fibreboards, oriented strand boards, and laminates (USB, 2015).

Bioadhesives are used widely in the paper, wood, and medical industries. Very little data are available on the bioadhesives market. However, two market research companies, MarketsandMarkets and Grand View Research, have published reports regarding estimates and forecasts of the bioadhesives market size. In 2014, the two largest consumers of bio-adhesives were Europe and North America, with Europe alone accounting for 43.2 percentage of global consumption. The global bioadhesive demand is expected to grow from 784 in 2014 to 2046 kilotonnes by the year 2022, which represents a monetary growth of approximately 2.5 billion USD. The use of inexpensive and renewable chemical feedstocks, such as starch and proteins, are poised to contribute to the global bioadhesive market as they provide an opportunity for the development of cost effective and environmentally friendly adhesives. The substantial potential for this market has inspired many major industrial players to become involved in the production, research, and development of bioadhesives, including Dow Chemical Company, 3M Company, and Henkel AG & Co., with Dow

Chemical Company being the major stakeholder (Grand View Research, 2015; MarketsandMarkets, 2015a).

An enormous amount of waste protein is generated globally as an agro-industrial waste. This material is currently available for a nominal cost and serves as a sustainable resource of proteins for the development of an array of materials for technical applications. Although proteinaceous materials are readily available, cost effective, and easily processed, proteins have some inherent limitations, mostly associated with poor water resistance. In recent years, attention has been focused on processing and chemically modifying proteins to impart properties required for specific applications. This chapter summarizes the potential application of proteins as wood adhesives and current trends of protein-based adhesive formulations from various protein sources.

Historical Aspects

According to Ebnesajjad (2010), the earliest use of glues—a type of resin—was by "prehistoric tribes" (Ebnesajjad, 2010). Although the tribes and the composition of the resin were not described, the resin is thought to be derived from tree sap, and was used to repair broken pottery vessels. On the other hand, Keimel (2003) and Brockmann et al. (2009), state that the origins of glue bonding technology developed in the times of the Egyptians, Greeks, Mesopotamians, Romans, and Sumerians circa 4000 B.C. (Brockmann et al., 2009; Keimel, 2003). With the exception of the Mesopotamians, who used asphalt in construction, these peoples used glues developed from proteinaceous materials such as animal hides and fish (Brockmann et al., 2009; Keimel, 2003).

The development of veneering and marquetry by the Romans and Greeks in 1–500 A.D. resulted in advances in refinement of adhesives from other proteinaceous materials, such as blood and milk protein (Ebnesajjad, 2010). Dioscorides, a Greek physician and botanist, described the use of whale intestines and bull hides to develop glues in around 50 A.D. (Fay, 2005). Unfortunately, it is thought that many of the adhesive technologies developed during this time period were lost after the fall of the Greek and Roman empires and did not resurface for centuries (Minford, 1991; Fay, 2005).

Although the first plant to manufacture glue from animal hides was established in Holland in 1690 (Keimel, 2003), the excitement and innovation during the industrial revolution instigated development of the adhesives sector and research into the use of novel materials for incorporation into adhesive formulations (Nicholson, 1991; Ebnesajjad, 2010). Radley (1976) noted that the development of photography, introduction of postage stamps, and the need for packaging materials were major drivers for the production of adhesives (Radley, 1976). The industrial revolution resulted in advancements in the adhesives sector, specifically relating to the use of proteinaceous materials (Ebnesajjad, 2010), though starches were also heavily employed (Radley, 1976). Several patents were filed during this period, primarily in the United States and Britain (Ebnesajjad, 2010; Keimel, 2003).

The world wars also played a major role in the development of the adhesive sector. For example, "exterior-grade adhesives" were needed for the military aircraft industry. This fueled research into developing better adhesives through chemical treatment and chemical synthesis. The properties of blood, casein, and soybean were improved with chemical denaturants and heat. However, as research into synthetic polymers gained prominence, the use of bio-based materials became limited (Bye, 1990; Fay, 2005; Keimel, 2003). For example, resin-modified phenolics started to be used to bond aluminum sheets for aeroplane propellers (Keimel, 2003).

Adhesive development continued after the world wars, though there continued to be decreasing interest in using bio-based feedstocks. Synthetic polymers based on epoxies and second generation acrylics were developed in the 1950s and 1970s, respectively (Fay, 2005; Licari and Swanson, 2011). Fast forward a few decades to the 21st century and the instability of fossil fuel reserves and concerns for the environment have provided the impetus for more research into the use of bio-based chemicals. This novel research includes the utilization of renewable chemical feedstocks such as protein and carbohydrates. Of significant interest is the chemical modification of renewable feedstocks to produce products that perform comparatively to their petroleum-based counterparts.

Protein as a Renewable Resource for Bioadhesives

Biopolymers are naturally occurring and predominantly organic materials that can adopt diverse structures and perform a variety of distinct functions. Because of these key features, biopolymers are being employed in an increasingly large number of industrial applications, particularly in the field of material sciences. In particular, proteins are extremely promising biomacromolecules for the production of an array of materials with many specific physicochemical properties that make them relevant for industrial utilization (Adebiyi and Aluko, 2011; Allen et al., 2014; Arthur, 1949; Kinsella and Melachouris, 1976). Therefore, before discussing the use of proteins as an industrial resource, we will first discuss the structure and fundamental physicochemical properties of proteins.

Structure of Proteins

A fundamental feature of all biological materials is that they are composed of basic molecular building blocks that are linked together to form a variety of large, complex structures. In proteins, the building blocks, known as amino acids, are linked together by peptide bonds to form a linear polypeptide chain (Fig. 1). The polypeptide chains can differ dramatically in chain length and structural complexity, ranging from simple dipeptides and oligopeptides to the largest known protein, titin, which consists of over 38,000 amino acid residues (Kruger and Linke, 2011). Although polypeptides and proteins are extremely diverse, they are all made of the same pool of twenty amino acids. Proteins can assume complex three-dimensional shapes, which are generally integral to the functions that they perform. The diversity in structure and function of proteins is due to the differences in number, type, and particular order of constituent amino acids in polypeptide chains (Cozzone, 2010). These attributes define the primary structure of proteins.

Theoretically, numerous conformations of similar energies are possible for each polypeptide chain due to the rotational freedom of C-C and C-N single bonds along the polypeptide chain, but in reality, these polypeptide chains adopt specific configurations due to secondary interactions. These secondary interactions include non-covalent interactions (i.e., hydrogen bonding, electrostatic forces, van der Waals forces, and hydrophobic interactions) as well as covalent interactions (i.e., disulphide linkages) between amino acids of the same or different polypeptide chains. Existence of these interactions leads to the formation of higher-level structures of proteins, known as secondary, tertiary, and quaternary level structures (Cozzone, 2010; Koshland, 2014).

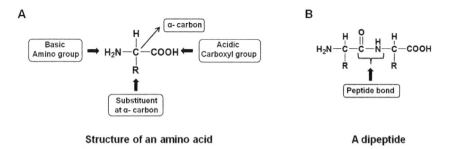

Structure of an amino acid   **A dipeptide**

Figure 1. Chemical structure of (A) an amino acid (building block of proteins) and (B) a dipeptide resulting from the formation of an amide linkage between two amino acids.

The secondary structure of proteins refers to localized structures adopted by a given stretch of amino acid residues. The two most common conformations that make up the secondary structure of proteins are the alpha helix and the beta sheet (Fig. 2). An alpha helix occurs through a regular pattern resulting from formation of several hydrogen bonds between certain amino acid residues in a localized region of a polypeptide chain. Specifically, regular hydrogen bonds between a C=O group of an amino acid residue and an NH group of the amino acid residue located four residues away in a polypeptide chain result in folding of the polypeptide chain in a helical shape (Fig. 2).

In a beta sheet conformation, a localized region of the polypeptide chain is folded in a zigzag orientation. The basic unit of this conformation is a beta strand of 5- to 10-amino acid residues. The sheet occurs due to hydrogen bonding interactions between adjacent beta strands. Beta sheets are stabilized by hydrogen bonds formed between the C=O groups of one polypeptide chain and the NH groups of another chain (Fig. 2).

The tertiary structure of proteins is the three-dimensional conformation of a protein, and refers to the spatial arrangement of a polypeptide chain through folding and coiling. Unlike secondary structures that are restricted to smaller, localized regions of the protein, tertiary structures result from

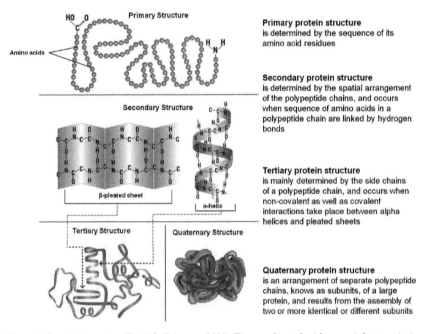

Figure 2. Protein structure (Particle Sciences, 2009). Figure adopted with copyright permission from Particle Sciences.

interactions between different regions within a polypeptide chain. These can involve interactions between secondary structures—alpha helices and beta sheets—and are stabilized by interactions between side chain functional groups that involve covalent disulfide bonds, electrostatic and hydrophobic interactions, and hydrogen bonding.

A protein's quaternary structure results from assembly of two or more separate polypeptide chains that are held together by noncovalent or, in some cases, covalent interactions. Many proteins are made up of two or more polypeptide chains, called subunits or monomers, which may have identical or different amino acid sequences. Protein complexes containing two polypeptides are called dimers or dimeric proteins. The prefixes homo- and hetero- are used to signify that the subunits are identical or different, respectively.

Physicochemical Properties of Proteins

The physicochemical properties of protein solutions and/or dispersions have been of major importance in determining their industrial applications. Such general properties include hydration/solvation, rheology, surface-activity, and structural properties, which comprise the specific functional properties of proteins as indicated in Table 1 (Adebiyi and Aluko, 2011; Gonza´lez-Pe´rez and Vereijken, 2007; Kinsella and Melachouris, 1976; Kinsella, 1979; Kinsella and Morr, 1984; Kristinsson and Rasco, 2000; Lestari et al., 2011; Morr and Ha, 1993; Tan et al., 2011).

The specific functional properties of proteins have been known for centuries, and proteins from various sources have been used historically in a number of non-food applications. For a protein to be suitable for a particular industrial application, there are some specific requirements for

Table 1. Specific functional properties of proteins. Table adapted with copyright permission from Taylor and Francis (for Kinsella and Melachouris, 1976).

General properties	Specific functional properties
Hydration/solvation	Wettability, water absorption and water holding capacity, dispersibility, swelling, gelation
Rheological	Viscosity, flowability, protein-solvent and protein-protein interaction, gelation, matrix formation, fluidity, elasticity
Surface	Emulsification, foaming, film formation and stabilization
Structural	Elasticity, grittiness, cohesion, adhesion, chewiness, viscosity, gelation, network formation, aggregation
Other	Compatibility with additives, diverse and hierarchical level of organization, molecular unfolding and denaturation, modification properties and design flexibility, biodegradability and potentially biocompatibility, renewability, sustainability

their physicochemical properties. For instance, for adhesive applications, protein dispersions and/or solutions must have good storage stability, flowability, wetting ability, and stickiness, and the resulting adhesive bonding should be strong and resistant to deterioration under various environmental conditions (Arthur, 1949; Frihart and Hunt, 2010). With the advent of modern science, it has been possible to modulate the properties of proteins and impart characteristic properties for specific applications by controlling the processing conditions or by chemical modifications.

There are four major approaches to achieve the necessary physicochemical properties of proteinaceous material (Audic et al., 2003; De Graaf, 2000). The first method involves dissolution of proteins in a suitable solvent that can make them easier to process using various physical and chemical treatments. Exposure of specific functional groups of a protein, such as polar and non-polar groups, is another approach that makes them more amenable to interactions with the substrate. This can be achieved through denaturation of proteins. A third method involves enhancing the entanglement of polypeptide chains, which can improve material strength by increasing cohesive strength. Finally, crosslinking of polypeptide chains can enhance the cohesive strength and water resistance. This method requires exposure of reactive functional groups such as carboxyl, amino, hydroxyl, and sulphahydroxyl groups, which can then be used to crosslink the polypeptide chains.

Industrial Resources of Proteins

Proteinaceous feedstocks for industrial applications can be obtained from various resources including meals and cakes of oil crops after extraction of vegetable oil, along with meat and dairy by-products. In addition, biomass of cultured microorganisms such as algae, yeasts, fungi, and bacteria can be potentially used as industrial sources of protein. Hence, industrial proteins can be sourced from single cell-based, plant-based, and animal-based proteins.

Single Cell Protein

The term single cell proteins (SCP) refers to the dried microbial cells or the proteins extracted from the biomass of cultivated microorganisms such as fungi, algae, yeast, and bacteria (Adedayo et al., 2011; Srividya et al., 2013). Proteins are involved in virtually every process occurring within microbial cells and are a key structural component. The following protein contents have been reported: 40–60 percentage for algae; 30–70 percentage for fungi; and 50–83 percentage for bacteria (Srividya et al., 2013). Consequently, compared to other cellular components, microorganisms generally produce significantly larger amounts of proteins.

For the production of SCP, microorganisms are grown on a suitable substrate through a fermentation process. A wide range of waste products such as agricultural and food processing wastes or industrial waste effluents, are potential carbon and nitrogen sources for SCP production (Bhalla et al., 2007). Production of SCP involves the following steps: (i) injection of selected microorganisms into a nutrient rich, sterilized carbon source in a fermentation tank; (ii) aeration and proper temperature maintenance of the fermentation tank for the growth of microorganisms; (iii) recovery of the microbial biomass from the growth medium by filtration; and (iv) drying and processing of the recovered biomass (Adedayo et al., 2011; Srividya et al., 2013).

One of the greatest advantages of producing SCP as an industrial protein source is that this bioconversion process utilizes a wide range of low-cost raw materials as nutrient sources to produce high-value products (Nasseri et al., 2011; Parashar et al., 2016; Srividya et al., 2013). Additionally, microorganisms have high rates of multiplication, and thus SCP organisms yield significant amounts of protein in a relatively small area and time. Furthermore, the growth of microorganisms in a fermenter is not affected by seasonal and/or climatic variations, and SCP can be produced throughout the year. However, production of SCP has high infrastructure and operating costs as it requires a sterilized fermenter with continuous aeration and proper control of nutrition and temperature. Moreover, the process requires a pure culture of microorganisms in the correct physiological state, which also makes the process costly.

Plant-Based Protein

Until the 1930s, vegetable proteins such as oilseed residues and starch were one of the most important constituents of woodworking glues (Davidson, 1929; Truax, 1929). The evolution of urea-formaldehyde and phenol-formaldehyde resins in the 1930s eroded the bioadhesive market due to the lower cost, better performance with regards to water resistance, and ease of processing and manufacturing of these petrochemical-based adhesives. Subsequently, urea-formaldehyde and phenol-formaldehyde resins largely replaced protein-based adhesives for the production of interior and exterior products, respectively (Wescott et al., 2006). However, recent fluctuations in the price of petrochemicals and the increasing desire for environmentally friendly, bio-based products have led to enhanced interest in adhesives manufactured from renewable resources. Thus, vegetable proteins and starches have regained popularity in the bioadhesive industry.

The main sources of plant-based protein are oilseeds and grains (Table 2). Generally, oilseeds contain a higher proportion of protein than grains (Table 3). With a protein content of 36% (USDA, 2015a), soybean has the largest protein content among plant-based resources, and is one of the most popular resources for industrial protein. While starch is the major constituent of grains, the second most abundant constituent is protein. Among the grains, wheat, which contains nearly 13% protein (USDA, 2015a), has also emerged as an attractive resource for industrial protein. The data presented in Tables 2 and 3 also provide an idea of the regional market potential for various plant-based protein applications.

Table 2. Production of grains from different parts of the world in 2013–2014 (million metric tons) (USDA, 2015b). Table adapted with copyright permission from USDA.

Region	Wheat	Corn	Sorghum	Major Oilseeds				
				Soybean	Sunflower-seed	Rape-seed	Cotton-seed	Peanuts
World	715.11	991.45	60.99	283.15	42.75	72.09	45.71	41.15
USA	58.11	351.27	9.97	91.39	0.92	1.00	3.81	1.89
China	121.93	218.49	2.70	12.20	2.42	14.46	12.84	16.97
EU	144.42	64.62	0.70	1.21	9.05	21.30	0.48	-
Brazil	5.30	80.00	1.89	86.70	0.23	-	2.67	0.32
Argentina	10.50	26.00	4.40	53.50	2.00	-	0.42	1.00
India	93.51	24.26	5.54	9.50	0.67	7.30	12.95	5.65
Canada	37.53	14.19	-	5.36	0.05	18.55	-	-
Russia	52.09	11.64	13.58	1.64	10.55	1.39	-	-
Ukraine	22.28	30.90	-	2.77	11.60	2.35	-	-
Australia	25.30	-	1.28	-	0.03	3.83	1.88	-

Table 3. Protein content of different grains (USDA, 2015a).

Grains	Form	Protein content (%)	Reference
Wheat	Flour, whole-grain	13	USDA basic report no. 20080
Corn	Grain, white/yellow	9	USDA basic report no. 20314
Sorghum	Grain	11	USDA basic report no. 20067
Soybean	Raw, mature seeds	36	USDA basic report no. 16108
Sunflowerseed	Seed kernels, dried	21	USDA basic report no. 12036
Cottonseed	Seed kernels, roasted	33	USDA basic report no. 12160
Peanut	Raw	26	USDA basic report no. 16087
Rapeseed (Canola)	Raw canola	20	(Barthet, 2012)

By 1950, oilseed proteins had largely replaced animal derived proteins, specifically casein, in non-food industrial applications due to unfavorable costs of the latter (USDA, 1951). Furthermore, soy protein was by far the leading protein used by the adhesive industry, followed by peanut and cottonseed meal protein (USDA, 1951). Soy protein is currently the most widely studied protein for the development of adhesives, plastics, composites, packaging, and coating materials. However, research into the use of peanut, canola, and cottonseed protein, as well as wheat gluten, for various technical applications has accelerated in recent years. Based on their worldwide production, protein content, and reports available on technical applications, soy protein, peanut protein, and wheat gluten are considered the most important resources of plant-based protein for development of wood adhesives. Accordingly, these proteins are discussed in subsequent sections from the perspective of production, market potential, and technical applications.

Soy Protein. Soybeans are one of the most abundant and widely cultivated high-protein agricultural products, and have become more popular as a renewable resource of protein. Soybeans yield a higher amount of protein than any other plant per unit area of cultivation (Potts et al., 2014). Soy products are important constituents of the global food chain system, serving as a food source for both humans and animals. Roughly 87 percentage of the globally produced soybean is utilized by soybean crushing industries to produce soy meal and soy oil, with the remaining 13 percentage consumed directly by humans and animals. In the soybean crushing process, roughly 80 percentage of the material is extracted as soy meal and 20 percentage is extracted as oil (MVO, 2011; Potts et al., 2014). According to a recent soy products guide released by the United Soybeans Board (USB), 88 percentage of the soy meal produced in soy crushing industries is used as animal feed, and 12 percentage is used to prepare food products and condiments for human consumption. Over the last 20 years, soy proteins have been increasingly used in human foods as protein supplements/substitutes, milk alternatives, cereal additives, and condiments (Webb, 2011), and this trend has been more pronounced in recent years (USB, 2015; 2016). Currently, only 0.3 percentage of soy meal produced annually is used in industrial production of adhesives, paper coatings, and fibers (USB, 2016).

Commercial soy products are available as defatted soy meal, soy flour, soy protein concentrate (SPC), and soy protein isolate (SPI). Defatted soy meal is the material remaining after solvent extraction of oil from soybean flakes, and usually contains 48 percentage protein and less than 1 percentage oil (Frihart and Birkeland, 2014; Kinsella, 1979). It is toasted and ground for the production of animal food. Soy flour is finely ground soy powder made from defatted soy meal or roasted and dehulled whole soybeans. It is manufactured with different fat levels—defatted, low-fat, high-fat, and full-fat. The first three types of soy flour are obtained from the defatted soy meal remaining after removing the oil from the crushed soybeans, while the fourth type of flour, full-fat soy flour, is produced from whole soy (Shurtleff and Aoyagi, 2004; USSEC, 2008; Webb, 2011).

Defatted soy flour contains less than 1 percentage fat and 53–56 percentage protein (Frihart and Birkeland, 2014; Kinsella, 1979) and is the most widely used type of soy flour. The low-fat soy flour contains about 46 percentage protein, and is commonly prepared by adding soy oil back to the defatted soy flour to a level of 5–6 percentage. The high-fat soy flour is similarly prepared by adding about 15 percentage soy oil back to the defatted soy flour, and contains about 41 percentage protein. The full-fat soy flour is the most natural soy product containing 18–20 percentage fat and 39–41 percentage protein (Shurtleff and Aoyagi, 2004). Defatted soy meal and soy flour are the starting materials for the production of high protein content value-added products such as SPC and SPI.

SPC is made by subjecting the defatted soy meal to aqueous alcohol extraction. This extraction process removes soluble carbohydrate, significantly lowers the antinutritional factors in regular soybean meal, and increases the protein content to nearly 70 percentage (USSEC, 2008). SPI is the most concentrated and refined form of commercially available soybean protein products and contains over 90 percentage protein (Frihart and Birkeland, 2014; Kinsella, 1979). SPI is produced by removing both insoluble and soluble carbohydrates, along with some protein, from the defatted soy flour or soy meal. In this case, protein from the starting material is first solubilized in water, the aqueous phase is separated from the solid residue, and finally the protein is precipitated from the solution, separated and dried. MarketsandMarkets has estimated that the 2015 market for soy protein ingredients (flours, protein concentrates, and isolates) will be 7.11 billion USD, and is projected to reach 10.12 billion USD by 2020, at a compound annual growth rate of 7.3 percentage (MarketsandMarkets, 2015c).

Technical Applications

Soy proteins were the typical proteins used in vegetable protein glues in the 1920s (Truax, 1929) and were widely used to produce glues for the coating and adhesive industries (Bradshaw and Dunham, 1929; Davidson, 1929; Fawthrop, 1933; Laucks and Cone, 1930). Soy proteins were used in paper coating industries mainly because they were cheaper, and soy protein glues had properties and characteristics similar to those of casein glues. In the following decades, soy proteins largely replaced animal-derived proteins from the softwood plywood adhesive market (USDA, 1951). The USDA technical bulletin in 1951 reported that the plywood industry used about 42 million pounds of soy meal and 3 million pounds each of casein and blood-based protein each year for the development of wood adhesives (USDA, 1951). However, soy protein glues subsequently lost ground to formaldehyde-based resins in the wood adhesive market due to their comparatively high cost and lower performance.

Li et al. in 2004 developed an adhesive system consisting of SPI and Kymene® 557H, a commercial polyamidoamine epichlorohydrin (PAE) resin (Li et al., 2004). This research attracted much interest in protein-based adhesive formulations, and a number of patented technologies have appeared recently in which soy protein has been used for development of bioadhesives (Allen et al., 2014; Brady et al., 2012; Li, 2007; Varnell et al., 2013; Wescott and Birkeland, 2008; Wescott and Birkeland, 2010). Columbia Forest Products, Ashland, Archer Daniels Midland, States Industries LLC, 9Wood, Solenis, Uniboard, and E2E are some of the companies that manufacture and market formaldehyde-free, soy-based industrial adhesives and/or use these resins in woodworking (USB, 2015).

Columbia Forest Products manufactures PureBond® Plywood Panels utilizing a formaldehyde-free soy protein-based adhesive. Reportedly, more than 50 million Purebond hardwood/plywood panels have been installed in interior products in the USA (USB, 2015; 2016). Ashland's Soyad™ co-adhesive is a soy-based, formaldehyde-free adhesive for internal wood products. Nutrasoy 7B, produced by Archer Daniels Midland, is a foamed plywood glue made using soy protein. States Industries LLC produces Elemental Hardwood Panel Products using soy-based adhesives. These products comply with California Air Resources Board regulation 93120, which is aimed at reducing formaldehyde emissions. 9Wood uses soy-based adhesives for making crosspiece backers in custom wood ceilings. Soyad adhesives developed by Solenis are formulated from soy flour and a proprietary crosslinking agent. Solenis uses this adhesive system in combination with isocyanate-based resins to produce composite panels such as particleboards, medium-density fibreboards and oriented strand boards. Uniboard's Nu Green® are particleboards produced using soy-based adhesive technology, and are claimed to be the lowest-priced environmentally friendly boards available in the market. Finally, E2E's Transform™ is a soy-based biocomposite used in fabricated furniture products (USB, 2015; 2016).

Peanut Protein. Processing of peanuts for production of edible oil generates a protein residue in the form of a cake, known as peanut meal, which is traditionally used for animal feed. Newer approaches for utilization of the peanut cake have been recently developed. In one such application, the cake is further processed to produce defatted peanut flour, as well as peanut protein concentrate and/or isolate, which can be used in a variety of food formulations. As in the case of soy, peanut protein isolates are the most refined form of peanut protein and contain the highest concentration of protein (Wu et al., 2009; Yu et al., 2007).

Defatted peanut meal consists of 48 percentage protein (Golden Peanut and Tree Nuts, 2015) whereas defatted peanut flour consists of 56 percentage protein (Wu et al., 2009). Similar to the defatted soy flour, the defatted peanut flour is obtained either from the peanut meal left after extraction of oil by solvent extraction processes or from roasted peanuts. In the first case, flakes left after solvent extraction of oils are dried to remove the solvent, toasted, and then finely ground. In the second case, the defatted peanut flour is obtained as a residue when the roasted peanuts are pressed for extraction of oil without the use of solvent. Peanut protein concentrates, which contain nearly 70 percentage protein, are obtained from defatted peanut flour through the combination of isolelectric precipitation and centrifugation (Wu et al., 2009; Yu et al., 2007). On the other hand, peanut protein isolates, which contain more than 90 percentage protein, are obtained by dissolving the peanut flour at high pH, separating the supernatant, and precipitating the protein at lower pH (Sibt-e-Abbas et al., 2015).

Technical Applications

Several aspects of the physicochemical properties of peanut proteins have been studied from the perspective of non-food applications (Arthur, 1949; Li et al., 2015; Liu et al., 2004). Peanut proteins possess certain functional properties such as good water solubility in certain pH ranges, water holding capacity, emulsifying capacity, foam forming capacity, and cohesive and adhesive properties (Li et al., 2015; Sibt-e-Abbas et al., 2015; Wu et al., 2009; Yu et al., 2007). Consequently, peanut proteins have been used in paper coatings (Arthur et al., 1948) and as adhesive glues (Hogan and Arthur, 1952; Truax, 1929). Patents were issued in the 1920s for making composite adhesives by mixing peanut protein with starch (Dunham and Lawrence, 1929) or casein (Bradshaw and Dunham, 1929). Recently, peanut protein has been explored for potential applications in making films (Liu et al., 2004) and wood adhesives (Li et al., 2015).

Wheat Gluten. Gluten is the cohesive and viscoelastic proteinaceous material in wheat. Typically, wheat flour contains 10–14 percentage gluten (Sarkki, 1979), which is obtained as a by-product from the processing of wheat flour for starch production. Although wheat gluten is often sold as a protein, it usually contains constituents other than protein. The approximate composition of dry gluten is 75 percentage protein, up to 8 percentage moisture, and varying amounts of starch, lipid, and fibre (Day et al., 2006; USDA, 2015a). World wheat production in 2013–14 was 715 million metric tons. Nearly 67 percentage of wheat produced worldwide in 2013–14 was used in food industries, mainly for making baked goods of various types.

The pet feed industry, consuming nearly 18 percentage of wheat worldwide, was the second largest user of wheat. Nearly 12 percentage of the wheat produced globally in 2013–14 was used in sectors other than food and feed (FAO, 2015a).

Technical Applications

The physicochemical properties of wheat gluten, such as thermoelasticity and good film-forming properties, have made this protein useful for technical applications. A number of scientific reports indicate that wheat gluten is a promising raw material for the paper coating (Kersting et al., 1994) and wood adhesive industries (D'Amico et al., 2013; Khosravi et al., 2014; Lei et al., 2010; Nordqvist et al., 2012a; Nordqvist et al., 2012b; Nordqvist et al., 2013). It has also been incorporated into the production of biodegradable films and coatings for food and non-food applications (Gontard and Guilbert, 1998; Hernández-Muñoz et al., 2003). With regard to wood adhesive applications, wheat gluten-based formulations prepared by simple treatments such as enzymatic or thermal treatment have demonstrated improved performance in comparison to the unmodified protein, and some of these adhesive systems have passed the European standard of durability of wood composites (Nordqvist et al., 2012a; Nordqvist et al., 2012b; Nordqvist et al., 2013). On the other hand, particleboards prepared using chemically cross-linked wheat gluten have passed the requirements for interior grade wood boards (Khosravi et al., 2014; Lei et al., 2010). Together, these reports indicate that wheat gluten has a great potential to be incorporated as a protein component in development of bio-based wood adhesive systems.

Key Points

- Even though the world production of wheat and corn is significantly higher than all of the other major farmed crops combined, they have received less attention in the past for the development of wood adhesive systems due to their low protein content. Recently, wheat gluten is becoming an attractive feedstock for the development of adhesives, due to overall abundance and cheap cost relative to other commonly used protein material, such as soy protein.
- Recent research and development on bio-based wood adhesives have led to the commercialization of soy protein-based, formaldehyde-free adhesive systems that have been used for commercial production of interior composite wood products such as hardwood plywood, medium density fibreboards and particleboards.

- Soy is an attractive source of protein due to its high protein content, and soy protein is increasingly being used in wood adhesive applications. However, soy protein is one of the major constituents of food and feed industries, and this protein is being increasingly used for human consumption. Diverting soy protein from food chains to industrial applications will lead to shortages in protein supply and consequently to increases in soy protein prices, which is likely to adversely affect the soy protein-based wood adhesive market. Thus, it is necessary to explore cheaper alternatives alongside soy protein in the development of bio-based adhesives.

Animal-Based Protein

Meat production for human consumption generates significant amounts of non-edible meat as a by-product, which is a potentially viable source of industrial proteins. In addition to the by-products from the meat industry, production of dairy products such as cheese generates substantial quantities of protein-enriched secondary products. Even though animal protein-based adhesives were the main adhesives used in the woodworking industries until 1920s (Truax, 1929), arrival of soy protein largely replaced other protein resources from technical industries (USDA, 1951). Nevertheless, 13 percentage of the recent global bioadhesives market is contributed by animal-based bioadhesives (Grand View Research, 2015) and the protein from non-edible animal by-products is expected to emerge as a sustainable protein resource for various industrial applications.

In addition to reducing dependency on petrochemical-based products and the associated environmental benefits of using animal-based proteins for technical applications, another major advantage lies in the utilization of waste materials, which are otherwise incinerated or landfilled, for the production of value-added materials. Generally, these waste materials are a complex mixture of different classes of compounds, and separation and purification of a particular class of compound becomes a major issue. Fortunately, improved methods for extraction/isolation of protein from animal-based by-products are being developed.

Meat By-products and/or Rendered products. The recent trend towards agro-industrialization and the expanding agro-food supply chain encompasses a number of processes such as agricultural production, food processing, and agro-food logistics. This generates enormous amounts of organic residues consisting of proteins, carbohydrates, and fatty acids (Lin et al., 2013). The renewability of these feedstocks, the increasing demand of environmentally benign products, and biodegradability of the end products are encouraging the efficient utilization of agro-industrial waste for the production of novel value-added materials. In this context, by-products from slaughter houses and rendering industries are a potential source of industrial proteins that have high potential to be converted to value-added materials.

Global meat production has risen by 23 percentage over the past decade—from 253.1 million tonnes in 2003 (FAO, 2004) to 311.1 million tonnes in 2013—and is expected to increase in upcoming years (FAO, 2015a). Consistent with this, the amount of meat by-products is also on the rise worldwide. By-products from the meat industry include all parts of a live animal that are not considered dressed meat. Based on live weight, an adult cow carcass consists of 45 percentage edible materials and 55 percentage by-products (Opio et al., 2013). In the case of hogs, the by-products constitute about 30 percentage of the live weight (Marti et al., 2011). The value of such by-products accounts for more than 6 and 10 percentage of the value for hogs and cattle, respectively (Marti et al., 2011). In modern times, by-products from meat production have been incorporated into the industrial, cosmetic, drug, and feed manufacturing sectors to recover value (Opio et al., 2013). In fact, the US rendering industry in 2012 was valued at 10 billion dollars (Swisher, 2013).

Currently, production of pet food and animal feed is the predominant route used by rendering industries for valorization of animal by-products generated in slaughter houses. However, this route typically produces low-value products such as meat and bone meal, blood meal, feather meal, and poultry by-product (Table 4). Despite being considered as low-value products, the total value of this feed industry exceeded $500 billion in 2013 with worldwide livestock feed production of 963 million metric tons (Alltech, 2014).

In addition to the rendered products discussed above, specified risk materials (SRM) also represent a significant proportion of slaughter house by-products. SRM includes the bovine tissues in which specific misfolded proteins, also known as prions, are likely to concentrate. Prions are known to cause a neurodegenerative disease called Bovine Spongiform

Table 4. Average annual price of some protein meals in 2013 (Swisher, 2014). Table adopted with copyright permission from Render Magazine.

Meal type	Price per metric ton (USD)
Blood meal, ruminant	1,118
Blood meal, porcine	1,187
Feather meal	636
Meat and bone meal, ruminant	421
Meat and bone meal, porcine	478
Poultry by-product meal (57% protein)	528
Poultry by-product meal (67% protein)	745

Encephalopathy (BSE), also known as "Mad Cow Disease", in cattle. While the risks are extremely low, humans who ingest such prions may develop Creutzfeldt-Jakob disease, which is the human equivalent of BSE (Collinge, 1997). This has led to the implementation of increasingly strict policies from the US Food and Drug Administration (FDA) and the Canadian Food Inspection Agency (CFIA) with regard to SRM. The tissues referred to as SRM constitute 10 percentage by live weight of a bovine (Opio et al., 2013). However, as per the strict policies of the CFIA and FDA, SRM and any tissues that come in contact with SRM needs to be discarded from food and feed applications as well. Furthermore, strategies to separate SRM from the surrounding tissues are not precise and thus some edible meat is discarded along with the SRM. Consequently, often in excess of 20 percentage of the live weight of cattle is being eliminated as waste. These tissues are either landfilled or incinerated, which has led to a declining supply of raw material for rendering companies even though the world's meat production is increasing (Alltech, 2014).

Currently, more than seven hundred thousand tonnes of SRM is being landfilled annually in the United States (Vonderwell et al., 2015). In Canada, an estimated three hundred thousand tonnes of SRM is generated annually, which is either landfilled or incinerated (Mekonnen et al., 2013a). Thus, the issue of SRM has become costly for the meat industry and can negatively impact the environment through release of GHGs during incineration. Because of the inherent fear that BSE may be transferred through these rendered products, the market for SRM is virtually non-existent. Fortunately, a government approved protocol for the thermal treatment of SRM has been shown to destroy any existing prions, and the peptides recovered from the SRM hydrolysates can be used in non-food applications. Appearance of a patented technology for the development of polymers and plastics from waste animal proteins has drawn significant attention for conversion of such waste materials to high-value products (Bressler and Choi, 2015). Thermally-treated SRM is likely to emerge as a viable industrial protein source for some technical applications such as for the development of wood adhesives and plastics (Bressler and Choi, 2015; Mekonnen et al., 2014; 2015).

Technical Applications

In addition to being used in feed industries, inedible animal by-products serve as raw materials for the manufacturing of a broad assortment of industrial, household, cosmetic, and pharmaceutical supplies (Opio et al., 2013; Park et al., 2000a). All these by-products and/or rendered products contain substantial amounts of protein, nearly 50 percentage in meat and bone meal, 60 percentage in poultry by-product meal, 80 percentage in feather meal, and 90 percentage in blood meal (NRC, 1994). The protein streams from these rendered co-products are well suited for technical applications. One potential way to create value from such products is to employ them in production of adhesives, biocomposites, and biodegradable plastics (Mekonnen et al., 2013b; 2013c; 2015; Park et al., 2000a).

Casein. Bovine milk typically contains about 3.5 percentage protein, out of which 80 percentage is casein and 20 percentage is whey. Hence, casein can be extracted commercially at a yield of 2.5–3.0 kg per 100 kg skim milk. Based on the coagulating agent used during production, there are two basic types of casein available in the market—acid and rennet casein. Acid casein is produced by the action of acid on milk and is of two types: (i) casein obtained as a precipitate by adjusting the pH of skim milk to 4.6 with mineral acids such as hydrochloric or sulphuric acid; and (ii) casein obtained as a precipitate of skim milk by acidification due to lactic acid produced *in situ* with lactic acid bacteria. In rennet casein, coagulation is obtained by the action of an enzyme called rennet on skim milk. Acid or enzyme treatment for precipitation of casein yields a mixture of casein curd and whey from which whey (the liquid product) is separated, and the casein curd is washed with water and then dried to produce casein (Audic et al., 2003; Southward, 1998). The pH of acid casein ranges from 4.6 to 5.4, and its protein content is nearly 85 percentage. The pH of rennet casein ranges from 7.3 to 7.7, and its protein content is nearly 80 percentage (Southward, 1998).

Technical Applications

Historically, casein was used in Europe as an adhesive binder in woodworking, and also as a flexible adhesive in bookbinding (Truax, 1929). It is believed that industrial production of casein glues started in the nineteenth century, with a US patent released for preparation of casein glue in 1876 (FPL, 1967; Ross and Ross, 1876). A patented technology for making plywood used casein as an ingredient for adhesive formulation (Henning, 1920). As casein glues had better color and remarkable water resistance compared to animal glues, there was enhanced interest in America in the 1920s to use casein to make water-resistant glues for the construction of military aircraft (FPL, 1967; Truax, 1929). Since then, casein was widely produced and used in various industrial applications such as in regenerated protein fibers, paper coatings, woodworking, paints, and plastics (Audic et al., 2003; USDA, 1951). The arrival of cheaper soy protein in the adhesive market, and the development of synthetic polymers/resins led to a decrease in the casein market for technical applications (USDA, 1951).

Whey Protein. Whey is the liquid by-product that remains after milk is curled and strained during the production of cheese or casein. Whey is found in two different forms: (i) sweet whey, obtained as a co-product during rennet-coagulated cheese production; and (ii) acid whey, generated during production of acid coagulated casein or cheese such as cottage cheese (Lagrange et al., 2015). Sweet and acid whey contain 6–10 and 6–8 percentage protein, respectively (Jelen, 2003).

In general, 10 L of milk used for cheese production yields 1 kg of cheese, with the rest being liquid whey (Božanić et al., 2014). Global cheese production in 2013 reached 2.4 million tonnes (FAO, 2015a), and thus resulted in the generation of an estimated 22 million tonnes of liquid whey from cheese making alone.

Because of its high biochemical oxygen demand (> 35,000 ppm) and chemical oxygen demand (> 60,000 ppm), whey is regarded as one of the most polluting by-product streams from food processing industries (Smithers, 2008). However, over the past decades, technologies have been developed for conversion of this unwanted by-product into valuable raw materials and a number of whey protein products such as whey protein concentrates (WPC) and whey protein isolates (WPI) (Table 5). WPC and WPI are produced after removing lactose and minerals from liquid whey using different types of membrane filtration such as ultrafiltration, microfiltration, electrodialysis, nanofiltration, and reverse osmosis, followed by drying (Tunick, 2008). WPC are marketed in different forms with protein content ranging between 25 and 89 percentage (Table 5), but the most common is WPC-35 with 35 percentage protein content (Jelen, 2003). Conversely, WPI contain at least 90 percentage protein and are obtained by removing almost all lactose through an array of microfiltration processing techniques.

Table 5. U.S. Production of whey-based products in 2013 (Lagrange et al., 2015). Table adapted with copyright permission from John Wiley and Sons.

Product	Production (kilotons)
Whey, dry	436
Whey, dry, animal feed	17
Whey, lactose	471
Whey, protein concentrate, 25% to 49.9%	118
Whey, protein concentrate, 50% to 89.9%	108
Whey, protein concentrate, animal feed	4
Whey, protein isolate, 90%	40
Whey, reduced lactose & mineral, animal feed	28

Technical Applications

Whey protein is mainly used in food applications due to its high nutritional value. As it possesses specific functional properties such as high solubility in water, as well as good gelation, foaming, emulsification, and adhesive bonding properties (Chobert et al., 1988; Wang et al., 2013), whey protein has many potential technical applications, such as production of paper glue (Wang et al., 2013). Similar to other proteins, whey protein can be cross-linked to form a network structure after curing. Consequently, they have potential application in adhesive formulation for woodworking. Recently, journal articles (Gao et al., 2011; Zhao et al., 2011) and a patent (Guo et al., 2012) have reported on the utilization of whey protein for the formulation of aqueous polymer-isocyanate wood adhesives by chemical crosslinking of whey protein with a diisocyanate crosslinking agent.

Key Points

- Through technological advancements, it has been possible to convert some of the undervalued food processing by-products, including SRM, casein, and whey, into value added materials.
- SRM constitute a significant amount of an animal by weight, and have negative value because of inherent fears regarding BSE transfer and costs associated with their disposal. Fortunately, these protein by-product streams can be hydrolyzed to small peptides that have great potential for the development of wood adhesives.
- Protein recovered from hydrolyzed SRM is likely to emerge as a sustainable resource of industrial protein for development of bio-based wood adhesives that do not disturb food chain systems and provides an opportunity to recover value.

Advantages and Disadvantages of Various Protein-based Adhesives

While wood adhesives vary depending on their use, the major factors affecting their suitability for a given application include:

i)   Rate of setting in a joint
ii) Strength and durability of the adhesive bond
iii) Resistance to moisture and heat
iv) Tendency to stain the wood
v)   Cost effectiveness
vi) Health and/or environmental concerns associated with the chemical constituents.

For instance, high adhesive strength and water resistance are required for wood composites for exterior use. Conversely, adhesives being used for interior wood products should be free from chemicals that are detrimental to health and can escape into the surrounding air. Table 6 summarizes some advantages and disadvantages of glues developed from various protein sources.

Current Progress in Formulating Protein-Based Wood Adhesives

There are both chemical and mechanical factors that determine the quality of a wood adhesive (Frihart and Satori, 2013; Wang et al., 2014). The ability of a protein to chemically interact with wood substrates depends on the quantity and type of functional groups exposed. Furthermore, to achieve effective mechanical interlocking between the substrate and adhesive, the latter needs to be able to penetrate the surface of the substrate. The ability to penetrate beyond the substrate surface is based on the flowability of the adhesive, and on how well its components are dispersed in a particular carrier medium, such as water.

Viscosity is an important physical property that governs the adhesive behavior of protein glues. Industrial formulations of adhesive resins and/or glues commonly require a viscosity of <5000 mPa·s at about 20 percentage solid content (Kalapathy et al., 1996). However, native protein dispersions have viscosities much higher than this industrial requirement. Hence, protein denaturation is important to achieve a suitable viscosity, flowability, and substrate penetration of adhesive formulations. Exposure of polar functional groups through unfolding of the protein molecules enhances its dispersion and solubility in water, thereby reducing the viscosity and increasing the flowability. Consequently, denaturation of proteins improves their ability to penetrate wood substrates, enhances mechanical interlocking, and augments the adhesive property. However, protein dispersions with very low viscosity can result in severe penetration through the bonding surfaces, which leads to thin bond lines and lower bond strengths (Nordqvist et al., 2012a).

Native proteins typically adopt highly folded structures in which the majority of their functional groups are not exposed to the substrate. Denaturation or unfolding of proteins exposes the reactive functional groups so that they are readily available to interact with the bonding substrate. This can be achieved through thermal treatment or hydrolysis at elevated temperatures, enzyme treatments, by treating native protein with protein disrupting chaotropic agents, or by increasing the pH of protein dispersions. These treatments destroy both covalent and non-covalent interactions that are responsible for holding the protein molecule in a given confirmation (Hettiarachchy et al., 1995).

Table 6. Advantages and disadvantages of various protein-based adhesives (Frihart and Hunt, 2010; Frihart and Birkeland, 2014; Henning, 1920; Li et al., 2015; Truax, 1929; USDA, 1951).

Adhesive	Advantages	Disadvantages
Animal-based glues	**Animal tissues and bones:** • Rapid rate of setting • Moderate to high dry shear strength • Not likely to stain wood • Inexpensive as it is produced from animal by-products • Non-toxic	• Moderate to low resistance to water and damp conditions • Does not produce clear (white) glue lines
	Albumin and blood: • Very rapid setting with heat • Moderate to high dry shear strength • Moderate to high water resistance • Moderate resistance to microorganisms • Does not stain wood	• Produces dark glue lines, and is thus not suitable for gluing thin veneer sheets
	Casein: • Moderate to high dry shear strength • Moderate resistance to water and damp environments • Moderate resistance to elevated temperatures	• Dissolves only at very high pH values which are caustic • Has a tendency to discolor certain woods • Not suitable for applications in exterior use • More expensive than animal waste-based and vegetable-based protein glues
Vegetable-based glues	**Soybean:** • A sustainable resource that is non-toxic • Good processability has resulted in adhesive formulations that have similar performance to commercial petrochemical-based adhesives • Adhesive formulations using crosslinked soy protein have good strength under dry conditions, moderate resistance to moisture and damp conditions, and good thermal resistance	• Native protein dispersions have high viscosity, and requires denaturation of proteins using an alkaline solution, which is caustic and can discolor wood • Formulations containing high soy protein are likely to be sensitive to microbial degradation • Not suitable for exterior applications
	Peanut: • A sustainable resource that is non-toxic • Superiority of color (produces white glue lines) • Less hygroscopic than animal glues	• Low to moderate to dry shear strengths • Produces blisters, small bubble-like voids developed at the adhesive joints • Severe peanut allergies may limit application of these adhesives
	Gluten: • A sustainable resource that is non-toxic	• Production of wheat gluten of high protein content is costly.

Thermally Treated Protein

Thermal treatment is one way to disrupt these internal bonds and uncoil the protein structure (Hettiarachchy et al., 1995; Kreibich, 2001; Pan et al., 2005). In theory, thermal treatment will allow the functional groups to become more accessible to interactions with the substrate, and the proteins will be more easily dispersed. Mekonnen et al. (2014) used the protein hydrolysates of thermally hydrolyzed SRM as a protein component for the development of proteinaceous plywood adhesives. They investigated the effect of thermal hydrolysis temperature on adhesive strength at 180, 200, and 220°C, and demonstrated that the temperature of protein hydrolysis had the most influential effect on dry as well as soaked shear strength when resulting peptides were used in peptide-glutaraldehyde-resorcinol adhesive formulations (Mekonnen et al., 2014). The dry shear strength increased almost linearly with increase in hydrolysis temperature and the adhesive formulation consisting of peptides hydrolyzed at 200 and 220°C passed the ASTM requirement of urea-formaldehyde resin wood adhesives. As the extent of protein hydrolysis is directly related to hydrolysis temperature, high temperature thermal treatments produced lower molecular weight protein fragments, resulting in protein adhesives with reduced viscosity, enhanced spreadability, and better penetration, all of which resulted in more effective bonding (Mekonnen et al., 2014). Conversely, adhesives made with smaller protein fragments were shown to have lower soaked shear strength, likely from the increase in exposed polar functional groups, which are hygroscopic in nature. Thus, experiments by Mekonnen et al. (2014) highlighted the fact that the relationship between thermal hydrolysis and adhesive strength is not that simple (Mekonnen et al., 2014).

A similar conclusion was drawn from experiments conducted using soy protein (Vnucec et al., 2015; Zhong et al., 2003). Vnucec et al. (2015) studied the effect of treating soy protein isolate in a vacuum at different temperatures (50, 100, 150, and 200°C) on adhesive properties. Use of soy protein isolate treated at 50°C in adhesive formulations resulted in better shear strength, though this was also dependent on the temperature at which the dispersions were prepared and the pH of the formulation. However, treatment of soy protein isolate at 150 and 200°C resulted in preparations with no adhesive properties indicating that there is a threshold level for thermal treatment.

Zhong et al. (2003) prepared adhesive formulations by drying the denatured soy protein isolate at 30, 50, 70, and 90°C. The dried materials were then used in the preparation of fibreboards. They found that although temperature and viscosity generally displayed a direct correlation, there were no significant changes in shear strength of the fibreboard samples (Zhong et al., 2003). Furthermore, the drying treatment increased the percentage of cohesive failures within the fibreboards. Higher temperature treatments of denatured soy protein isolate (70 and 90°C) resulted in adhesive formulations that displayed 100% cohesive failure within the fibreboards in dry, soaked, and wet conditions. The main difference between soaked and wet specimen testing is that in the former, samples are dried and conditioned prior to shear strength testing, while in the latter, shear strength analysis is performed on wet specimens. Taken together, these experiments suggest that although exposing proteins to high temperatures may create smaller molecules, enhance protein dispersion in water, and/ or expose functional groups, additional treatments are likely necessary to produce superior adhesives.

Enzymatically Treated Protein

Treating proteins with enzymes (i.e., proteases) can also result in fragmentation of polypeptide chains and exposure of functional groups that are normally embedded on the interior of the protein. The effects of enzymatic hydrolysis on the adhesive properties of proteins sourced from wheat gluten (Nordqvist et al., 2012a) and soy protein (Kalapathy et al., 1995; Kumar et al., 2004; Qin et al., 2013) have been studied. Using low levels of the serine protease alcalase (0.3–0.6 percentage) improved the adhesive properties, including bond strength and moisture resistance, of wheat gluten (Nordqvist et al., 2012a). This was attributed to denaturation of the proteins and enhanced penetration into the substrate, resulting from decreased viscosity. However, excessive hydrolysis led to adhesive formulations with very low viscosities that led to over-penetration of the wood substrate, resulting in poor adhesive strength.

Adhesive formulations using soybean protein treated with various enzymes have also been explored. The use of papain and urease to enzymatically modify soy protein led to an increase in shear strength by 22 and 28 percentage, respectively (Kumar et al., 2004). Conversely, the use of trypsin resulted in adhesive performance comparable to unmodified soy protein, and treatments using chymotrypsin and pepsin led to significant decreases in the adhesive strength. Reduced adhesive strength was attributed to excessive hydrolysis that lowered viscosity, perhaps leading to thin bond lines due to over-penetration of the substrate.

The effect of enzymatic hydrolysis with trypsin on the viscosity and adhesive strength of soy protein has been studied (Kalapathy et al., 1995). One hour hydrolysis of soy protein with trypsin led to a decrease in the viscosity of the soy protein dispersion from 240 to 70 cP, with concomitant increase in adhesive strength from 300 (unreacted control) to 700 N. However, when the hydrolysis time was increased to six hours, viscosity was further reduced leading to a decreased adhesive strength of 280 N. Even though the relationship between enzymatic hydrolysis and adhesive strength seems to be quite complex, what is generally observed is that mild hydrolysis results in protein dispersions of reasonable viscosity and enhances the adhesive strength, but excessive hydrolysis leads to protein dispersions of too low viscosity and reduces the adhesive strength.

Chemically Treated Protein

Many chemical treatments have been employed on proteinaceous material to overcome the main challenges of high viscosity, short pot-life, and low water resistance of native protein-based adhesives. For example, various protein feedstocks have been treated with alkali (Kalapathy et al., 1995; Kalapathy et al., 1996), urea (Sun and Bian, 1999; Zhang and Hua, 2007), sodium dodecyl sulfate (SDS) and sodium dodecylbenzene sulphonate (SDBS) (Huang and Sun, 2000a), and guanidine hydrochloride (Huang and Sun, 2000b).

Sodium hydroxide (NaOH) has been used extensively to denature various sources of proteins including soy protein (Hettiarachchy et al., 1995), wheat gluten (Nordqvist et al., 2010), and protein obtained from jatropha, a tropical flowering plant (Zhang et al., 2011). The denaturation of proteins with NaOH has been primarily used to improve adhesive strength and water resistance. NaOH treatment of protein enhances the adhesive property by (i) exposing specific functional groups as a result of unfolding of protein molecules, and (ii) increasing the intermolecular interactions with the solvent/medium and reducing the viscosity. For example, the modification of soy protein with moderate alkali treatment (pH 10, 50°C) increased the adhesive strength from 340 to 730 N (Hettiarachchy et al., 1995). Mo et al. (2001) extended work on alkali treatment by incorporating an oxidant in methodologies for processing of wheat straw proteins. Proteins treated with 1 M NaOH and 0.2 percentage H_2O_2 resulted in adhesives with tensile strength and compression strength values that were significantly higher than those obtained using untreated proteins (Mo et al., 2001).

The presence of disulphide bonds in native protein molecules affects their flexibility and unfolding properties. Disrupting the disulphide linkages by the use of proper chemicals is another approach for enhancing adhesive properties of proteins. Reducing agents, such as sulphites (Kalapathy et al., 1996; Li et al., 2012b; Qi et al., 2013a), and thiols (Kawamura et al., 1985; Kim and Kinsella, 1986) cleave the inter- and intramolecular disulphide bonds in protein molecules. Cleavage of disulphide bonds by reducing agents generally leads to an improvement in adhesive properties due to a reduction in viscosity. For example, the dry strength increased from 5.2 to 6.2 MPa and the wet strength increased from 1.63 to 1.9 MPa for adhesives developed from glycinin rich soy protein treated with sodium bisulphite (Qi et al., 2012).

Sodium dodecyl sulfate and sodium dodecyl benzene sulfonate are both detergents and denaturing agents, and they have been used to chemically treat various protein sources. Sodium dodecyl sulfate binds with protein molecules disrupting secondary, tertiary, and quaternary structures. This disruption is believed to expose functional groups due to rapid changes in the conformation of proteins (Tanford, 1968). It is expected that sodium dodecyl benzene sulfonate will have the same effect. The use of sodium dodecyl sulfate and sodium dodecyl benzene sulfonate generally led to an improvement in the adhesive properties of the proteins (Cheng et al., 2013; Huang and Sun, 2000a; Zhong et al., 2003). For example, the use of sodium dodecyl sulfate led to an increase in tensile strength from 130 to 200 lb/in^2 for adhesives developed from cottonseed protein (Cheng et al., 2013). Huang and Sun (2000) studied the effect of the concentration of sodium dodecyl benzene sulfonate (0–3 percentage) on the dry shear strength and soaked shear strength of adhesives developed from soy protein. Concentrations of sodium dodecyl benzene sulfonate from 0.5–1 percentage improved both the dry and soaked shear strengths of walnut specimens bonded with the sodium dodecyl benzene sulfonate-treated soy protein. The use of 1 percentage sodium dodecyl benzene sulfonate resulted in an increase in dry shear strength of walnut wood specimens from 25 to 48 kg/cm^2. However, the use of 3 percentage concentrations of sodium dodecyl benzene sulfonate generally did not impact dry and soaked shear strengths (Huang and Sun, 2000a).

Guanidine hydrochloride, a known chaotrope, has also been studied as a denaturing agent to improve the adhesive properties of proteins (Cheng et al., 2013; Zhong et al., 2002; Zhong et al., 2003). Denaturing agents help improve the adhesive properties of proteins by unfolding protein molecules, resulting in random structures that have more exposed functional groups. This enhances the interactions between the adhesive and the substrate, resulting in improved strength properties (Cheng et al., 2013; Zhong et al., 2002; Zhong et al., 2003). Reportedly, treatment of soy protein isolate with 1 M guanidine hydrochloride led to improved adhesive strength, as well as water resistance, of bonded fibreboards (Zhong et al., 2002). However, higher concentrations of such chaotropic agents have been reported to negatively affect protein adhesion. A general understanding is that secondary structures are desirable for protein adhesion as they allow for better interactions between protein molecules, as well as with the substrate, during the curing process, which results in stronger bonding. Destruction of secondary structure reduces the number of intermolecular interactions, resulting in poor mechanical strength and water resistance of the resulting formulation.

Urea is another effective denaturant that can unfold proteins by disrupting the hydrogen bonds that maintain protein structure. This leads to unfolded structures of varying degrees. Partly unfolded protein molecules with a certain amount of secondary structure have been found to exhibit good water resistance (Huang and Sun, 2000b). Furthermore, the use of urea has been exploited to improve the adhesive properties of soy protein (Huang and Sun, 2000b; Xu et al., 2012), soy flour (Wang et al., 2010), and cottonseed (Cheng et al., 2013). Huang and Sun (2000b) studied the effect of concentration of urea (1, 3, 5, and 8 M) on the adhesive properties of soy protein isolate. The increase in concentration of urea from 1 to 3 M improved both the dry and soaked shear strengths of walnut specimens. For instance, modification with 3 M urea resulted in an increase in dry shear strength from 25 to 49 kg/cm². Conversely, the use of higher concentrations of urea (5 and 8 M) resulted in decreasing dry and soaked shear strengths. Similar trends were observed for cherry and pine wood specimens. As was the case with guanidine hydrochloride, it is believed that some secondary structure was maintained at low concentrations of urea, which improved adhesive qualities, but at higher concentrations, secondary structure was completely destroyed leading to poor mechanical strength and lower water resistance (Huang and Sun, 2000b).

Key Points

- The two most important parameters that determine the adhesive strength of protein-based adhesives are: (i) the ability of a protein solution/dispersion to penetrate beyond the surface debris and damaged fibers into sound wood to create effective mechanical interlocking; and (ii) the extent of exposure of specific functional groups of the adhesive formulation for bond formation with the substrate.
- Thermal, enzymatic, and chemical treatments of proteins destroy native structures and produce smaller peptide fragments with more exposed functional groups. All these treatments help reduce the viscosity of protein dispersions and enhance the penetration ability of the adhesive formulation resulting in enhanced adhesive performance. However, severe treatment leads to excessive breakdown of peptides producing fragments that are too small and lead to undesirable adhesive performance.
- Even though thermal, enzymatic or chemical treatments alone have not resulted in protein-based adhesive systems that meet industrial requirements, protein denaturation using these treatments has emerged as an essential pretreatment step in more advanced approaches of adhesive formulation that employ chemical modification or chemical crosslinking.

Trends for the Future

There are several methods available for the thermal, enzymatic, and chemical treatment of proteins for the production of adhesives. While many of these methods have been shown to generally improve dry shear strengths—depending on other conditions and treatments examined, most did not produce adhesives with sufficient soaked and/or wet shear strengths (i.e., sufficient moisture resistance). The inherently low water resistance of protein-based adhesives is one of the major barriers preventing widespread application of proteins in adhesives for the wood composite market. To address this issue, many of the ongoing studies surrounding the development of protein-based adhesives are focused on chemical modification and chemical crosslinking of the protein feedstock and blending of protein with resins.

Current research on development and commercialization of protein-based wood adhesives focuses on two different type of adhesive systems: (i) a protein-phenol-formaldehyde adhesive system; and (ii) a formaldehyde-free adhesive system to replace the formaldehyde-based resins. For the latter, different approaches for denaturation of native proteins, modification of end functional groups, and crosslinking of polypeptide chains to overcome problems with viscosity and poor water resistance are being explored for the development of various adhesive systems. Thus far, greater success has been achieved for interior wood products as formaldehyde-free adhesives have not yet been able to outcompete phenol-formaldehyde-based commercial resins with regards to cost effectiveness and performance in wood products to be used in exterior applications. Ongoing research may be able to improve formaldehyde-free adhesive systems such that they can replace phenol-formaldehyde-based resins currently used in exterior wood products.

Chemical Modification of Protein

There is a significant amount of research being conducted that focuses on improving adhesive strength of protein-based systems by chemically modifying the proteins. In this approach proteins are modified by means of chemical reactions with multifunctional compounds that possess functional groups capable of reacting with active groups of protein, and also contribute to adhesive strength and/or water resistance enhancing properties. The main reactive functional groups of proteins are amines, carboxyls, and hydroxyls. These moieties can react with other functional groups, such as amines, imines, and carboxyls, to produce substances with different properties. The multifunctional compounds used for such studies include dopamine (Li and Liu, 2008; Liu and Li, 2002), dicarboxylic acid anhydrides such as maleic anhydride, succinic anhydride, and phthalic anhydride (Coco et al., 1984; Huang and Li, 2008; Liu and Li, 2007), as well as polyglycidyl methacrylate (Wang et al., 2014). Reportedly, chemical modification of various proteins has resulted in improved bond strength and water resistance of the resulting formulations used in wood adhesion (Table 7), some of which performed comparably to commercial phenol-formaldehyde and urea-formaldehyde adhesives (Liu and Li, 2002).

Mussels secrete a unique protein that helps them to adhere to the substrate on which they reside. It is believed that the compound 3,4-dihydroxyphenylalanine (DOPA) present in mussel proteins elicits adhesive and crosslinking properties (Silverman and Roberto, 2007). In an approach to impart DOPA like properties to soy protein adhesives, Li and Liu (2008) reacted alkali denatured soy protein isolate (SPI) with dopamine. Compared to the alkali denatured SPI, dopamine-grafted SPI exhibited remarkably high adhesive strength both in dry and wet conditions (Li and Liu, 2008; Liu and Li, 2002). The maple veneer plywood specimens bonded with dopamine modified SPI yielded very good shear strengths that did not drop significantly even after three soaking-and-drying cycles. The shear strengths increased remarkably when the dopamine content of modified SPI increased from 4.12 to 8.95 percentage (wt/wt), indicating that the phenolic hydroxyl groups of dopamine had substantial contribution to adhesive strength and water resistance of the resulting wood adhesive. The favorable level of dopamine residue for adhesive strength and water resistance was around 9 percentage by weight as further increasing the dopamine content to 14.7 percentage (wt/wt) led to decreased shear strengths (Liu and Li, 2002). It is possible that this is due to excessive crosslinking of SPI, which would cause the formulation to become rigid and thus lead to lower dry shear strength. Furthermore, increasing dopamine content to 14.7 percentage (wt/wt) likely resulted in a greater number of polar phenolic groups, which could potentially decrease soaked shear strength.

In another approach of chemical modification, maleic anhydride (MA) was grafted on SPI through reactions with amino groups and hydroxyl groups of protein (Liu and Li, 2007). In this modification, maleyl groups were added to SPI through amide and ester linkages. Grafting of maleic anhydride decreased the adhesive performance of SPI, but the addition of polyethylene imine (PEI) to maleic anhydride grafted SPI dramatically enhanced the adhesive strength due to the formation of highly crosslinked networks of modified SPI and PEI during hot pressing (Liu and Li, 2007). A number of polar groups such as amide, amine, and hydroxyl of the modified proteins formed covalent linkages with wood substrates during hot pressing. As a result, maple veneer plywood composites bonded with a combination of a maleic anhydride grafted SPI and PEI had dry shear strengths even higher than those bonded with commercially used phenol-formaldehyde resins. The combination of SPI, 10 percentage (wt/wt) maleic anhydride, and 20 percentage (wt/wt) PEI yielded the highest shear strength. However, the

Table 7. Current approaches for enhancing adhesive performance of protein-based wood adhesives developed from various protein sources.

Protein source	Treatment/formulation	Gluing condition	Test method/standard	Adhesive strength (Mpa)		Application	Reference
				Dry	Soaked		
Canola protein	Native	170°C, 3.57 MPa load, 10 min	ASTM std. D2339-98 (for dry strength) and D1183-96 (for wet strength)	5.15 ± 0.33[1]	2.23 ± 0.45	Cherry wood adhesive for making plywood	(Li et al., 2012a)
	30% NaHSO₃ treated			5.08 ± 0.35[1]	2.86 ± 0.10		
	60% NaHSO₃ treated			5.15 ± 0.21[1]	2.78 ± 0.12		
Canola protein isolate (CPI)	Alkali denatured	110°C, 3.1 Mpa load, 4 min	Automated bonding evaluation system (ABES)	3.72 ± 0.21	1.02 ± 0.14	Adhesive for birch wood veneer	(Wang et al., 2014)
	Polyglycidyl methacrylate grafted (82% grafting)			8.25 ± 0.12	3.68 ± 0.29		
Sorghum protein	Acetic acid extracted sorghum protein from DDGS (12% protein)	170°C, 3.57 MPa load, 10 min	ASTM std. D2339-98 and D1183-96	4.78 ± 0.39	1.32 ± 0.10	Cherry wood adhesive for making plywood	(Li et al., 2011)
Whey protein isolate (WPI)	API adhesive containing 55.4% WPI, 11.1% PVA, 30% PVAc and 3.5% CaCO3 crosslinked with 15% pMDI	20°C, 1.5 kN load, 2 hr	Japanese Industrial Standards K6806-2003	13.38 ± 1.42	6.81 ± 0.79	Adhesive for birch blocks (30 × 20 × 10 mm size)	(Zhao et al., 2011)

Table 7. contd....

Table 7. contd.

Protein source	Treatment/ formulation	Gluing condition	Test method/standard	Adhesive strength (Mpa)		Application	Reference
Thermally denatured WPI	API adhesive containing 58.3% WPI, 11.7% PVA and 30% PVAc crosslinked with 15% pMDI	20°C, 1.5 kN load, 2 hr	Japanese Industrial Standards K6806-2003	10.56	5.65	Adhesive for birch blocks (30 × 25 × 10 mm size)	(Gao et al., 2011)
Soy protein (SP)	β-conglycinin rich SP at pH 5.6 (33% solid content)	150°C, 10 min, 1.4 MPa load	ASTM std. D2339-98, D1183-96 and D1151-00 (wet)	6.82 ± 0.35[1]	2.84 ± 0.39	Adhesive for cherry wood veneers	(Qi et al., 2012)
	Glycinin rich SP at pH 5.4 (33% solid content)			6.21 ± 0.69[1]	1.92 ± 0.29		
	NaHSO₃ modified SP blended with sorghum lignin (80:20)	170°C, 10 min, 2 MPa load	ASTM std. D2339-98	6.40 ± 0.53[1]	2.27 ± 0.36	Adhesive for cherry wood veneers	(Xiao et al., 2013)
	Glycin dominated SP modified by 0.4% NaHSO₃	170°C, 10 min, 2.0 MPa load	ASTM std. D2339-98 and D1151-00	5.5 ± 0.10[1]	2.7 ± 0.40	Adhesive for cherry wood veneers	(Qi et al., 2013a)
	β-conglycin dominated SP modified by 0.4% NaHSO₃			6.8 ± 0.54[1]	3.3 ± 0.20		

	NaHSO$_3$ modified SP blended with urea-formaldehyde-based resin (40:60)	170°C, 10 min, 2.0 MPa load	ASTM std. D2339-98, D1183-96 and D1151-00	6.13 ± 0.28[1]	6.42 ± 0.49[1]	Adhesive for cherry wood veneers	(Qi and Sun, 2011)
	NaHSO$_3$ modified SP blended with PVA-based resin (40:60)			7.26 ± 0.32[1]	2.01 ± 0.39		
Soy protein isolate (SPI)	33% SPI	150°C, 10 min, 1.4 MPa load	ASTM std. D2339-98 and D1183-96	5.29 ± 0.16[1]	1.63 ± 0.20	Adhesive for cherry wood veneers	(Qi et al., 2012)
	12% SPI (pH 4.5)	170°C, 5 min, 1.4 MPa load	ASTM std. D2339-98, D1151-00 and D1183-96	5.36 ± 0.21[1]	2.84 ± 0.22[2]	Adhesive for cherry wood veneers	(Zhong et al., 2007)
	SPI modified with 5% PAE (pH 5.5)			6.36 ± 0.40[1]	3.90 ± 0.17[3]		
	28% SPI (pH 5.5)	190°C, 10 min, 4.9 MPa load	ASTM std. D2339-98 and D1151-00	6.52	3.67	Plywood adhesive for cherry wood samples	(Mo and Sun, 2013)
	28% SPI and 0.5% NaCl (pH 5.5)			6.57	3.55		
	SPI blended with sorghum lignin (80:20)	170°C, 10 min, 2.0 MPa load	ASTM std. D2339-98	5.64 ± 0.72[1]	2.41 ± 0.13	Adhesive for cherry wood veneers	(Xiao et al., 2013)
Soy flour (SF)	30% SF (pH 6.2)	120°C, 2 min, 0.2 MPa load	Automated bonding evaluation system (ABES)	6.2 ± 1.1	0.8 ± 0.1	Adhesive for hard maple veneer	(Frihart and Satori, 2013)
	30% reaction mixture of SF and PAE resin containing 5% PAE to SF (pH 5.9)			6.9 ± 0.70[1]	2.4 ± 0.20		

Table 7. contd....

Table 7. contd.

Protein source	Treatment/formulation	Gluing condition	Test method/standard	Adhesive strength (Mpa)		Application	Reference
	Defatted SF	150°C, 10 min, 1.4 MPa load	ASTM std. D2339-98 and D1183-96	5.67 ± 0.46[1]	1.81 ± 0.41	Adhesive for two ply-plywood from cherry wood veneers	(Qi et al., 2013b)
	3.5% 2-octen-1-ylsuccinic anhydride grafted SF			5.82 ± 0.86[1]	3.13 ± 0.26[4]		
	Defatted SF	150°C, 10 min, 1.03 MPa load	ASTM std. D906-98	0.90 ± 0.10[5]	-	Adhesive for three ply-plywood from yellow poplar veneers	
	3.5% 2-octen-1-ylsuccinic anhydride grafted soy flour			0.94 ± 0.09[1]	0.28 ± 0.11[5]		
	Defatted SF			1.56 ± 0.29[5]	1.27 ± 0.33[5]	Adhesive for three ply-plywood from maple veneers	
	3.5% 2-octen-1-ylsuccinic anhydride grafted soy flour			2.33 ± 0.21[6]	2.26 ± 0.34[7]		

[1] 100% cohesive wood failure
[2] 50% cohesive wood failure
[3] 72% cohesive wood failure
[4] fiber pulled out
[5] 5% cohesive wood failure
[6] 80% cohesive wood failure
[7] 60% cohesive wood failure

wet shear strength of the wood composites bonded with this adhesive was lower than that of wood composites bonded with a phenol-formaldehyde resin, and was attributed to the presence of a number of polar functional groups (Liu and Li, 2007).

The adhesive system based on soy protein isolate, maleic anhydride, and polyethylene imine developed by Liu and Li had good adhesive strength (Liu and Li, 2007), but had no potential for industrial application because SPI is too expensive as a raw material. Additionally, the SPI-MA-PEI adhesive system had some drawbacks such as: (i) the extended reaction time and elevated temperature needed for chemical modification of SPI; and (ii) the time-consuming and labor-intensive process of drying modified SPI (Huang and Li, 2008). Subsequently, Li's group investigated the use of soy flour to replace SPI from their protein-maleic anhydride-polyethylene imine-based adhesive system, and also simplified the adhesive formulation protocol.

Three-ply plywood samples were prepared from yellow-poplar veneer, and the water resistance of plywood panels was determined using a three-cycle soak test. Each component of the SF-MA-PEI adhesive system had a remarkable effect on bond strength and water resistance. With 80 percentage of the test panels passing the three-cycle soak test, the three-component formulation consisting of SF-MA-PEI performed better than the two-component formulations consisting of SF-MA or SF-PEI, which had pass rates of 0 and 60 percentage, respectively. Addition of 0.39 percentage (wt/wt) of NaOH to the SF-MA-PEI formulation remarkably affected the water resistance of the formulated adhesive, and all panels passed the three-cycle soak test. Reportedly, addition of NaOH helped expose the functional groups by unfolding the native structure of soy protein, and the exposed functional groups were readily available for curing reactions during hot pressing of test specimens. Based on their studies, a 7:1 weight ratio of SF:PEI and 32 percentage (wt/wt) of MA produced the best adhesive formulation, as all test panels passed the three cycle soak test and no specimens were delaminated even after the boiling water test (Huang and Li, 2008).

Grafting hydrophobic functions such as long alkyl chains is an alternative approach to impart water resistance to protein-based adhesives. Qi et al. (2013b) grafted alkyl chains on to soy protein by reacting defatted soy flour with 2-octen-1-ylsuccinic anhydride (OSA). Grafting of OSA did not significantly enhance the dry shear strength of the soy protein adhesive, but the wet shear strength of the OSA-modified soy protein adhesive was remarkably improved through OSA grafting (Qi et al., 2013b). In the two-ply plywood system using cherry wood veneers, the wet strength of soy protein adhesive was considerably improved from 1.8 MPa using unmodified soy protein to 3.1 MPa using soy protein modified with OSA. Protein modification using OSA introduced hydrocarbon chains and enhanced the hydrophobic property of the protein resulting in enhanced water resistance (Qi et al., 2013b).

Reaction of polyglycidyl methacrylate (pGMA) and canola protein isolate (CPI) in different combinations yielded CPI-GMA conjugates with various grafting degrees of GMA (Wang et al., 2014). The dry, wet, and soaked shear strengths of glued birch wood veneers using all CPI-GMA conjugates were remarkably higher than that of CPI alone, indicating that grafting of GMA on canola protein yielded an adhesive with good adhesive strength and water resistance. A conjugate with a GMA grafting degree of 82.0 percentage exhibited the best performance with enhancement of dry shear strength by 121 percentage and wet shear strength by 276 percentage, compared to that of alkali denatured CPI (Table 7). As pure GMA itself does not possess adhesive property, the enhancement in adhesive strength of the protein isolate after modification was due to the effect of grafting GMA on to the protein. GMA grafting led to enhanced interactions, such as covalent bonding, hydrogen bonding, and hydrophobic interactions, between the modified CPI and wood substrate, thus resulting in improved adhesive strength (Wang et al., 2014).

Key Points

- Chemical modification of proteins, which can be achieved by numerous mechanisms, is a highly relevant method to enhance adhesive strength and/or water resistance of protein-based adhesive formulations.
- Chemical modification of proteins based on adhesive systems found in nature can potentially result in wood adhesive systems that perform similarly to petrochemical-based wood adhesive resins currently being used by woodworking industries.
- Grafting excess polar functional groups on to proteins is detrimental for adhesive applications, as the resulting adhesive bonding does not survive moist conditions.
- If the grafted functional groups are capable of producing stronger crosslinks with a curing agent through covalent interactions, chemical modification followed by crosslinking dramatically enhances the adhesive strength as well as moisture resistance of the final adhesive formulation.

Chemical Crosslinking of Protein

In this approach, the polypeptide chains of protein are crosslinked with a crosslinking reagent resulting in the formation of three-dimensional networks of polypeptide chains *via* covalent linkages. The formation of a rigid network of polypeptide chains prevents movement of individual polymers, which helps to maintain structural integrity. Additionally, crosslinking of polypeptide chains decreases the ability of water to permeate through the polypeptide network. Together, the phenomenon of crosslinking enhances the mechanical strength and water resistance of protein adhesives (Bowes and Cater, 1966; Park et al., 2000b; Wang et al., 2007; Zhang et al., 2006).

Generally, compounds possessing two or more reactive groups capable of reacting with the functional groups of polypeptide chains are used as polypeptide crosslinking agents. Since aldehydes and isocyanates readily react with amine functions, dialdehydes and diisocyanates are typical polypeptide crosslinking agents. As the reaction of a carbonyl compound with an amine produces a Schiff's base, a dialdehyde crosslinks the polypeptide chains through aldimine linkages by reacting with the amine groups of the polypeptide chain (Fig. 3A; Cuesta-Garrote et al., 2014; Marquie, 2001). A diisocyanate, on the other hand, crosslinks the polypeptide chains through terminal amine groups by the formation of urea linkages (Fig. 3B; Guo et al., 2012; Zhao et al., 2011).

Compounds possessing nitrogen or oxygen-containing heterocyclic functional groups are also prone to react with amine or carboxyl groups of polypeptide chains. Hence, macromolecules containing these functional groups, such as epoxy-based resins and cationic polyamidoamine-epichlorohydrin (PAE) resins, are also used as polypeptide crosslinking agents. It is well known that a curing agent is necessary for curing (hardening) of epoxy and/or PAE resins, and the most commonly used curing agents are (poly)amines. In case of crosslinking of peptides with epoxy or PAE resins, the polypeptides act as the curing agent. Both the amine group and carboxyl group of polypeptides react with azetidinium group of the PAE resin (Fig. 3C) and epoxy group of the epoxide resin (Fig. 3D) leading to the formation of highly crosslinked, three dimensional, and compact networks of peptides and the resin (El-Thaher et al., 2014; Lei et al., 2014; Li et al., 2004).

Crosslinking with a Dialdehyde

Among dialdehydes, glutaraldehyde is a widely studied compound for crosslinking of peptides due to the following advantages: (i) it is more reactive than glyoxal, malonaldehyde, succinaldehyde, and adipaldehyde; (ii) it has appropriate chain length for bridging the gap between the amino groups of the peptide chains; and (iii) it produces more crosslinks than other dialdehydes (Bowes and Cater, 1966). Even though it can react with a large number of available amino groups of polypeptides, it readily does so with the amino groups of lysine residues (Marquie, 2001). This reaction introduces three-dimensional crosslinks between polypeptide chains by

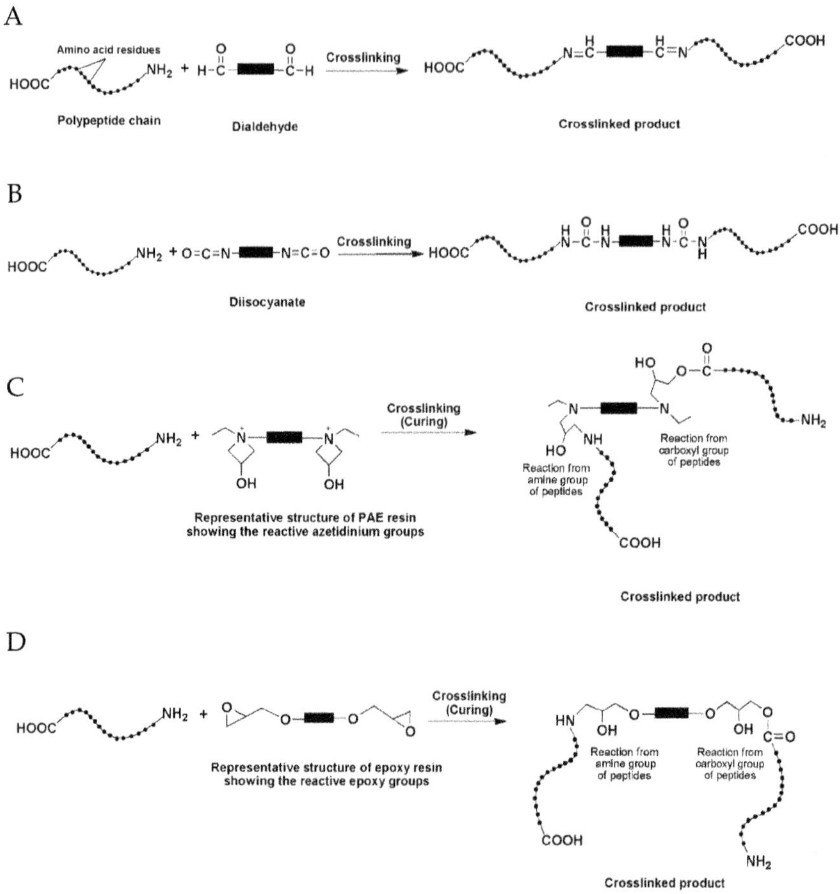

A

B

C

D

Figure 3. Chemical crosslinking of polypeptides with (A) dialdehyde, (B) diisocyanate, (C) PAE resin, and (D) epoxy resin.

inter- and intramolecular crosslinking, increases the molecular weight, hinders the formation of crystalline structures of protein, and causes some structural and/or morphological changes making the crosslinked product more suitable as an adhesive (Cuesta-Garrote et al., 2014; Ghosh et al., 2009; Marquie, 2001; Park et al., 2000b; Wang et al., 2007). Accordingly, the glutaraldehyde crosslinked soy protein had a higher tensile strength and greater elongation property than the native soy protein (Park et al., 2000b). Additionally, glutaraldehyde crosslinking decreases the hydrophilic property by reducing the number of amino groups and N-terminals in polypeptide chains, and increases the hydrophobic property by adding the hydrocarbon part of glutaraldehyde to the peptide chain. The glutaraldehyde crosslinking induced structural changes, reduced the number of hydrophilic groups, and made the crosslinked product more water resistant due to addition of hydrophobic propylene bridges (Wang et al., 2007).

In addition to creating compact three-dimensional crosslinks of polypeptide chains, glutaraldehyde crosslinking of SPI also led to conformational changes of protein molecules and reduced the viscosity of the soy protein suspension (Wang et al., 2007). The dry, wet, and soaked shear strengths of cherry wood veneer specimens prepared by using glutaraldehyde-crosslinked SPI (20 µM glutaraldehyde) displayed enhancements of 31.5, 115, and 29.7 percentage, respectively, from those of wood specimens prepared by using unmodified SPI (Wang et al., 2007). According to Wang et al. (2007), glutaraldehyde crosslinking induces more entanglements of polymers during hot pressing, resulting in the improved integrity of polypeptide chains. Furthermore, such entanglements are more resistant to damage by water than that of the unmodified SPI in wet and soak tests (Wang et al., 2007). Adhesive performance of the crosslinked product also benefited from the mild conformational changes that resulted from crosslinking. However, a higher degree of crosslinking resulted in further conformational and structural changes that were not beneficial to adhesive performance, and thus the adhesive strength decreased at higher glutaraldehyde concentrations (Wang et al., 2007).

Crosslinking with a Diisocyanate

Methylene diphenyl isocyanate (MDI) or polymeric methylene diphenyl isocyanate (pMDI) are the most widely used diisocyanate crosslinking agents for formulation of protein-based adhesives. Similar to glutaraldehyde crosslinking, MDI and pMDI crosslinking of polypeptide chains enhances the crosslink density of proteins by connecting the polypeptide chains, reduces the motional freedom of individual polypeptide chains, and creates a three-dimensional network of peptides. In some approaches for the development of protein-based adhesives, MDI and pMDI were used as crosslinking agents to prepare aqueous polymer-isocyanate (API) adhesives. Typically, API adhesives are formulated by crosslinking the components of water-based glue with an isocyanate crosslinking agent. The water-based glue is a mixture of natural polymer (starch, protein, etc.) with soluble synthetic polymer such as polyvinyl alcohol (PVA) or a mixture of PVA with emulsions of polyvinyl acetate (PVAc), styrene-co-butadiene rubber (SBR), ethylene-co-vinyl acetate, or their mixtures (Gao et al., 2011; Guo et al., 2012; Zhao et al., 2011).

Whey protein is composed of smaller molecular weight polypeptides, and has relatively high water solubility. As a result, whey protein exhibited poor performance as a wood adhesive in terms of bond strength and bond durability. The dry strength of whey protein adhesives was much less than the required value for structural applications, and most samples yielded very low wet strength in 28 h boiling-drying cycles indicating very poor bond durability (Gao et al., 2011; Zhao et al., 2011). Crosslinking of whey protein with pMDI dramatically enhanced the strength of API adhesives by producing three-dimensional networks of polypeptide chains and by forming covalent bonds *via* urethane linkages with the functional groups of wood in bondlines (Gao et al., 2011; Guo et al., 2012; Zhao et al., 2011). An adhesive formulation containing 15 percentage (wt/wt) pMDI in a 40 percentage aqueous solution of whey protein isolate exhibited much better adhesive properties and bond strength (5.78 MPa dry shear strength and 2.64 MPa wet shear strength) as compared to a 40 percentage aqueous solution of whey protein isolate (2.06 MPa dry shear strength). Further improvement in dry as well as wet shear strengths of bonded wood specimens was achieved through the addition of polyvinyl alcohol to the mixture of whey protein and diisocyanate crosslinker. The added polyvinyl alcohol further enhanced the adhesive strength *via* increased covalent linkages among polyvinyl alcohol, whey protein, and the crosslinker in the curing process. Addition of 3.5 percentage (wt/wt) of nano-scale $CaCO_3$ powder as a filler augmented the phenomenon of mechanical interlocking and resulted in the further increase of both dry and wet shear strengths. A water-based adhesive consisting of 55.4 percentage thermally denatured whey protein isolate, 30 percentage polyvinyl acetate, 11.1 percentage polyvinyl alcohol, and 3.5 percentage $CaCO_3$, crosslinked with 15 percentage (wt/wt) pMDI exhibited the highest dry shear strength (13.38 MPa) and wet shear strength (6.81 MPa), with values comparable to that of commercial API adhesives being used for structural applications (Gao et al., 2011; Zhao et al., 2011).

Lei et al. (2010) developed adhesive resins from alkali hydrolyzed wheat gluten using different crosslinking agents, and the resulting adhesive formulations were used for the preparation of one-layer particleboard panels. The combination of pMDI and hydrolyzed gluten (20:80) alone gave such poor particleboards that they had no measurable internal bond strength. When the hydrolyzed gluten was first reacted with formaldehyde or glyoxal, prior to crosslinking with pMDI, the obtained particleboards had an internal bond strength that was almost double the strength requirement (>0.35 MPa) of the European Norm EN 312 for interior grade panels (Lei et al., 2010). The formulations consisting of formaldehyde or glyoxal-reacted gluten protein and 20–30 percentage pMDI yielded good bond strength and bond durability as a wood adhesive, satisfying the standard specifications for interior grade particleboards (Lei et al., 2010).

Recently, Mekonnen et al. (2014) developed waste animal protein-based wood adhesives by crosslinking thermally hydrolyzed protein (specific risk material) with MDI, which was then used to make oriented strandboard (OSB) panels. Different adhesive formulations were developed by varying the weight proportions of the protein hydrolysates and MDI, and the performance of each formulation was evaluated in terms of static bending, internal bond strength, and bond durability of the OSB panels. The static bending and internal bond strength values of the panels fabricated using adhesive formulations containing 40 or 50 percentage hydrolyzed protein satisfied the CSA 0437 requirement in moisture free conditions. However, the developed OSB panels had intermediate water resistance, and they did not meet the CSA 0437 requirement of bond durability for structural applications (Mekonnen et al., 2014).

Crosslinking with Epoxy Resin

US patent no. 2,882,250 (Baker, 1959) disclosed a method of preparing epoxy resin modified protein and utilization of the modified protein as a wood adhesive for the production of three-ply panels of yellow birch veneer. The water resistance of plywood panels made with epoxy resin crosslinked soy protein was superior to those made from unmodified soy protein (Baker, 1959). Recently, Lei et al. (2014) used a reaction product of soy protein with an epoxy-based resin as an adhesive for production of plywood panels, and the water resistance of the formulated adhesive was evaluated by soaking the test samples in 63°C water for 3 h, as well as by employing a boiling water test (Lei et al., 2014). When using soy protein without crosslinking for adhesion of wood specimens, all the samples were delaminated in the soaked test indicating that the soy protein alone had no water resistance. Crosslinking the soy protein with different amounts of epoxy resin yielded plywood specimens that had measurable soaked shear strength. However, more than 50 percentage of the test samples did not pass the standard requirement of Chinese national standard (GB/T 9846.3-2004, ≥ 0.70 MPa). On the other hand, all the test specimens prepared by using epoxy resin crosslinked soy protein were delaminated in the boiling water test. The limited water resistance of the soy protein-epoxy resin adhesive was due to the inability of the epoxy resin to form extensive crosslinks with soy protein, which in turn was due to the limited water solubility of the crosslinking agent (Lei et al., 2014). Hence, in order to get strong, water resistant crosslinks after curing, it is necessary for a crosslinking agent to be miscible in the adhesive system.

Crosslinking with PAE Resin

Polyamidoamine-epichlorohydrin (PAE) resin is available commercially as Kymene® 557H wet strength resin. Sun et al. (2008) studied the interaction of PAE resin with SPI at different pH conditions. When the protein is in an isoelectric condition, the macromolecules of PAE and the polypeptide chains form reversible protein-PAE complexes at room temperature through ionic interactions between the cationic azetidinium groups of the PAE resin and the anionic carboxyl groups of the protein. However, the protein-PAE complexes disintegrate at extremely acidic or basic conditions. Interaction of hydroxyl ions with cationic PAE releases the PAE resin from the complex at higher pH, and protonation of the carboxylate anion destroys the protein-PAE interactions at lower pH (Sun et al., 2008).

The ionic interactions between PAE and the polypeptide chains act as a physical crosslinking at room temperature. Chemical crosslinking of the PAE resin and the polypeptide chains, however, takes place at elevated temperatures through chemical reactions between the azetidinium groups of PAE and the active hydrogen containing functional groups of polypeptides (Li et al., 2004; Sun et al., 2008). Reportedly, the following reactions are possible between protein and PAE at higher temperatures: (i) reactions between the azetidinium groups and the secondary amines of PAE (self-crosslinking reaction of PAE); (ii) reactions between the azetidinium groups and the primary and secondary amino groups of protein; and (iii) reactions between the azetidinium groups of PAE and the carboxyl groups of protein. Reactions between the azetidinium groups of the resin and the carboxyl and/or hydroxyl groups of wood fibers are also possible during hot pressing (Li et al., 2004; Sun et al., 2008).

The effect of crosslinking SPI with PAE on adhesive strength at high temperatures (i.e., hot pressing conditions) was demonstrated by Sun et al. (2008). SPI alone exhibited low bond strength and poor water resistance as an adhesive for bonding of cherry wood veneers (Sun et al., 2008). The SPI-PAE adhesive system, prepared by crosslinking of SPI with PAE resin, demonstrated great improvement in adhesion properties and water resistance. This improvement was thought to have resulted from the formation of a three-dimensional network of PAE and polypeptides, which resulted from the chemical reactions between PAE and SPI, as well as self-crosslinking reactions of the PAE resin. Additionally, the reactions between the crosslinked product and wood substrate were confirmed to contribute to the improvement in adhesive strength (Sun et al., 2008).

In another experiment, the adhesive developed by the reaction of alkaline SPI with Kymene was used for gluing of maple veneers. The resulting wood composites had dramatically high shear strengths and water resistance. The SPI-Kymene adhesive yielded higher dry shear strength than the alkaline SPI, Kymene alone, and a commercial PF resin (Li et al., 2004).

All the specimens bonded with alkaline SPI alone underwent delamination, and some specimens bonded with Kymene alone were also delaminated in the boiling water test. However, the specimens bonded with SPI-Kymene adhesive retained relatively high strength even after the boiling-water test (Li et al., 2004).

Key Points

- Chemical crosslinking of peptides enhances the mechanical strength and water resistance of protein-based adhesives by creating compact, three-dimensional networked structures of peptides connected by covalent linkages.
- Protein-based adhesive formulations using an isocyanate or PAE crosslinking agent have demonstrated an adhesive performance comparable to urea-formaldehyde and phenol-formaldehyde type wood adhesives.
- Low molecular weight protein/peptide fragments typically demonstrate poor performance as wood adhesive. However, using suitable crosslinking agents, it has been possible to formulate water-based adhesives from those low molecular weight protein fragments, and the resulting formulations have demonstrated comparable performance to that of commercial water-based adhesives.
- The development of an environmentally friendly wood adhesive system that is competitive with urea-formaldehyde and phenol-formaldehyde resins will likely require a combination of technologies, such as protein denaturation, followed by chemical modification of denatured protein, and finally chemical crosslinking.

Co-polymerization and/or Blending of Protein with Formaldehyde-based Resins

In an alternative approach for formulating proteinaceous wood adhesives, co-reaction of protein with the constituents of formaldehyde-based resins has led to the development of adhesive systems containing 30 to 70 percentage by weight of protein. Reportedly, adhesives prepared by proper denaturation and stabilization of protein, followed by its co-reaction with commercial resin constituents were shown to be very promising as wood adhesives for exterior applications (Frihart and Wescott, 2004; Hse et al., 2000; Kuo et al., 2003; Lorenz et al., 2006; Qu et al., 2015; Yang et al., 2006). In this approach, the denatured protein is first reacted with formaldehyde to make a stabilized protein, and the resulting material is then reacted with phenol and formaldehyde to produce protein-phenol-formaldehyde adhesives. The final adhesive contains substantial amounts of amalgamated protein, either by co-polymerization with phenol and formaldehyde or by irreversible incorporation in the phenol-formaldehyde network (Frihart and Wescott, 2004; Lorenz et al., 2006). A protein content of as high as 70 percentage by weight in the adhesive formulation dramatically reduces the use of petroleum-based phenol and formaldehyde. These adhesive formulations have bonding strength and water-resistance values similar to those obtained using commercial phenol-formaldehyde resins, and have been used to make wood composites such as fibreboards, flakeboards, and strandboards (Frihart and Wescott, 2004; Hse et al., 2000; Kuo et al., 2003; Lorenz et al., 2006; Qu et al., 2015; Yang et al., 2006).

US patent no. 6,518,387 (Kuo et al., 2003) described a method for preparing protein-phenol-formaldehyde adhesives and their application in the fabrication of fibreboards, hardboards, and flakeboards. The soy protein-based adhesive resin was prepared by reacting seven parts of soy flour with three parts of phenol-formaldehyde prepolymer in an aqueous solution. The mechanical properties and dimensional stability of fiberboards and flakeboards produced using this adhesive were comparable to that of boards formed from fibers bonded with urea-formaldehyde and phenol-formaldehyde resins (Kuo et al., 2003). Consequently, the soy protein-based adhesive resin was recommended not only for the replacement of urea-formaldehyde resin in production of fiberboard panels for interior applications, but also as an exterior adhesive resin for the production of construction grade wood composites (Kuo et al., 2003).

Bryer et al. (2006) found that addition of an effective amount of a modified protein to urea-formaldehyde prepolymer yields wood composites having improved internal bond strength and stability as compared to the wood composites made using urea-formaldehyde resin alone (Bryer et al., 2006). Adhesives developed by co-polymerization exhibited higher adhesive strength than those developed by simple blending (Qu et al., 2015). An adhesive developed either by blending or co-reacting hydrolyzed soy protein isolate (HSPI) with urea-formaldehyde was employed for making three-layer plywood panels. The bond strength of plywood made using HSPI alone was 0.58 MPa, and improved to 0.89 MPa when an HSPI-UF blend was employed. Co-polymerization of HSPI, urea, and formaldehyde further increased the bonding strength to 1.31 MPa. Addition of HSPI at the beginning of adhesive formulation through the co-polymerization process resulted in better bonding strength and low formaldehyde emissions from the final adhesive product. In addition, the HSPI-UF co-polymers were more biodegradable than the urea-formaldehyde resin (Qu et al., 2015).

Hse et al. (2000) developed a soy protein-based adhesive through the co-reaction of soy flour hydrolysates with a phenol-formaldehyde mixture in alkaline conditions, and used the adhesive for bonding of exterior grade flakeboards (Hse et al., 2000). In those soy protein-phenol-formaldehyde co-reacted adhesives, 30 percentage of the phenol from conventional phenol-formaldehyde resins was substituted with the soy protein hydrolysates. The internal bond strength of the flakeboard panels made with the co-reacted adhesive was higher than those of conventional phenol-formaldehyde resins. Additionally, substitution with soy flour hydrolysates lowered the materials cost by 20 percentage (Hse et al., 2000). The strandboards prepared by Lorenz et al. (2006) using a soy flour-phenol-formaldehyde co-reacted resin with 40 percentage protein content had qualities equivalent to that of boards produced with the commercial adhesive (Lorenz et al., 2006).

The blend of melamine-urea-formaldehyde resin and denatured soy flour had remarkably lower viscosity than that of soy flour dispersions. Blending of soy flour and melamine-urea-formaldehyde also increased the water resistance and wet shear strength of plywood panels made from poplar veneers. The soy flour-melamine-urea-formaldehyde blend containing 20 percentage melamine-urea-formaldehyde resin had a reasonable viscosity and yielded good bond strength in plywood panels, satisfying the Chinese Industrial Standard requirement (≥ 0.7 MPa) for interior plywood panels (Fan et al., 2011).

Blending of $NaHSO_3$-modified soy protein with some synthetic latex adhesives such as urea-formaldehyde-based and polyvinyl alcohol-based adhesives significantly reduced viscosity. Addition of 40 to 60 percentage modified soy protein in the synthetic latex adhesives also led to improved water resistance (Qi and Sun, 2011). For example, the wet adhesion strength of the 40 percentage modified soy protein blend with the urea-formaldehyde-based resin was 6.41 MPa, which was higher than the 4.66 MPa of the pure urea-formaldehyde-based resin. According to Qi and Sun (2011), chemical crosslinking between the functional groups of protein and hydroxymethyl groups of the urea-formaldehyde resin occurs in such blends, which leads to the enhancement of adhesive performance. Additionally, protein components within these blends act as an acidic catalyst for the self-polymerization of the urea-formaldehyde resin (Qi and Sun, 2011).

In addition to soy protein, the protein hydrolysates of specified risk materials (SRM) were recently used for the formulation of proteinaceous adhesives by co-reacting the hydrolyzed protein with resorcinol-glutaraldehyde prepolymer (Mekonnen et al., 2015). The adhesives were formulated by varying protein concentration and the mole ratio of resorcinol and glutaraldehyde using the Taguchi method. When tested as an adhesive for birch veneers, 88 percentage of the formulations passed the ASTM requirement (ASTM D4690) of minimum dry shear strength (2.344 MPa) and one-third of the formulations passed the wet shear strength (1.93 MPa) for urea-formaldehyde adhesives. Studies on the effect of temperature on protein hydrolysis, as well as the mole ratio of glutaraldehyde-resorcinol in the prepolymer on final adhesive strength revealed that the highest lap shear strength in bonded wood specimens resulted from the use of an adhesive formulation having 20 percentage (wt/wt) protein hydrolysates (hydrolyzed at 220°C) crosslinked with 40 percentage (wt/wt) of glutaraldehyde:resorcinol (mole ratio of 1:0.5) prepolymer (Mekonnen et al., 2015).

Applied Protein Systems and Eka chemicals have developed and marketed some adhesive systems using co-polymerization technology. ProSoy 4315—a soy-based additive for phenol formaldehyde-based wood adhesives—is a product of Applied Protein Systems. Reportedly, this product is a sustainable and economical partial replacement for the phenol component of the adhesive. Developed by Eka Chemicals, PRF Soy/2000 is a two-component adhesive system consisting of a soy-based adhesive and a phenol-resorcinol-formaldehyde resin. This formulation is developed for bonding of green lumber during finger-jointing applications (USB, 2015; USB, 2016).

Key Points

- The adhesive systems developed by blending and/or co-reacting protein with phenol-formaldehyde or resorcinol-formaldehyde-based prepolymers are two-component adhesives in which the petroleum-based component is largely replaced with a bio-based component, thus substantially reducing the hazard associated with formaldehyde emission from wood products.
- Thus far, soy protein has found a market as a protein component of such formulations, and commercial wood adhesive products employing a soy-based additive and phenol-formaldehyde prepolymers are available.

Conclusions

The inherent adhesive properties of proteins have been recognized for millennia, and proteinaceous substances have been used for making wood adhesives since early civilization. Even though protein-based glues lost ground in the marketplace after the arrival of petroleum-based resins, they have recently been gaining in popularity in the wood adhesive sector. Development of newer and better technologies for efficient valorization of proteinaceous materials, and a growing interest in the use of bio-based materials to replace toxic formaldehyde-based products are major drivers for the revival of interest in protein-based adhesives. Additionally, their abundance, easy of processing, renewability, and sustainability are key factors that make protein-based adhesives more appealing for woodworking industries.

A review of recent scientific reports on protein-based wood adhesives indicates that denaturation, chemical modification, chemical crosslinking, and/or co-reacting or blending protein with a prepolymer are the current methods employed for improving adhesive properties of these bio-based adhesives. Utilizing these approaches, ongoing research focuses on developing two different protein-based adhesive products for wood adhesive market: (i) a formaldehyde-free protein-based adhesive to completely replace urea-formaldehyde resins in interior wood products such as hardwood plywood, medium density fibreboard, and particleboard; and (ii) blending and/or co-reacting protein with synthetic resins, such as polyvinyl acetate or phenol-formaldehyde resins, for use in exterior wood products such as oriented strand board, plywood, and engineered wood for exterior applications. It is hoped that protein-based adhesives are developed in the near future that can be incorporated into such exterior wood products without the use of synthetic resins.

Denaturation of protein followed by chemical modification using adhesive enhancing functional groups and/or water resistance enhancing moieties has greatly enhanced the adhesive strength and water resistance, respectively, of modified protein-based adhesives when compared to the native protein. On the other hand, chemical crosslinking of denatured protein fragments has also been shown to improve the mechanical strength as well as water resistance of adhesive formulations to such an extent that some formulations of this method have resulted in protein-based adhesive systems performing comparable to urea-formaldehyde- and phenol-formaldehyde-based wood adhesives. Consequently, formaldehyde-free protein-based adhesive technologies have been developed and commercialized for production of plywood and various composite wood products.

Blending or co-reacting protein with synthetic resins is an alternative solution for reducing dependency on petrochemicals and lowering the potential health hazard due to formaldehyde emission from interior wood products. Protein blends with polyvinyl acetate or phenol-formaldehyde resins have resulted in adhesive systems containing up to 70 percentage (wt/wt) protein that have comparable performance to that of synthetic resins currently used in wood adhesive applications. Adhesive systems developed by proper denaturation and stabilization of protein, followed by its co-reaction with phenol-formaldehyde resin constituents, have shown even higher adhesive strength than the phenol-formaldehyde resins alone, and are very promising as wood adhesives for exterior applications.

Industrial protein can be obtained from a number of sources. Among the various proteins, soy proteins are widely studied for their potential application in wood adhesives, and some soy protein-based adhesive products have seen commercial success as formaldehyde-free adhesives. Currently, raw protein from a variety of sources is being predominantly utilized in animal feed and pet food industries. Processed proteins from soy, peanut, whey, and casein are also consumed as protein supplements and meat substitutes. Diverting these edible proteins to large scale industrial applications may create protein shortages in food/feed industries. Accordingly, it is necessary to examine alternative protein sources for adhesive development. In this context, waste proteins that have little or even negative-value, such as specified risk materials from the rendering industry, may provide a cost-effective and sustainable industrial protein source for the development of wood adhesives that meet the ASTM requirement for making composite wood products. Use of such waste material to produce adhesives would have a significant advantage over traditional products and other protein-based adhesives.

Keywords: Proteins, Bioadhesives, Wood adhesives, Formaldehyde-free adhesives, Chemical modification, Chemical crosslinking, Adhesive strength, Water resistance

References

Adebiyi, A.P. and R.E. Aluko. 2011. Functional properties of protein fractions obtained from commercial yellow field pea (*pisum sativum* L.) seed protein isolate. Food Chem. 128: 902–908.

Adedayo, M.R., E.A. Ajiboye, J.K. Akintunde and A. Odaibo. 2011. Single cell proteins: As nutritional enhancer. Adv. Appl. Sci. Res. 2: 396–409.

Allen, A.J., J.M. Wescott, D.F. Varnell and M.A. Evans. 2014. Protein adhesives formulations with amine-epichlorohydrin and isocyanate additives. U.S. Patent No. 8845851B2.

Alltech. 2014. Feed industry worth $500 billion. Alltech global feed survey summary. Alltech, Kentucky, USA.

Arthur, J.C.J. 1949. Evaluation of peanut protein for industrial utilization. A review. J. Am. Oil Chem. Soc. 26: 668–671.

Arthur, J.C., Jr., T.W. Mason, Jr. and M.E. Adams. 1948. Peanut protein paper coatings. J. Am. Oil Chem. Soc. 25: 338–340.

Audic, J.-L., B. Chaufer and G. Daufin. 2003. Non-food applications of milk components and dairy co-products: A review. Le Lait 83: 417–438.

Baker, E.B. 1959. Process for preparing epoxy resin modified proteins and compositions resulting therefrom. U.S. Patent No. 2882250A.

Barthet, V.J. 2012. Quality of western canadian canola 2011. Grain research laboratory. Canadian Grain Commission, Canada.

Bhalla, T.C., N.N. Sharma and M. Sharma. 2007. Production of metabolites, industrial enzymes, amino acid, organic acids, antibiotics, vitamins and single cell proteins. Food and Industrial Microbiology. pp. 1–46. National Science Digital Library, India.

Bowes, J.H. and C.W. Cater. 1966. The reaction of glutaraldehyde with proteins and other biological materials. J. Royal Microscop. Soc. 85: 193–203.

Božanić, R., I. Barukčić, K.L. Jakopović and L. Tratnik. 2014. Possibilities of whey utilisation. Austin J. Nutri. Food Sci. 2: 1036–1042.

Bradshaw, L. and H.V. Dunham. 1929. Adhesive material and process of making. U.S. Patent No. 1703134A.

Brady, L.R., Q.-M. Gu and R.R. Staib. 2012. Diluents for crosslinked-containing adhesive compositions. U.S. Patent No. 8147968B2.

Bressler, D.C. and P. Choi. 2015. Polymers and plastics derived from animal proteins. U.S. Patent No. 9120845B2.

Brockmann, W., P.L. Geiss, J. Klingen and K.B. Schroder. 2009. Adhesive bonding: Adhesives, applications and processes. Wiley-VCH, Weinheim, Germany.

Bryer, R.A., R.H. Carey, X.S. Sun, E.-M. Cheng and J.D. Rivers. 2006. Wood composites bonded with soy protein-modified urea-formaldehyde resin adhesive binder. U.S. Patent No. 20060234077A1.

Bye, C.N. 1990. Casein and mixed protein adhesives. pp. 135–152. *In:* I. Skeist (ed.). Handbook of Adhesives. Springer, New York, USA.

Cheng, H.N., M.K. Dowd and Z. He. 2013. Investigation of modified cottonseed protein adhesives for wood composites. Ind. Crops Prod. 46: 399–403.

Chobert, J.-M., C. Bertrand-Harb and M.-G. Nicolas. 1988. Solubility and emulsifying properties of caseins and whey proteins modified enzymatically by trypsin. J. Agric. Food Chem. 36: 883–892.

Coco, C.E., P.M. Graham and T.L. Krinski. 1984. Modified protein adhesive binder and method of producing. U.S. Patent No. 4474694A.

Collinge, J. 1997. Human prion diseases and bovine spongiform encephalopathy (BSE). Hum. Mol. Genet. 6: 1699–1705.

Cozzone, A.J. 2010. Proteins: Fundamental chemical properties. pp. 1–10. *In:* Encyclopedia of Life Sciences (ELS). John Wiley & Sons Ltd., Chichester, UK.

Cuesta-Garrote, N., M.J. Escoto-Palacios, F. Arán-Ais and C. Orgile´s-Barcelo. 2014. Structural changes in the crosslinking process of a protein bioadhesive. Proceedings of the Institution of Mechanical Engineers, Part L: Journal of Materials: Design and Applications 228: 115–124.

D'Amico, S., U. Müller and E. Berghofer. 2013. Effect of hydrolysis and denaturation of wheat gluten on adhesive bond strength of wood joints. J. Appl. Polym. Sci. 129: 2429–2434.

Davidson, G. 1929. Process of preparing substances composed in part of protein containing cells for the manufacture of adhesives. U.S. Patent No. 1724695A.

Day, L., M.A. Augustin, I.L. Batey and C.W. Wrigley. 2006. Wheat-gluten uses and industry needs. Trends Food Sci. Technol. 17: 82–90.

De Graaf, L.A. 2000. Denaturation of proteins from a non-food perspective. J. Biotechnol. 79: 299–306.

Dunham, H.V. and B. Lawrence. 1929. Vegetable glue and process of making same. U.S. Patent No. 1703133A.

Ebnesajjad, S. 2010. Characteristics of adhesive materials. pp. 137–184. *In:* S. Ebnesajjad (ed.). Handbook of Adhesives and Surface Preparation: Technology, Applications and Manufacturing. Elsevier Inc., Oxford, UK.

Eggersdorfer, M., J. Meyer and P. Eckes. 1992. Use of renewable resources for non-food materials. FEMS Microbiol. Rev. 103: 355–364.

El-Thaher, N., P. Mussone, D. Bressler and P. Choi. 2014. Kinetics study of curing epoxy resins with hydrolyzed proteins and the effect of denaturants urea and sodium dodecyl sulfate. ACS Sustainable Chem. Eng. 2: 282–287.

Fan, D.-B., T.-F. Qin and F.-X. Chu. 2011. A soy flour-based adhesive reinforced by low addition of MUF resin. J. Adhes. Sci. Technol. 25: 323–333.

[FAO] Food and Agriculture Organization of the United Nations. 2015a. Food outlook. Biannual report on global food markets. Food and Agriculture Organization of the United Nations, Rome, Italy.

[FAO] Food and Agriculture organization of the United Nations. 2015b. 2014 Global Forest Products Facts and Figures. Forest Product Statistics. Food and Agriculture organization of the United Nations, Rome, Italy.

[FAO]. Food and Agriculture Organization of the United Nations 2004. Food outlook no. 4. Global information and early warning system on food and agriculture. Food and Agricultural Organization of the United Nations, Rome, Italy.

Fawthrop, W.D. 1933. Adhesive and method of making the same. U.S. Patent No. 1897469A.

Fay, P.A. 2005. History of adhesive bonding. pp. 3–22. *In*: R.D. Adams (ed.). Adhesive Bonding: Science, Technology and Applications. CRC, Cambridge, England.

[FPL] Forest Product Laboratory. 1967. Casein glues: Their manufacture, preparation, and application. Forest Product Laboratory, U.S. Department of Agriculture, Madison, WIS.

Frihart, C.J. and J.M. Wescott. 2004. Improved water resistance of bio-based adhesives for wood bonding. pp. 293–302. 1st international conference on environmentally-compatible forest products. September 22–24, Oporto, Portugal.

Frihart, C.R. and M.J. Birkeland. 2014. Soy properties and soy wood adhesives. pp. 167–192. *In*: R.P. Brentnin (ed.). Soy-based Chemicals and Materials. ACS Symposium Series. American Chemical Society, Washington, D.C., USA.

Frihart, C.R. and C.G. Hunt. 2010. Adhesives with wood materials: Bond formation and performance. pp. 1–24. *In*: Anonymous (ed.). Wood Handbook : Wood as an Engineering Material. General Technical Report FPL; GTR-190. U.S. Dept. of Agriculture, Forest Service, Forest Products Laboratory, Madison, WI, USA.

Frihart, C.R. and H. Satori. 2013. Soy flour dispersibility and performance as wood adhesive. J. Adhes. Sci. Technol. 27: 2043–2052.

Gao, Z., W. Wang, Z. Zhao and M. Guo. 2011. Novel whey protein-based aqueous polymer-isocyanate adhesive for glulam. J. App. Polym. Sci. 120: 220–225.

Ghosh, A., M.A. Ali and G.J. Dias. 2009. Effect of crosslinking on microstructure and physical performance of casein protein. Biomacromolecules 10: 1681–1688.

Golden Peanut and Tree Nuts. 2015. Peanut meal. Golden Peanuts and Tree Nuts. Accessed 12/03/2015. http: //www.goldenpeanut.com/PeanutMeal.aspx.

Gontard, N. and S. Guilbert. 1998. Edible and/or biodegradable wheat gluten films and coatings. pp. 324–328. *In*: J. Guéguen and Y. Popineau (eds.). Plant Proteins from European Crops. Springer-Verlag, Berlin Heidelberg.

González-Pe´rez, S. and J.M. Vereijken. 2007. Sunflower proteins: Overview of their physicochemical, structural and functional properties. J. Sci. Food Agric. 87: 2173–2191.

Grand View Research. 2015. Bioadhesives market analysis by source (plant based & animal based), by application (packaging & paper, construction, wood, personal care, medical) expected to reach USD 2,549.2 million by 2022. Renewable chemicals report. Grand View Research, USA.

Guo, M., M.E. Vayda and Z. Gao. 2012. Whey-protein based environmentally friendly wood adhesives and methods of producing and using the same. U.S. Patent No. 20120183794A1.

Henning, S.B. 1920. Process of manufacturing plywood. U.S. Patent No. 1329402A.

Hernández-Muñoz, P., A. Kanavouras, P.K. Ng and R. Gavara. 2003. Development and characterization of biodegradable films made from wheat gluten protein fractions. J. Agric. Food Chem. 51: 7647–7654.

Hettiarachchy, N.S., U. Kalapathy and J.D. Myers. 1995. Alkali-modified soy protein with improved adhesive and hydrophobic properties. J. Am. Oil Chem. Soc. 72: 1461–1464.

Hogan, J.T. and J.C.J. Arthur. 1952. Cottonseed and peanut meal glues. Resistance of plywood bonds to chemical reagents. J. Am. Oil Chem. Soc. 29: 16–18.

Hse, C.Y., F. Fu and B.S. Bryant. 2000. Development of formaldehyde based wood adhesives with co-reacted phenol/soybean flour. pp. 13–9. Proc. wood adhesives conference. 22–23 June, South Lake Tahoe, NV, USA.

Huang, J. and K. Li. 2008. A new soy flour-based adhesive for making interior type II plywood. J. Am. Oil Chem. Soc. 85: 63–70.

Huang, W. and X. Sun. 2000a. Adhesive properties of soy proteins modified by Sodium Dodecyl sulfate and sodium dodecylbenzene sulfonate. J. Am. Oil Chem. Soc. 77: 705–708.

Huang, W. and X. Sun. 2000b. Adhesive properties of soy proteins modified by urea and guanidine hydrochloride. J. Am. Oil Chem. Soc. 77: 101–104.

IARC. 2006. Formaldehyde, 2-butoxyethanol and 1-tert-butoxypropan-2-ol. IARC monographs on the evaluation of carcinogenic risks to humans. 88: 1–497.

Jelen, P. 2003. Whey processing: Utilization and products. pp. 2739–2745. *In*: H. Roginski, J.W. Fuquay and P.F. Fox (eds.). Encyclopedia of Dairy Sciences. Academic Press, London, UK.

Kalapathy, U., N.S. Hettiarachchy, D. Myers and M.A. Hanna. 1995. Modification of soy proteins and their adhesive properties on woods. J. Am. Oil Chem. Soc. 72: 507–510.

Kalapathy, U., N.S. Hettiarachchy, D. Myers and K.C. Rhee. 1996. Alkali-modified soy proteins: Effect of salts and disulfide bond cleavage on adhesion and viscosity. J. Am. Oil Chem. Soc. 73: 1063–1066.

Kawamura, Y., Y. Matsumura, T. Matoba, D. Yonezawa and M. Kito. 1985. Selective redution of interpolypeptide and intrapolypeptide disulfide bonds of wheat gluten from defatted flour. Cereal Chem. 62: 279–283.

Keimel, F.A. 2003. Historical development of adhesives and adhesive bonding. pp. 1–12. *In*: A. Pizzi and K.L. Mittal (eds.). Handbook of Adhesive Technology, 2nd edition. CRC Press, New York, USA.

Kersting, H.-J., M.G. Lindhauer and W. Bergthaller. 1994. Application of wheat gluten in non-food industry—wheat gluten as a natural cobinder in paper coating. Ind. Crops Prod. 3: 121–128.

Khosravi, S., F. Khabbaz, P. Nordqvist and P. Johansson. 2014. Wheat-gluten-based adhesives for particle boards: Effect of crosslinking agents. Macromol. Mater. Eng. 299: 116–124.

Kim, S.H. and J.E. Kinsella. 1986. Effects of reduction with dithiothreitol on some molecular properties of soy glycinin. J. Agric. Food Chem. 34: 623–627.

Kinsella, J.E. 1979. Functional properties of soy proteins. J. Am. Oil Chem. Soc. 56: 242–258.

Kinsella, J.E. and N. Melachouris. 1976. Functional properties of proteins in foods: A survey. CRC Crit. Rev. Food Sci. 7: 219–280.

Kinsella, J.E. and C.V. Morr. 1984. Milk proteins: Physicochemical and functional properties. CRC Crit. Rev. Food Sci. 21: 197–262.

Koshland, D.E., Jr. 2014. Protein. *In*: Encyclopedia Britannica. Encyclopedia Britannica Inc. Accessed 01/07/2016. http: //www.britannica.com/science/protein.

Kreibich, R.E. 2001. New adhesives based on soybean proteins. American soybean association. Weyerhaeuser Technology Centre, Tacoma, WA, USA.

Kristinsson, H.G. and B.A. Rasco. 2000. Fish protein hydrolysates: Production, biochemical, and functional properties. CRC Crit. Rev. Food Sci. 40: 43–81.

Kruger, M. and W. Linke. 2011. The giant protein titin: A regulatory node that integrates myocyte signalling pathways. J. Biol. Chem. 25: 9905–9912.

Kumar, R., V. Choudhary, S. Mishra and I.K. Varma. 2004. Enzymatically-modified soy protein part 2: Adhesion behaviour. J. Adhes. Sci. Technol. 18: 261–273.

Kumar, R., V. Choudhary, S. Mishra, I.K. Varma and B. Mattiason. 2002. Adhesives and plastics based on soy protein products. Ind. Crops Prod. 16: 155–172.

Kuo, M., D.J. Myers, H. Heemstra, D. Curry, D.O. Adams and D.D. Stokke. 2003. Soybean-based adhesive resins and composite products utilizing such adhesives. U.S. Patent No. 6518387B2.

Lagrange, V., D. Whitsett and C. Burris. 2015. Global market for dairy proteins. J. Food Sci. 80: 16–22.

Laucks, I.F. and C.N. Cone. 1930. Process of manufacture of glue and the product thereof. U.S. Patent No. 1757805A.

Lei, H., G. Du, Z. Wu, X. Xi and Z. Dong. 2014. Cross-linked soy-based wood adhesives for plywood. Int. J. Adhes. Adhes. 50: 199–203.

Lei, H., A. Pizzi, P. Navarrete, S. Rigolet, A. Redl and A. Wagner. 2010. Gluten protein adhesives for wood panels. J. Adhes. Sci. Technol. 24: 1583–1596.

Lestari, D., W.J. Mulderb and J.P.M. Sandersa. 2011. Jatropha seed protein functional properties for technical applications. Biochem. Eng. J. 53: 297–304.

Li, J., X. Li, J. Li and Q. Gao. 2015. Investigating the use of peanut meal: A potential new resource for wood adhesives. RSC Adv. 5: 80136–80141.

Li, K. 2007. Formaldehyde free lignocellulosic adhesives and composites made from the adhesives. U.S. Patent No. 7252735B2.

Li, K. and Y. Liu. 2008. Modified protein adhesives and the lignocellulosic composites made from the adhesives. U.S. Patent No. 7393930B2.

Li, K., S. Peshkova and X. Geng. 2004. Investigation of soy protein-kymene® adhesive systems for wood composites. J. Am. Oil Chem. Soc. 81: 487–491.

Li, N., G. Qi, X.S. Sun, M.J. Stamm and D. Wang. 2012a. Physicochemical properties and adhesion performance of canola protein modified with sodium bisulfite. J. Am. Oil Chem. Soc. 89: 897–908.

Li, N., G. Qi, Z.X. Sun and D. Wang. 2012b. Effects of sodium bisulfite on the physicochemical and adhesion properties of canola protein fractions. J. Polym. Environ. 20: 905–915.

Li, N., Y. Wang, M. Tilley, S.R. Bean, X. Wu, X.S. Sun and D. Wang. 2011. Adhesive performance of sorghum protein extracted from sorghum DDGS and flour. J. Polym. Environ. 19: 755–765.

Licari, J.L. and D.W. Swanson. 2011. Adhesives technology for electronic applications: Materials, processing, reliability. Elsevier, MA, USA.

Lin, C.S.K., L.A. Pfaltzgraff, L. Herrero-Davila, E.B. Mubofu, S. Abderrahim, J.S. Clark, A.A. Koutinas, N. Kopsahelis, K. Stamatelatou, F. Dickson, S. Thankappan, Z. Mohamed, R. Brocklesby and R. Luque. 2013. Food waste as a valuable resource for the production of chemicals, materials and fuels. Current situation and global perspective. Energy Environ. Sci. 6: 426–464.

Lipinsky, E.S. 1981. Chemicals from biomass: Petrochemical substitution options. Science 212: 1465–1471.

Liu, C.-C., A.M. Tellez-Garay and M.E. Castell-Perez. 2004. Physical and mechanical properties of peanut protein films. LWT—Food Sci. Technol. 37: 731–738.

Liu, Y. and K. Li. 2007. Development and characterization of adhesives from soy protein for bonding wood. Int. J. Adhes. Adhes. 27: 59–67.

Liu, Y. and K. Li. 2002. Chemical modification of soy protein for wood adhesives. Macromol. Rapid Commun. 23: 739–742.

Lorenz, L., C.R. Frihart and J.M. Wescott. 2006. Analysis of soy flour/phenol-formaldehyde adhesives for bonding wood. pp. 501. Wood adhesives 2005. Nov 2–5, 2005, San Diego, California, USA.

MarketsandMarkets. 2015a. Bioadhesive market by type (plant based, and animal based), by application (packaging & paper, construction, wood, personal care, medical, and others) - global forecast to 2019. MarketsandMarkets report. marketsandmarkets.com.

MarketsandMarkets. 2015b. Renewable chemicals market - alcohols (ethanol, methanol), biopolymers (starch blends, regenerated cellulose, PBS, bio-PET, PLA, PHA, bio-PE, and others), platform chemicals & others - global trends & forecast to 2020. MarketsandMarkets report. marketsandmarkets.com.

MarketsandMarkets. 2015c. Soy protein ingredients market by type, application, and by region - global trends & forecast to 2020. MarketsandMarkets report. marketsandmarkets.com.

Marquie, C. 2001. Chemical reactions in cottonseed protein cross-linking by formaldehyde, glutaraldehyde, and glyoxal for the formation of protein films with enhanced mechanical properties. J. Agric. Food Chem. 49: 4676–4681.

Marti, D.L., R.J. Johnson and K.H. Mathews. 2011. Where's the (not) meat? - byproducts from beef and pork production. LDPM: Livestock, dairy, and poultry outlook (LDP-M-209-01). Economic Research Service, United States Department of Agriculture, Washington, DC, USA.

Mecking, S. 2004. Nature or petrochemistry?—Biologically degradable materials. Angew. Chem. Int. Ed. 43: 1078–1085.

Mekonnen, T.H., P.G. Mussone, P. Choi and D.C. Bressler. 2015. Development of proteinaceous plywood adhesive and optimization of its lap shear strength. Macromol. Mater. Eng. 300: 198–209.

Mekonnen, T.H., P.G. Mussone, P. Choi and D.C. Bressler. 2014. Adhesives from waste protein biomass for oriented strand board composites: Development and performance. Macromol. Mater. Eng. 299: 1003–1012.

Mekonnen, T.H., P.G. Mussone, N. Stashko, P.Y. Choi and D.C. Bressler. 2013a. Recovery and characterization of proteinaceous material recovered from thermal and alkaline hydrolyzed specified risk materials. Process Biochem. 48: 885–892.

Mekonnen, T., P. Mussone, K. Alemaskin, L. Sopkow, J. Wolodko, P. Choi and D. Bressler. 2013b. Biocomposites from hydrolyzed waste proteinaceous biomass: Mechanical, thermal and moisture absorption performances. J. Mater. Chem. A 1: 13186–13196.

Mekonnen, T., P. Mussone, N. El-Thaher, P.Y.K. Choi and D.C. Bressler. 2013c. Thermosetting proteinaceous plastics from hydrolyzed specified risk material. Macromol. Mater. Eng. 298: 1294–1303.

Minford, J.D. 1991. Treatise on adhesion and adhesives. CRC Press, Boca Raton, FL, USA.

Mo, X., J. Hu, X.S. Sun and J.A. Ratto. 2001. Compression and tensile strength of low-density straw-protein particle board. Ind. Crops Prod. 14: 1–9.

Mo, X. and X.S. Sun. 2013. Soy proteins as plywood adhesives: Formulation and characterization. J. Adhes. Sci. Technol. 27: 2014–2026.

Montano-Leyva, B., G.G.D. da Silva, E. Gastaldi, P. Torres-Chávez, N. Gontarda and H. Angellier-Coussy. 2013. Biocomposites from wheat proteins and fibers: Structure/mechanical properties relationships. Ind. Crops Prod. 43: 545–555.

Morr, C.V. and E.Y.W. Ha. 1993. Whey protein concentrates and isolates: Processing and functional properties. CRC Crit. Rev. Food Sci. 33: 431–476.

MVO. 2011. Fact sheet soy. Product Board MVO, Netherlands.

Nasseri, A.T., S. Rasoul-Amini, M.H. Morowvat and Y. Ghasemi. 2011. Single cell proteins: Production and processes. J. Am. Food Technol. 6: 103–116.

Nicholson, C. 1991. History of adhesives. Educational services committee (ESC) report. Bearing Specialists Association (BSA), IL, USA.

Nordqvist, P., E. Johansson, F. Khabbaz and E. Malmström. 2013. Characterization of hydrolyzed or heat treated wheat gluten by SE-HPLC and ^{13}C NMR: Correlation with wood bonding performance. Ind. Crops Prod. 51: 51–61.

Nordqvist, P., F. Khabbaz and E. Malmströma. 2010. Comparing bond strength and water resistance of alkali-modified soy protein isolate and wheat gluten adhesives. Int. J. Adhes. Adhes. 30: 72–79.

Nordqvist, P., M. Lawther, E. Malmström and F. Khabbaz. 2012a. Adhesive properties of wheat gluten after enzymatic hydrolysis or heat treatment—A comparative study. Ind. Crops Prod. 38: 139–145.

Nordqvist, P., D. Thedjil, S. Khosravi, M. Lawther, E. Malmström and F. Khabbaz. 2012b. Wheat gluten fractions as wood adhesives—glutenins versus gliadins. J. App. Polym. Sci. 123: 1530–1538.

[NRC] National Research Council. 1994. Nutrient requirements of poultry: Nineth revised edition. National Academy Press, Washington, D.C., USA.

Opio, C., P. Gerber, A. Mottet, A. Falcucci, G. Tempio, M. MacLeod, T. Vellinga, B. Henderson and H. Steinfeld. 2013. Greenhouse gas emissions from ruminant supply chains—A global life cycle assessment. Food and Agriculture Organization of the United Nations (FAO), Rome, Italy.

Pan, Z., A. Cathcart and D. Wang. 2005. Thermal and chemical treatments to improve adhesive property of rice bran. Ind. Crops Prod. 22: 233–240.

Parashar, A., Y. Jin, B. Mason, M. Chae and D.C. Bressler. 2016. Incorporation of whey permeate, a dairy effluent, in ethanol fermentation to provide a zero waste solution for the dairy industry. J. Dairy Sci. 99: 1859–1867.

Park, S.K., D.H. Bae and N.S. Hettiarachchy. 2000a. Protein concentrate and adhesives from meat and bone meal. J. Am. Oil Chem. Soc. 77: 1223–1227.

Park, S.K., D.H. Bae and K.C. Rhee. 2000b. Soy protein biopolymers crosslinked with glutaraldehyde. J. Am. Oil Chem. Soc. 77: 879–884.

Particle Sciences. 2009. Protein structure. Particle Sciences—Technical Brief. Volume 8. Particle Sciences, Bethlehem, PA, USA.

Potts, J., M. Lynch, A. Wilkings, G. Huppé, M. Cunningham and V. Voora. 2014. Soybean market. pp. 253. *In*: R. Ilnyckyj, D. Holmes and E. Rickert (eds.). The state of sustainability initiatives review 2014. International Institute for Sustainable Development, Winnipeg, MB, Canada.

Qi, G., N. Li, D. Wang and X.S. Sun. 2013a. Adhesion and physicochemical properties of soy protein modified by sodium bisulfite. J. Am. Oil Chem. Soc. 90: 1917–1926.

Qi, G., N. Li, D. Wang and X.S. Sun. 2012. Physicochemical properties of soy protein adhesives obtained by *in situ* sodium bisulfite modification during acid precipitation. J. Am. Oil Chem. Soc. 89: 301–312.

Qi, G., N. Li, D. Wang and X.S. Sun. 2013b. Physicochemical properties of soy protein adhesives modified by 2-octen-1-ylsuccinic anhydride. Ind. Crops Prod. 46: 165–172.

Qi, G. and X.S. Sun. 2011. Soy protein adhesive blends with synthetic latex on wood veneer. J. Am. Oil Chem. Soc. 88: 271–281.

Qin, Z., Q. Gao, S. Zhang and J. Li. 2013. Glycidyl methacrylate grafted onto enzyme-treated soybean meal adhesive with improved wet shear strength. BioResources 8: 5369–5379.

Qu, P., H. Huang, G. Wu, E. Sun and Z. Chang. 2015. The effect of hydrolyzed soy protein isolate on the structure and biodegradability of Urea–Formaldehyde adhesives. J. Adhes. Sci. Technol. 29: 502–517.

Radley, J.A. 1976. Industrial uses of starch and its derivatives. Springer, Dordrecht, Netherlands.

Ross, J.H. and C.D. Ross. 1876. Improvement in process of preparing glue. U.S. Patent No. 183024A.

Sarkki, M.-L. 1979. Food uses of wheat gluten. J. Am. Oil Chem. Soc. 56: 443–446.

Schnepp, Z. 2013. Biopolymers as a flexible resource for nanochemistry. Angew. Chem. Int. Ed. 52: 1096–1108.

Shurtleff, W. and A. Aoyagi. 2004. History of soy flour, grits, flakes, and cereals- soy blends- part 1. Anonymous (eds.). History of soybeans and soyfoods, 1100 B.C. to the 1980s. Soyinfo Center, California, USA.

Sibt-e-Abbas, M., M.S. Butt, M.T. Sultan, M.K. Sharif, A.N. Ahmad and R. Batool. 2015. Nutritional and functional properties of protein isolates extracted from defatted peanut flour. Int. Food Res. J. 22: 1533–1537.

Silverman, H.G. and F.F. Roberto. 2007. Understanding marine mussel adhesion. Marine Biotechnol. 9: 661–681.

Smithers, G.W. 2008. Whey and whey proteins—from 'Gutter-to-gold'. Int. Dairy J. 18: 695–704.

Southward, C.R. 1998. Casein products. J.E. Packer, J. Robertson and H. Wansbrough (eds.). Chemical processes in New Zealand. New Zealand Institute of Chemistry, New Zealand.

Srividya, A.R., V.J. Vishnuvarthan, M. Murugappan and P.G. Dahake. 2013. Single cell protein- A review. International Journal for Pharmaceutical Research Scholars 2: 472–485.

Sun, S., D. Wang, Z. Zhong and G. Yang. 2008. Adhesives from modified soy protein. U.S. Patent No. 7416598B2.

Sun, X. and K. Bian. 1999. Shear strength and water resistance of modified soy protein adhesives. J. Am. Oil Chem. Soc. 76: 977–980.

Swisher, K. 2014. Pressure is on US supplies, prices, and exports. Render. National Renders Association, Camino, CA, USA.

Swisher, K. 2013. US rendering: A $10 billion industry. Render. National Renders Association, Camino, CA, USA.

Tan, S.H., R.J. Mailer, C.L. Blanchard and S.O. Agboola. 2011. Canola proteins for human consumption: Extraction, profile, and functional properties. J. Food Sci. 76: 16–28.

Tanford, C. 1968. Protein denaturation. pp. 121–283. *In:* C.B. Anfinsen, M.L. Anson, J.T. Edsall and F.M. Richard (eds.). Advances in Protein Chemistry. Academic Press, Inc, New York, USA.

Truax, T.R. 1929. The gluing of wood. US Department of Agriculture, Department Bulletin No. 1500 1–78.

Tunick, M.H. 2008. Whey protein production and utilization: A brief history. pp. 1–13. *In:* C.I. Onwulata and P.J. Huth (eds.). Whey pProcessing, Functionality and Health Benefits. Wiley-Backwell, Iowa, USA.

[UNEP] United Nations Environment Programme. 2011. Global partnership on waste management. Waste agricultural biomass to energy. United Nations Environment Programme, Division of Technology, Industry and Economics. International Environmental Technology Centre, Osaka, Japan.

[USB] United Soybean Board. 2016. Soy products guide. United Soybean Board, Chesterfield, MO, USA.

[USB] United Soybean Board. 2015. Soy products guide. United Soybean Board, Chesterfield, MO, USA.

[USB] United Soybean Board. 2012. Soy-based wood adhesives. United Soybean Board, Chesterfield, MO, USA.

[USDA] United States Department of Agriculture. 2015a. National nutrient database for standard reference release 28. National Agriculture Library, Agricultural Research Services, United States Department of Agriculture. Accessed 05/02/2016. https://ndb.nal.usda.gov/.

[USDA] United States Department of Agriculture. 2015b. World agriculture production. Circular series WAP. Foreign Agricultural Service, United States Department of Agriculture, Washington, D.C., USA.

[USDA] United States Department of Agriculture. 1951. Marketing potential for oilseed protein materials in industrial uses. United States Department of Agriculture, Washington, D.C., USA.

[USSEC] United States Soybean Export Council. 2008. Soy protein concentrate for aquaculture feeds. US Soybean Export Council, St. Louis, MO, USA.

van der Leeden, M.C., A.A.C.M. Rutten and G. Frens. 2000. How to develop globular proteins into adhesives. J. Biotechnol. 79: 211–221.

Varnell, D.F., B.K. Spraul and M.A. Evans. 2013. Adhesive compositions. U.S. Patent No. 8399544B2.

Vnucec, D., A. Gorsek, A. Kutnar and M. Mikuljan. 2015. Thermal modification of soy proteins in the vacuum chamber and wood adhesion. Wood Sci. Technol. 49: 225–239.

Vonderwell, R., E. Fritz and N. Metz. 2015. Alkaline hydrolysis of organic waste including specified risk materials and effluent disposal by mixing with manure slurry. U.S. Patent No. 20150107319A1.

Wang, C., J. Wu and G.M. Bernard. 2014. Preparation and characterization of canola protein isolate–poly(glycidyl methacrylate) conjugates: A bio-based adhesive. Ind. Crops Prod. 57: 124–131.

Wang, G., T. Zhang, S. Ahmad, J. Cheng and M. Guo. 2013. Physicochemical and adhesive properties, microstructure and storage stability of whey protein-based paper glue. Int. J. Adhes. Adhes. 41: 198–205.

Wang, Y., X. Mo, X.S. Sun and D. Wang. 2007. Soy protein adhesion enhanced by glutaraldehyde crosslink. J. App. Polym. Sci. 104: 130–136.

Wang, W.H., X.P. Li and X.Q. Zhang. 2010. Soy flour adhesive modified with urea, citric acid and boric acid. Pigm. Resin Technol. 39: 223–227.

Webb, D. 2011. Soyfoods made easy–A soy primer. Today's dietitian. Great Valley Publishing Company Inc., Spring City, PA, USA.

Wescott, J.M. and M.J. Birkeland. 2010. Stable soy/urea adhesives and methods of making same. U.S. Patent No. 20100069534A1.

Wescott, J.M. and M.J. Birkeland. 2008. Stable adhesives from urea-denatured soy flour. U.S. Patent No. 20080021187A1.

Wescott, J.M., C.R. Frihart and A.E. Traska. 2006. High soy-containing water-durable adhesives. J. Adhes. Sci. Technol. 20: 859–873.

Wu, H., Q. Wang, T. Ma and J. Ren. 2009. Comparative studies on the functional properties of various protein concentrate preparations of peanut protein. Food Res. Int. 42: 343–348.

Xiao, Z., Y. Li, X. Wu, G. Qi, N. Li, K. Zhang, D. Wang and X.S. Sun. 2013. Utilization of sorghum lignin to improve adhesion strength of soy protein adhesives on wood veneer. Ind. Crops Prod. 50: 501–509.

Xu, H.N., Q.Y. Shen, X.K. Ouyang and L.Y. Yaing. 2012. Wetting of soy protein adhesives modified by urea on wood surfaces. Eur. J. Wood Wood Prod. 70: 11–16.

Yang, I., M. Kuo and D.J. Myers. 2006. Bond quality of soy-based phenolic adhesives in southern pine plywood. J. Am. Oil Chem. Soc. 83: 231–237.

Yu, J., M. Ahmedna and I. Goktepe. 2007. Peanut protein concentrate: Production and functional properties as affected by processing. Food Chem. 103: 121–129.

Zhang, X., P. Hoobin, I. Burgar and M.D. Do. 2006. Chemical modification of wheat protein-based natural polymers: Cross-linking effect on mechanical properties and phase structures. J. Agric. Food Chem. 54: 9858–9865.

Zhang, Z. and Y. Hua. 2007. Urea-modified soy globulin proteins (7S and 11S): Effect of wettability and secondary structure on adhesion. J. Am. Oil Chem. Soc. 84: 853–857.

Zhang, S., X. Liu, J. Zhang and J. Li. 2011. A novel formaldehyde-free adhesive from jatropha curcas press-cake. Adv. Mat. Res. 236-238: 1549–1553.

Zhao, Z., Z. Gao, W. Wang and M. Guo. 2011. Formulation designs and characterisations of whey-protein based API adhesives. Pigm. Resin Technol. 40: 410–417.

Zhong, Z., X.S. Sun, X. Fanga and J.A. Ratto. 2002. Adhesive strength of guanidine hydrochloride modified soy protein for fiberboard application. Int. J. Adhes. Adhes. 22: 267–272.

Zhong, Z., X.S. Sun and D. Wang. 2007. Isoelectric pH of Polyamide–Epichlorohydrin modified soy protein improved water resistance and adhesion properties. J. App. Polym. Sci. 103: 2261–2270.

Zhong, Z., X.S. Sun, D. Wang and J.A. Ratto. 2003. Wet strength and water resistance of modified soy Protein Adhesives and effects of drying treatment. J. Polym. Environ. 11: 137–144.

2

Adhesion Properties of Soy Protein Subunits and Protein Adhesive Modification

Guangyan Qi,[1] Ningbo Li,[2] Xiuzhi Susan Sun[1] and Donghai Wang[3,*]

ABSTRACT

Environmental concerns and the need for sustainable development have stimulated investigations of plant protein-based adhesives. Soy protein, which has been used for wood adhesives for centuries, is a promising biodegradable adhesive with great potential as an alternative to current synthetic petroleum-based adhesives. This chapter contains a review of soy protein subunit characterization and their adhesion properties when used as wood adhesives. Glycinin was found to be the main contributor to adhesion strength of soy protein. The basic and β subunits with larger portions of hydrophobic amino acid were shown to have greater adhesion properties than their respective subunits, acidic and $\alpha' \alpha$. Soy latex adhesives with high protein content (33%–38%), good flowability, long storage life and good adhesion performance can be produced by *in situ* sodium bisulfite modification in soy flour-water extract. Protein modification uses physical and chemical methods to improve adhesion

[1] Biomaterials and Technology Laboratory, Department of Grain Science and Industry, Kansas State University, Manhattan, KS 66506.

[2] Sunhai Bioadhesive Technologies LLC, 2005 Research Park Cir, Manhattan, KS 66502.

[3] Department of Biological and Agricultural Engineering, Kansas State University, Manhattan, KS 66506.

* Corresponding author: dwang@ksu.edu

properties by altering protein molecular structure and conformation. This chapter also describes current technologies for soy protein adhesion improvement, including cross-linking, protein blends (synthetic adhesives, lignin, wet-strength resin), enzymatic treatment, and nano technology.

Introduction

Soy protein-based adhesives have been used as wood adhesives for more than a century. The first commercial soy protein adhesive for plywood production was used in the 1930s (Lambuth, 1989). Due to many unique features such as low cost, ease of handling, cured by either hot or cold pressing conditions, and easily modified functional groups, soy protein based adhesives were widely used for wood composite production in the United States from the 1930s to the 1960s. However, the bonding strength and water resistance of soy adhesives are much lower than synthetic urea formaldehyde (UF) and phenol formaldehyde (PF) resins. Besides, the low cost of synthetic resins have caused them to dominate the wood adhesive market since World War II. Now with the increased environmental concerns and regulations, there has been a resurgence of interest in the development of biodegradable soy wood adhesives in the last two decades, especially with the emphasis on environmentally acceptable emissions with reduced formaldehyde.

The United States is the world's leading soybean producer and exporter, accounting for 34% of worldwide soybean production with 91.4 million metric tons in 2013 and 108.0 million metric tons in 2014 (Soystats, 2015). Farmers in more than 30 states in the United States grow soybeans, and they are the second largest crop in cash sales and foremost value crop export (Fig. 1) (Soystats, 2015). Bulk soybean flour, which contains 44%–50% protein, is typically the source of wood adhesives. Soy protein in the form

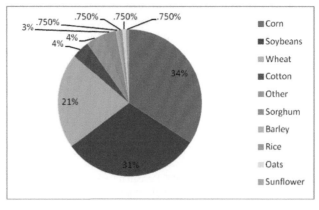

Figure 1. U.S. land use for crop production in 2014 (Soystats, 2015).

of soy flour is used primarily for animal feed and food applications; only a small portion of soy protein is currently used for non-food applications. Wood adhesives are potentially a huge market for the oversupplied soybean (Ren and Soucek, 2014).

Soybean protein is primarily composed of 7S and 11S storage protein, and each fraction has distinct subunits: α' α, β subunit for 7S; acidic, basic subunits for 11S. Each subunit demonstrates unique native structures and molecular properties due to variation in amino acid composition and polypeptide chain arrangement, resulting in unique contributions to soy protein adhesiveness. This chapter reviews the effects of soy protein subunits on adhesion properties of soy protein adhesives. An innovative soy protein latex adhesive with high protein content (33%–38%), good flowability, long storage life, and good adhesion strength could be produced by manipulating the $NaHSO_3$ concentration in soy flour-water extract and adjusting the fraction pH. In addition, protein modification is designed to improve its adhesion properties by altering protein molecular structure or conformation using physical and chemical methods. This chapter also summarizes current technologies that improve adhesion properties of soy protein adhesive, including cross-linking, protein blends (synthetic adhesives, lignin, and wet-strength resin), enzymatic treatments, and nano technology, etc. Soy protein has great potential to produce adhesives with high bonding strength and improved water resistance.

Soy Protein Chemistry and Protein Subunits Isolation

Soy Protein Chemistry

A mature soybean, when dry, consists of 38% protein, 30% carbohydrate, 18% oil, and 14% moisture, ash and hull. Soybeans are typically cleaned, cracked, dehulled, and rolled into soy flakes, during which the oil cells are ruptured for efficient oil extraction using hexane solvents. After oil removal, the remaining defatted soy flakes are processed into a variety of soy protein products, such as soy flour, soy protein isolates, soy protein concentrates, etc. Besides the food and feed applications, soy protein is also considered to be one of the most promising adhesive alternatives to petroleum-based adhesives because soy protein is abundant, renewable, easy to handle, and suitable for cold or hot press.

Approximately 90% of seed proteins in a soybean are globulins that can be extracted with dilute salt solutions or water (Nielsen, 1974). Based on sedimentation coefficients obtained by ultracentrifuge or gel filtration, soy protein contains four major components, designated as 2S, 7S, 11S, and 15S (Hou and Chang, 2004). The 7S (conglycinin) and 11S (glycinin) fractions represent the dominant proteins, accounting for 30%–50% and 20%–30% of total protein, respectively. The 7S globulins contain three major fractions: β-conglycinin, γ-conglycinin, and basic 7S globulin, of which β-conglycinin is the main component, with 30%–50% of the total seed protein (Hou and Chang, 2004). Soy protein subunits are often determined using a gel electrophoresis method (sodium dodecyl sulfate-polyacrylamide gel electrophoresis). An SDS-PAGE picture of soy protein, β-conglycinin, and glycinin is shown in Fig. 2 (Mo et al., 2006; 2011).

β-conglycinin (140–170 kDa) is a trimeric glycoprotein consisting of three types of subunits: α' (MW 57–72 kDa), α (57–68 kDa), and β (42–52 kDa) (Thanh and Shibasaki, 1977). All three subunits are rich in glutamate, aspirate, leucine, and arginine but very low in methionine and cystine. The β subunit is reported to have higher contents of hydrophobic amino acids than α α'. The β-conglycinin is a trimeric structure in which subunits are associated by noncovalent bonds such as hydrophobic interaction and hydrogen bonding without any disulfide bonds (Thanh and Shibasaki, 1977; 1978). The 11S globulin is a very heterogeneous oligomeric protein (340–375 kDa) (Utsumi et al., 1981) consisting of six subunits formed into a hexamer structure. Each subunit contains an acidic polypeptide (37–45 kDa; pI = 4.2–4.8) and basic polypeptide (18–20 kDa; pI = 8.0–8.5) that are linked by disulfide bonds ($A_{1a}B_{1b}$, A_2B_{1a}, $A_{1b}B_2$, $A_5A_4B_3$, and A_3B_4) (Badley et al., 1975). The acidic and basic polypeptides alternate in the hexagonal layer and are held together by hydrophobic and disulfide bonds, forming three subunits. The two identical hexagonal layers are held together by electronic interaction and hydrogen bonding to form one glycinin molecule (Peng et al., 1984). Acidic subunits have high contents of glutamic acid, proline, and half cystine, while basic subunits have high

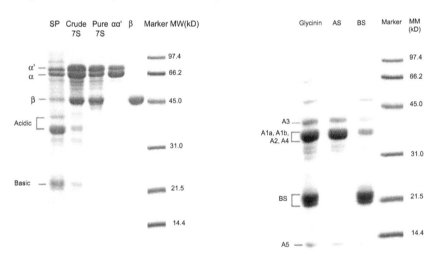

Figure 2. Soy protein fractions determined using SDS-PAGE method. Adapted from Mo et al. (2006; 2011).

hydrophobic amino acid content (Leu, Ala, Val, Tyr, and Phe) (Catsimpoolas et al., 1971). The inherent difference in native structure and molecular properties of soy protein subunits contribute to variations in their functional properties such as thermal stability, water-holding capacity, hydrophobicity, gelling properties, and adhesion strength, etc. Therefore, fractionation of soy protein subunits with high purity and yield is of interest to researchers.

Soy Protein Isolation and Fractionation

Soy proteins are typically presented as soy protein concentrates and soy protein isolates (SPI), and they are traditionally extracted from soy flour using acid precipitation (pI = 4.5–4.8). Soy protein concentrates are isolated by removing soluble carbohydrates and low molecular weight proteins using an ethanol/water wash (Hettiarachchy and Kalapathy, 1999). SPI, the most refined forms of soy proteins, are isolated by removing insoluble and soluble carbohydrates along with some protein through a series of steps, containing 90% or more protein (Wolf, 1970). Defatted soy flakes or flours are extracted with water at pH of 7 or 8.5, insoluble polysaccharides and fibers are centrifuged out, and then pH of supernatant is adjusted to about pH 4.5 to predicate the protein. Discarded supernatant contains sugars, ash, and minor 2S and 7S soluble proteins (Iwabuchi and Yamauchi, 1987).

Thanh and Shibasaki (1976) developed a straightforward fractionation of 7S and 11S globulins at pH 6.1–6.6 based on their different solubility. 11S globulin was precipitated when the pH of the extract of soybean meal was adjusted to 6.4; the 7S globulin was separated from whey protein by adjusting the pH to 4.8. Factors affecting precipitation (i.e., pH, Tris concentration, protein concentration, and ionic strength) were also optimized for simultaneous fractionation of 7S and 11S globulins. However, cross contamination of proteins in both globulin fractions was a hindrance to protein purification. The glycinin fraction was 79% glycinin, 6% conglycinin, and 15% other components. The purity of the β-conglycinin fraction was only 52%; it contained 3% glycinin, and 45% other components. Gel filtration and affinity chromatography were required to purify these two fractions, which are costly and difficult to scale up. Results of previous studies have shown that reducing agents prevented co-precipitation of glycinin and β-conglycinin, thereby improving the fraction's purity (Thiering et al., 2001; Wolf, 1993). Reducing agents for protein fractionation aids are any sulfite compound that yields SO_2 in solution, such as $NaHSO_3$, glutathione, cysteine, and β-mercaptoethanol, believed to reduce disulfide bonds and avoid co-precipitation (Deak et al., 2006).

In order to improve the purity of each globulin protein fraction, several researchers have investigated three pI precipitations. O'Keefe et al. (1991) modified the conventional method by replacing the two isoelectric precipitations at pH 6.4 and 4.8 with three precipitations at pH 6.4, 5.3, and 4.8. Nagano et al. (1992) and Bogracheva et al. (1996) also obtained three fractions at isoelectric pH 6.4, 5.0 (5.5), and 4.8 instead of two. Therefore, one more soy protein fraction composed of a mixture of glycinin and conglycinin was obtained in the pH 5.5, 5.3, or 5.0. Nagano et al. (1992) claimed that the purities of crude glycinin and β-conglycinin were >90% but that those purities came at the expense of yield. Wu et al. (1999) and Rickert et al. (2004) utilized three-step fractionation in the pilot scale and optimized the processing parameters (pH, temperature, water-to-flake ratio) to yield high purity conglycinin fractions; the purities of both globulins were comparable to those of the laboratory-scale process. Qi et al. (2011) obtained the β-conglycinin fraction with purity of 88.5% using a simple extraction procedure. The pH of the soy flour-water extract was adjusted to 9.5 and stabilized for 2 hrs. The slurry pH was then adjusted to a pH series (5.1, 5.4, 5.6, 5.8, 6.0, and 6.4), and carbohydrates and some glycinin proteins were removed by centrifuge. The supernatant pH was then adjusted to 4.8 in order to acidify the soy protein. The authors obtained pure β-conglycinin with fractionation pH of 5.1–5.4; the purity can reach 86%–88% but with only approximately 8% yield. This pure β-conglycinin demonstrated unique viscoelastic properties with potential for applications such as gluten-free baking products, chewing gum, and edible film.

Adhesive Performance of Soy Protein-based Adhesives

Protein Composition and Adhesion Properties

Adhesion properties of several types of soy protein subunits have been investigated, including 7S globulin (conglycinin), 11S globulin (glycinin), a mixture of 7S and 11S globulins with various ratios, or 7S subunits (α' α, and β) and 11S subunits (acidic, basic). Mo et al. (2004) isolated 7S and 11S globulins according to the method of Thanh and Shibasaki (1976) and then studied their adhesion properties on cherry wood. Results showed that glycinin had greater adhesion performance than conglycinin. For example, wet strength from glycinin was about 50% higher than the wet strength of the SPI and about 35% higher than the wet strength of conglycinin. Because glycinin is known to have a higher molecular weight (M.W. 350 kDa) than conglycinin (150–200 kDa), and native glycinin potentially contains more active chemical groups available for bonding, resulting in high shear strength. However, for 7S and 11S mixture adhesives, gluing strength and water resistance were significantly improved for sodium dodecyl sulfate (SDS) and urea-modified proteins with higher 7S globulin amounts. The opposite behavior was observed for proteins with an increased portion of 11S globulins. Adhesion performance of soybean protein adhesives is determined primarily by their structure. High degrees of denaturation usually lead to high adhesive strength. Soy proteins with high 11S content have a more ordered structure; therefore, they showed a lower degree of denaturation and gave relatively limited improvements in adhesive strength for modified proteins. Therefore, the conclusion was made that chemical modification with denaturants can significantly improve adhesive strength and water resistance of soy proteins rich in 7S globulins.

The authors further fractionated glycinin into acidic and basic subunits with estimated purity of 90 and 85%, respectively (Mo et al., 2006). Dry strength of the basic subunits was higher than the acidic subunits throughout a pH range from 2 to 12. Wet strength of basic subunits was 160% more than that of acidic subunits prepared at their respective pIs. Basic subunits have similar wet and soaked strength to those of glycinin protein, suggesting that water resistance of glycinin is mainly contributed by basic subunits. High hydrophobic amino acid content (47.1%) found in basic subunits in glycinin (36.6%) was considered to be the primary contributor to water resistance of the wood adhesive. The a′a and β subunits were also separated from purified β-conglycinin protein (Mo et al., 2011). According to the hydrophobic amino acid composition, the β subunit is more hydrophobic than the a′a subunit. Transmission electron microscopy showed that the β subunits exist as spherical hydrophobic clusters, whereas a′a subunits exist as uniformly discrete particles at pH 5.0. Adhesives made from β subunits showed greater water resistance compared to adhesives from a′a subunits and β-conglycinin. Both a′a and β subunits showed similar dry adhesion strength, but their cohesive wood failures were 70% and 95%, respectively, suggesting that the β subunit has better adhesive cohesiveness. Results indicated that proteins with higher hydrophobic amino acid content can promote protein-protein associations and could be favorable for producing adhesives with increased water resistance.

Latex-like Adhesives

Several studies have revealed that sodium bisulfite (NaHSO$_3$) modification has negative or insignificant effects on adhesion properties of soy protein adhesives (SPI, glycinin, conglycinin) (Kalapathy et al., 1996; Zhang and Sun, 2008; Zhang and Sun, 2010). It is reported that the presence of RS-SO$_3^-$ groups, converted from R-SH groups on protein molecules, could decrease the effective interfacial area and enhance electrostatic repulsion between the protein and a wood surface, resulting in decreased adhesive strength (Kalapathy et al., 1996). However, Sun et al. (2008) exploited the *in situ* NaHSO$_3$ modification on soy protein in soy flour-water extracts and then precipitated the modified soy protein with high solid content at the pI. This innovative viscous cohesive soy protein adhesive system advantageously provides high solid protein content, good flowability, long storage life, and high water resistance (Sun et al., 2008; Qi et al., 2012; 2013a).

Extraction conditions such as the $NaHSO_3$ amount and fractionation pH were determined (Qi et al., 2012; 2013a). First, $NaHSO_3$ concentrations from 0 g/L to 16 g/L were used to modify the soy protein in the soy flour-water extracts at pH 9.5, and then the modified soy protein was precipitated at pH 5.4 and pH 4.5 in order to obtain 11S rich fractions (SP 5.4) and 7S rich (SP 4.5) fractions, respectively (Qi et al., 2013a). $NaHSO_3$ introduced additional negative charges on the protein surface and then lowered the isoelectric point of 7S and 11S. The reduced soy protein pI caused the protein composition (7S/11S ratios) of SP 4.5 altered with $NaHSO_3$ amount, as shown in Table 1. Unmodified samples SP 5.4 and SP 4.5 possessed clay-like properties and viscoelastic properties, respectively. The addition of $NaHSO_3$ ranging from 4 to 8 g/L significantly enhanced the rheological properties of the soy protein adhesive: SP 5.4 and SP 4.5 had the viscous cohesive phase with good hand-ability and flowability. Results showed that the balance between hydrophobic interaction and electrostatic force among proteins is critical to the continuous phase of SP 5.4 (11S rich) and that the 7S/11S ratios and the high molecular weight protein aggregates connected by disulfide bonds are closely related to the continuous phase of SP 4.5 (7S rich). The adhesion performance of SP 4.5 was better than SP 5.4, and the wet strength of these two fractions was in the range of 2.5–3.2 MPa compared to 1.6 MPa of the control SPI.

Second, different fractionation pHs between 4.8 and 6.0 were also selected in order to obtain the mixture of 7S and 11S globulins at various ratios. They were extracted directly from soy protein slurries modified with 6 g/L $NaHSO_3$ using the acid precipitation method, which is based on the different solubility of 7S and 11S globulins as mentioned previously (Qi et al., 2012). The flowchart in Fig. 3 illustrates the procedure using the pH 6.0 sample as an example. External morphology of the samples changed from the viscous cohesive phase to the clay-like phase without cohesiveness (Table 2). The 11S rich soy protein displayed viscous cohesive properties in the pH range of 5.4 to 5.8, while the soy protein became a clay-state phase of non-cohesiveness when the pH was less than 5.4. The 7S rich soy protein fractions exhibited viscous cohesive properties

Table 1. Estimated content of polypeptides of soy protein (SP) adhesives modified with $NaHSO_3$ at different concentrations. Adapted from Qi et al. (2013a).

Protein subunits	SP fraction distribution (%)								
	SP 5.4 ($NaHSO_3$ g/L)					SP 4.5 ($NaHSO_3$ g/L)				
	0	2	6	8	16	0	2	6	8	16
7S	19.7	11.8	11.2	10.2	13.7	87.5	66.8	57.7	39.5	15.8
11S	75.9	85.7	85.8	87.7	76.3	3	23.5	34.2	55.9	82.9

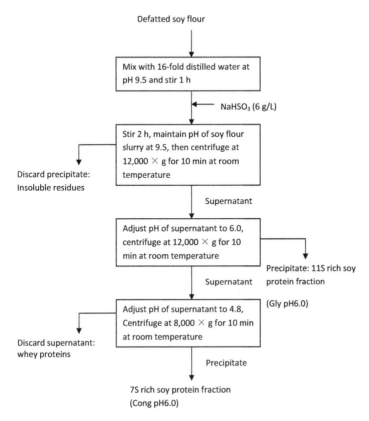

Figure 3. Procedures of NaHSO$_3$-modified soy protein adhesive extraction from soy flour. Adapted from Qi et al. (2012).

for samples Cong pH 5.0, Cong pH 5.2, and Cong pH 5.4, whereas other sample fractionations obtained at pH > 5.4 lost the continuous phase. The soy protein structure was unfolded, and the protein conformation was altered by NaHSO$_3$ modification through the ion pair shielding effect and reducing effects (Babajimopoulos et al., 1983). The re-associated protein could be coalesced upon centrifugation to form a homogeneous and continuous protein phase, exhibited as viscous cohesive substances when the re-formed protein network had the ability to retain the proper amount of water. Results suggested that proper protein-protein interaction, hydration capacity (11S soy protein fractions), and certain ratios of 7S and 11S (7S rich soy protein fractions) in the soy protein sample are crucial to continuous protein phase formation. Hydrogen bonding, electrostatic forces, and hydrophobic interactions are involved in maintaining the protein viscous cohesive network, whereas disulfide bonds do not exert significant effects. The NaHSO$_3$-modified 7S rich adhesives with pH close to pI (4.5) and 5%

Table 2. NaHSO$_3$-modified soy protein adhesive product yield, water content, and external morphology. Adapted from Qi et al. (2012).

Number	Soy products	Yields	Solid content	External morphology
11S rich soy protein adhesives				
1	Gly pH 6.0	9.10%	22%	Diluted, no adhesion
2	Gly pH 5.8	23.00%	28%	Viscous cohesive
3	Gly pH 5.6	43.80%	27%	Viscous cohesive
4	Gly pH 5.4	62.50%	29%	Somewhat clay
5	Gly pH 5.2	67.70%	31%	Clay-state
6	Gly pH 5.0	69.30%	33%	Clay-state
7	Gly pH 4.8	70.00%	34%	Clay-state
7S rich soy protein adhesives				
8	Cong pH 6.0	64%	34%	Clay-state
9	Cong pH 5.8	56.90%	33%	Clay-state
10	Cong pH 5.6	39.20%	32%	Somewhat clay
11	Cong pH 5.4	14.90%	33%	Viscous cohesive
12	Cong pH 5.2	6.54%	33%	Viscous cohesive
13	Cong pH 5.0	1.70%	34%	Viscous cohesive

higher in protein concentration than the 11S-rich adhesive, contributed to better adhesion strength in 7S rich samples.

To date, soy protein adhesives under general investigation have been in the form of protein suspension (freeze-dried soy protein powder mixed with water) with protein concentration ranging from 10% to 15%. Limited protein concentration is attributed to extremely high viscosity with high protein content. However, *in situ* NaHSO$_3$ modification produces the latex-like soy protein adhesives, having high solid content (28%–39%) but good flowability. The high protein content phases were stable for up to several months without phase separation when stored at room temperature (Qi et al., 2012; Qi et al., 2013a).

Effect of pH

The protein pH value has been shown to significantly affect adhesion properties of soy protein, especially wet adhesion strength. Wang et al. (2008) reported that SPI with pH levels at or close to its pI (e.g., pH = 4.5) could result in maximum shear strengths for dry and wet adhesive strengths. For example, wet strength at the pI was approximately 33% greater than those at neutral pH 7.6 (3.45 MPa versus 2.59 MPa). Dry and soaked adhesive strengths were higher at pH levels ranging from 3.6 to 7.6 (7.01–7.41 MPa) than those at pH 8.6 and above (5.04–6.24 MPa). Glycinin and its subunits (acidic and basic subunits) also demonstrated maximum adhesion strength at pH values close to their pI values (Mo et al., 2006). Both β-conglycinin and a'a subunits exhibited maximum wet adhesion strength of 2.0 MPa at pH 5.0, whereas β subunits exhibited maximum wet adhesion strength of 2.4 MPa at pH 8.0. In general, maximum wet adhesion strength was observed at pH levels close to those at which proteins have minimum solubility and are strongly associated with each other.

Proteins commonly exhibit different surface charges, charge distributions, surface structures, and conformations at various pH values, and proteins typically have charged functional groups on their surfaces that make energetically favorable polar interactions with the surrounding water. A protein surface becomes negatively charged at pH levels above its pI and positively charged at pH levels below its pI. At the pI, the net charge on the protein surface is zero, the uncharged molecules will not migrate in an imposed electric field, and hydrophobic interactions increase. Protein tends to have the minimum solubility at the pI (Fennema, 1996; Renkema et al., 2002). Increased hydrophobic interactions and decreased solubility can promote protein associations and are favorable for further protein-protein interactions during the curing process and cause high adhesion strength. At extreme pH levels, such as pH 2.0 or 12.0, increased electrostatic repulsions between proteins weaken protein association. Such changes might increase water-protein interaction instead of protein-protein and protein-wood interaction, leading to reduced adhesive strength.

The adhesion strength of basic subunits was affected by pH to a less degree than the acidic subunits, and the β subunits were affected by pH to a lesser degree than the a'a subunits (Mo et al., 2006; 2011). Because basic and β subunits are more hydrophobic than their respective subunits, at extreme pH, strong hydrophobic interactions may act against electrostatic repulsions and favor protein associations, resulting in stable adhesion strength. In contrast, acidic and a'a subunits have weak protein associations due to dominant electrostatic repulsions and reduced adhesion strength.

Combined with isoelectric pH technology as above, adhesive strength and water resistance of modified soy proteins are significantly improved. Zhong et al. (2007) studied interactions between polyamide-epichlorohydrin (PAE) and soy protein as affected by pH. The isoelectric pH of the PAE-modified soy protein shifted from 4.5 to 5.5, at which point the highest adhesive strength was also observed (3.9 MPa wet strength with 72% wood failure compared to 2.8 MPa with 50% wood failure). PAE and soy protein molecules formed reversible ionic complexation interactions at a pH range of 4–9, which can stabilize the soy protein structure and improve adhesion properties. At pH 5.5, complexation interactions were greatly enhanced, as shown in absorbance profiles and conductivity titration curves. Combined with the fact that PAE-modified SPI also had the lowest net charge at pH 5.5; therefore, the water resistance of soy protein was greatly improved.

Previous studies have used detergent SDS to increase the protein unfolding degree and improve adhesion strength and water resistance of SPI at pH 7.6 (Zhong et al., 2001; Huang and Sun, 2000). However, in the study by Wang et al. (2009), SPI with neutral surface charge demonstrated greater water resistance than SDS-modified SPI at neutral pH and pI, possibly due to extra charges protein complex induced by SDS. Adjusting the protein surface charge was suggested to be a more powerful method of enhancing adhesive performance of soy protein.

Soy Protein Modification

Protein modification is designed to improve its adhesion properties by altering protein molecular structure or conformation using physical and chemical methods. Modification strategies such as cross-linking, denaturation, protein blends (synthetic adhesives, lignin, wet strength resin), enzymatic treatments, and nano technology have been investigated for improving the strength and water resistance of soy protein used for wood adhesives.

Cross-linking

Cross-linking of protein involves joining the two molecular components by a covalent bond between protein and the cross-linking agent. A compact protein complex would be formed, and induce more entanglements and cross-linking during the thermal setting, thereby maintaining their structure better than the unmodified adhesive after water soaking and improving soy protein water resistance. A number of functional groups on amino acid side chains are available for the chemical/cross-linking reaction, such as carboxyl, hydroxyl, amino, disulfide, imidazole, indole, phenolic, and sulfhydryl groups (Feeney and Whitaker, 1985). Use of cross-linking technology to improve water resistance has been demonstrated in previous studies. Rogers et al. (2004) reported that 1, 3-dichloro-2-propanol could induce cross-linking of soy protein through the reaction among functional groups and improve soy protein adhesive performance. Liu and Li (2007) developed modified soy protein adhesives using two-step modification. SPI was first modified by maleic anhydride (MA) to form MA-grafted SPI (MPSI), and then polyethylenimine (PEI) was used to modify MSPI. The optimum formula of modified SPI was made from 20% PEI and 80% MSPI, resulting in a dry strength 6.8 MPa and boiling strength of 1.5 MPa. Qi et al. (2013b) studied the effect of 2-octen-1-ylsuccinic anhydride (OSA) on the adhesion performance of soy protein-based adhesive and found that wet strength increased to 3.2 MPa at 3.5% OSA compared to 1.8 MPa for adhesive without OSA; the wet strength leveled off as OSA concentration increased further. The improvement of water resistance of soy protein-OSA blends was attributed to the fact that OSA was grafted to soy protein molecules through a reaction between amine, hydroxyl groups of protein, and anhydride groups in OSA molecules. 3-aminopropyltriethoxysilane (APTES) with amino-propyl group was reported to react with side chains, such as carboxylic acid and hydroxyl groups, on soy protein molecules, forming a cross-linked structure (Kim and Sun, 2014). Adding 20% APTES greatly improved the dry and wet shear strength on wood due to the cross-linking effects of the soy protein-APTES mixture. The entangled and interwoven polymeric structure based on the cross-linked interface promoted the protein's attachment to the wood surface, consequently leading to improved bonding strength among protein molecules compared to unmodified soy protein-based adhesives. Liu et al. (2015) reported that wet strengths of undecylenic acid (UA)-modified soy protein adhesives improved from 2.04 MPa to 3.03 MPa at UA amounts up to 10% due to chemical grafting between amine groups from protein and carboxyl groups from UA.

Epoxies, which are active cross-linking agents for alkaline soy adhesives, improve adhesive strength and durability (Lambuth, 1989; Huang, 2007). A mixture of ethylene glycol diglycidyl ether (EGDE) and diethylenetriamine (DETA) as the cross-linker was shown to improve pure soy meal-based adhesive's wet bonding strength to plywood from 0.38 MPa to 1.15 MPa (Li et al., 2015). This mechanism proposed by the authors was that DETA reacted with EGDE to form a long chain structure with epoxy groups, which cross-linked the soy protein molecules to form a denser cured adhesive layer and combined with the soy protein molecules to form an interpenetrating network, and consequently improved the water resistance of the adhesive. In the study by Luo et al. (2015a), 5,5-Dimethyl hydantoin polyepoxide (DMHP) contained epoxy groups that reacted with hydroxyl, amino, and carboxyl groups on protein to form three-dimensional cross-linked networks. Adding 13% of DMHP with SDS and polyacrylamide to soy meal significantly improved the bonding strength and reduced the viscosity but the solid content of soy meal-based adhesives increased. DMHP loading level higher than 13% reduced the bonding strength of soy meal-based adhesives due to excessive penetration of adhesives into the wood. Lei et al. (2014) asserted that epoxyresin (EPR), melamine–formaldehyde (MF), and their mixture (EPR+MF) could enhance the water resistance of soy protein-based adhesives due to the chemical reaction between epoxy and –OH and between MF and –NH on the protein. In addition, the cross-linking reaction between soy-based adhesive and formaldehyde or its derivatives was also studied (Wang et al., 2007a; Qi and Sun, 2010). Wang et al. (2007a) reported that wet strength of soy protein improved by 115% at optimum concentration of glutaraldehyde (20 mM).

Zhu and Damodaran (2014) reported that phosphorus oxychloride (POCl$_3$) improved soy flour's bonding strength and water resistance via chemical phosphorylation of soy flour. Particularly, POCl$_3$ reacts with amino and hydroxyl groups on protein or polysaccharides chains. The attached phosphate groups acted as cross-linking agents via covalent esterification with hydroxyl groups on wood chips or via ionic and hydrogen-bonding interactions with functional groups in wood chips. To phosphorylate the soy flour, the soy flour slurry with 15% solid content (w/w) in water at pH 10.5 was stirred for 60 min at 60°C, and then different amount of POCl$_3$ was added under vigorous mixing. The pH of slurry was maintained at 10–10.5 for 1 h until the reaction was completed. Until the temperature decreased to room temperature, the soy flour solution was adjusted to pH 8.0 and then freeze-dried. Phosphorylation of soy meal using POCl$_3$ significantly increased the wet bond strength, and the optimum POCl$_3$: Soy flour ratio was about 0.15 (g/g) to produce maximum wet bond strength. At hot-press temperature above 160°C the wet bond strength of PSF was 2.6 MPa, a level that might be acceptable for interior-used hardwood plywood and particleboard.

Denaturation

The native state of soy protein is typically a folded and coiled biologically active structure, meaning that many hydrophobic and potentially reactive groups are buried inside the protein molecules. In order to improve the water resistance of soy protein-based adhesives, denaturation agents are often used to open the compact structure of protein and expose more hydrophobic functional groups. Urea is a commonly used protein denaturing agent because it interacts actively with hydroxyl groups of soy protein and then breaks down hydrogen bonding, resulting in an unfolded protein structure. Zhang and Hua (2007) reported that wettability and adhesive properties of 7S and 11S improved under 1 M urea modification. Similar to urea, guanidine hydrochloride enhanced protein hydrophobicity and adhesion strength (Huang and Sun, 2000a; Zhong et al., 2002). SDS is the anionic detergent that dissociates the protein by disrupting hydrophobic and electrostatic bonds, potentially moving various interior hydrophobic side chains outward where they can interact with hydrophobic moieties of detergent molecules and form micelle-like regions. Previous studies have shown that 0.5% concentration of SDS could enhance water resistance and adhesion strength of soy protein adhesive (Huang and Sun, 2000b; Mo et al., 2004; Xu et al., 2014).

The glycinin component in soy protein contains 18–20 intermolecular and intramolecular disulfide bonds, while only two disulfide bonds per mole exist in β-conglycinin (Koshiyama, 1972; Kella et al., 1986). The presence of disulfide bonds in glycinin protein molecules significantly contributes to differences in functional properties between glycinin and conglycinin, including structural integrity, protein stability, and thermal properties. Because high viscosity is a drawback that limits application of soy protein-based adhesives, reducing agents, such as sulfites, bisulfites, and sulfates, cleave inter- and intra-disulfide bonds in protein molecules, thus increasing protein molecule flexibility, solubility, and surface hydrophobicity and decreasing viscosity (Babajimopoulos et al., 1983; Kella et al., 1986; Kalapathy et al., 1996). Many previous studies have revealed insignificant or negative effects of Na_2SO_4, Na_2SO_3, and $NaHSO_3$ modification on adhesion properties of soy protein adhesives (Kalapathy et al., 1996; Zhang and Sun, 2008; 2010). Kalapathy et al. (1996) showed that counteracting effects exist between increased hydrophobicity of soy protein and decreased effective wood-protein interfacial area due to the conversion of SH to $-SSO_3$ groups in soy protein consequently adversely affecting adhesion strength and water resistance of soy protein. Zhang and Sun (2008; 2010) also reported that negative effects of $NaHSO_3$ exert adhesiveness of pure glycinin and β-conglycinin globulin, respectively. In the study by Qi et al. (2012; 2013a), *in situ* $NaHSO_3$ modification on protein in soy flour-water slurries produced a latex-like soy protein adhesive that demonstrated slightly decreased adhesion strength as the amount of $NaHSO_3$ increased because negative effects of $NaHSO_3$ (decreased effective wood-protein interfacial area) exceeded the positive effects (increased hydrophobicity) on adhesion strength. However, the latex-like soy protein adhesive had high solid protein content (33%–38%), good flowability, long storage life, and good adhesion strength compared to the traditional SPI adhesive.

Blends with Synthetic Adhesives

Partial replacement of synthetic adhesives with biodegradable adhesives is one way to decrease dependence on fossil resins and improve adhesion performance of soy-based adhesives (Steele et al., 1998; Qi et al., 2011; Qu et al., 2015). Soy protein-based adhesive showed good compatibility with many commercial adhesives and synthetic resins. Qi and Sun (2011) investigated the effect of six commercial synthetic glues on the adhesion performance of soy protein. Cherry wood veneers were bonded with soy protein-commercial mix adhesives in order to evaluate their adhesion performances. By mixing with UF at a ratio of 40:60 (soy protein/UF), the wet strength increased to 6.4 MPa with 100% wood cohesive failure from 4.7 MPa of pure UF and 3.6 MPa of pure soy protein-based adhesive. Loading level of UF less than 60% also improved the wet shear strength compared to pure soy protein-based adhesive. Increasing the loading level of UF higher than 60% did not result in higher wet shear strength due to significant increasing of adhesive blend viscosity, leading to poor diffusion on wood substrates and resulting in decreased adhesion strength. Chemical reaction between soy protein and UF occurred, and a new ester bond was formed and proven by FTIR. Blending soy protein-based adhesive with other commercial adhesives such as Veneer Glue (Constantine's Wood Center, Fort Lauderdale, FL, USA), Heat Lock Iron Veneer Glue and Cold Press Veneer Glue (Atlanta, GA, USA), and Cold Press Light Veneer Glue and Flex-Pro Veneer Adhesive (Veneer Supplies Company, Louisville, KY, USA) showed higher wet shear strength than pure commercial glues, indicating that soy protein-based adhesive has good compatibility with commercial adhesives.

In order to obtain more active groups on protein, soy protein was initially hydrolyzed by sodium hydroxide (0.5 wt%) and then synthesized into UF-soy protein resins (Qu et al., 2015). After sodium hydroxide hydrolyzation, soy protein was added to the UF resins using a series of steps. In the first step, methylolation, formaldehyde was mixed with urea and slowly heated to 90°C at a rate of 1°C. In the second step, polycodensation, ammonium chloride was used to adjust and maintain a pH of 4.5–5.0 for the mixture obtained from the first step until the reaction was complete. In the third step, post-treatment, the mixture harvested from step 2 was adjusted to pH 7.8 using sodium hydroxide, and the urea was loaded after the temperature decreased to 75–80°C and was maintained for 1 h. The hydrolyzed soy protein was added to the mixture at steps 1, 2, and 3 and designated as 5I, 5II, and 5III of adhesives, respectively. Viscosity of 5I, 5II, and 5III significantly decreased, indicating good compatibility of soy protein and UF resins. Three-layer plywood panels were bonded with the adhesives at 1.0 MPa for 20 min or 135°C for 6 min. The blends of soy protein and UF adhesives showed improved bonding strength compared to pure UF or soy protein adhesives. Results showed that soy protein could partially substitute UF-based adhesives in industrial applications. Investigation of blending soy meal with melamine-urea-formaldehyde (MUF) was conducted by Gao et al. (2012). The soy meal-MUF blends showed decreased viscosity, which enhanced the spreading and wetting ability of the adhesives. Wet shear strength of the blends improved due to the cross-linking interaction between soy meal and MUF.

Blends of soy-based protein with blood, casein, polyvinyl alcohol, polyvinyl acetate resin, or phenol–formaldehyde (PF) have also shown improved water resistance for wood applications (Kumar et al., 2002). Steele et al. (1998) developed blends of soy protein and phenolic resins used for finger jointing of green lumber that cured rapidly at room temperature, demonstrated excellent water resistance, and reduced formaldehyde emissions. As much as 70% of PF can be replaced by soy protein-based adhesive with comparable physical properties for oriented and random strand board (Hse et al., 2001; Wescott and Frihart, 2004). In a study by Zhong and Sun (2007), blended adhesives (SPI/PF = 100/20) showed the same level of adhesion strength as commercial PF adhesive, as well as reduced formaldehyde usage and economic advantages. Because polyamide-epichlorohydrin (PAE) is an excellent curing agent for soy protein, Zhong et al. (2007) showed that the complexation interaction for PAE–soy protein blends significantly enhances their adhesion properties.

Blend with Lignin

Together with cellulose, and hemicellulose, lignin is the principle component of plants. Lignin, one of the most available polysaccharides, is the unwanted by-product of the pulping process for papermaking or biofuel production. Lignin is composed of highly cross-linked phenolic C_6C_3 units; these phenylpropane units are connected by a series of carbon-oxygen (ether) and carbon-carbon linkages (Heitner et al., 2010). Lignin is defined as a network-polymer comprised of three main lignin building units: p-hydroxyphenyl, guaiacyl, and syringyl units, linked by carbon–carbon and ether bonds. Because lignin is an aromatic compound available from renewable resources, depolymerization of lignin can potentially produce bio-based polymers (Luo et al., 2015; Yang et al., 2015). Lignin is a hydrophobic polymer that offers mechanical support, acts as a decay retarder, forms a barrier for evaporation, and helps channel water to critical areas of the plant as a part of its cellular structure (Hemmila et al., 2013). Inspired by the natural function to glue and hold together cellulose fibers in plants, lignin has been investigated for industrial applications such as adhesives (Aracri et al., 2014). The crowded macromolecule structure of lignin inevitably leads to less reactive systems in adhesive formulations (Tejado et al., 2008). Although limited commercial success has been achieved due to the intrinsic brittleness and low chemical reactivity of lignin (Aracri et al., 2014), many studies have partially replaced thermosetting resins such as polyurethane, PF, and epoxy resins with lignin (Danielson and Simonsom, 1998; Feldman, 2002; Hatakeyama, 2002) or have chemically treated lignin to develop adhesives that were able to be used directly.

Lignin has also been blended with soy protein-based adhesives in order to improve adhesive bonding performance (Xiao et al., 2013; Luo et al., 2015). Lignin and extruded lignin from sorghum were blended with $NaHSO_3$-modified soy protein at various ratios to test the effects of lignin on the physicochemical and bonding properties of soy protein-based adhesives (Xiao et al., 2013). $NaHSO_3$ was used as the modifier to partly unfold the compact structure of protein, releasing more reactive groups from the protein that could chemically bond with wood substrates

(Qi et al., 2013a). Dry shear strength did not increase when 10% to 40% lignin (based on soy protein, dry basis) was blended with soy protein. However, wet shear strength increased significantly when 50% lignin was added to soy protein (Xiao et al., 2013).

Because some part of lignin has similar structure to phenol, therefore it is a potential substitute for phenol in PF resin synthesis (Luo et al., 2015; Yang et al., 2015). Luo et al. (2015b) copolymerized lignin, phenol, and formaldehyde in a weight ratio of 150:140:132 (100% solid content) into lignin resins, and the lignin resins were blended with soy protein at 0%–25% (based on soy protein, dry basis) for making adhesives. Incorporating 10 wt% lignin resins effectively improved adhesive wet shear strength by 200% when tested on plywood. However, further increase of the lignin resin level did not result in higher shear strength because excess NaOH in lignin severely hydrolyzed the soy protein molecules and reduced the bond strength. Low viscosity of adhesive with high lignin resin loading levels that cause over-penetration could be another cause of decreased wet bonding strength (Luo et al., 2015).

Blends with Wet Strength Resin

Water resistance of soy protein-based adhesives may also be improved by blending with commercially available wet strength resin. In previous studies, researchers added PAE, a wet-strength resin, to a soy protein-based adhesive in order to improve the wet bonding strength of adhesive blends. Major mechanisms involved in the synthesis of SPI-PAE adhesives are illustrated in Fig. 4. First, the azetidium group in PAE resins initially reacts with the remaining secondary amines in the PAE resin, causing a cross-linking reaction (reaction A in Fig. 4). Second, the azetidium group may also react with carboxylic acid groups, such as glutamic acid and aspartic acid in SPI (reaction B in Fig. 4). Third, various amino groups in SPI can also react with the azetidium group (reaction C in Fig. 4). All reactions result in a water-insoluble three-dimensional network (Li et al., 2004). Both the PAE and soy protein have functional groups that can react to the wood. Figure 4 also proposes the possible reaction between soy protein-PAE adhesive and wood substrate (Frihart and Birkeland, 2014).

The soy protein-PAE adhesive was prepared by mixing soy protein with PAE at 1/6 g/ml for 60 min before use. The adhesive was brushed onto the maple veneer with a spread rate of 10 mg/cm^2, and the two wood veneers were stacked together at the glued area and pressed at 120°C with a pressure of 200 psi for 5 min. Bonding test results showed that soy protein-PAE exhibited higher dry shear strength than pure soy protein adhesive, PAE, and PF adhesives (Li et al., 2004). Gui et al. (2013) investigated the effect of itaconic acid-based polyamidoamine-epichlorohydrin (IA-PAE) on the

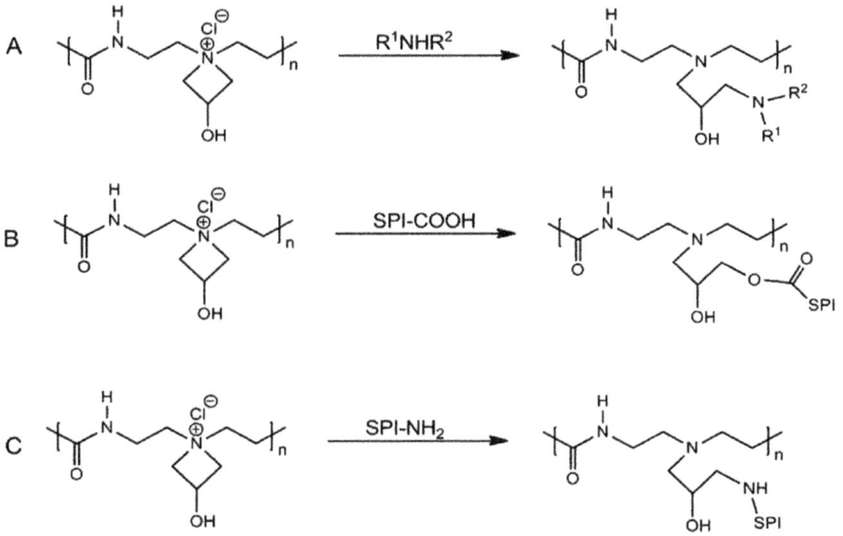

Figure 4. Proposed mechanism of soy protein-PAE (polyamide-epichlorohydrin) adhesive. Adapted from Gui et al. (2013), Ren and Soucek (2014) and Li et al. (2004).

adhesion performance of soy protein adhesives. IA-PAE was synthesized with readily available renewable itaconic acid and PAE at 170°C for 1.5 h, and then the mixture reacted with epichlorohydrin in aqueous solution at 70°C for 1 h. The IA-PAE had a solid content of 50 wt% and viscosity of 144 cp measured at 100 rpm shear rate and a much lower molecular weight compared to commercial PAE. IA-PAE acted as the cross-linking agent when blended with soy protein to improve the water resistance of soy protein-based adhesives, resulting in significantly improved wet shear strength compared to soy protein adhesive and comparable wet shear strength of commercial PAE. Besides soy protein, PAE has also been used with soy flour to improve the bonding strength of soy flour-based adhesives (Frihart and Satori, 2013; Xu et al., 2012).

Nano Technology

Nanoparticle-reinforced adhesive has recently attracted research attention because the environmentally friendly nanocomposite shows promise as a wood adhesive enhancer and provides a new opportunity for the wood composite industry (Gao et al., 2011). Several studies have confirmed that mechanical properties of adhesive are significantly improved with the addition of nanoscale fillers (Gilbert et al., 2003; Chen et al., 2004; Jang and Li, 2015) due to the large surface area of the nanoscale filler and its ability to mechanically interlock with polymer (Hussain et al., 1996). Liu et al. (2010) reported that the wettability, affinity, adhesion strength, and water resistance of soy protein/$CaCO_3$ glue showed great advantages over pure soy protein adhesives; the hybrid adhesive demonstrated stable wet adhesion strength greater than 6 MPa. The researchers attributed this improvement to the compact rivets, interlocking links, and ion cross-linking reaction induced by calcium, carbonate, and hydroxyl ions in the adhesives. Gao et al. (2011) also proved that the addition of 5%–10% nanoscale $CaCO_3$ significantly improved the bond strength and bond durability of whey protein-based adhesives.

Nano-SiO_2 can also enhance the bonding strength of soy protein-based adhesives. Because SiO_2 nanoparticles have large specific surface areas and high surface activity, more reacting sites are available on SiO_2, causing the interaction between the adhesive and wood to increase, leading to enhanced shear strength (Xu et al., 2011). Nanoscale montmorillonite (MMT) served as a filler and effectively prolonged the work life of a modified soybean protein adhesive but reduced the bond strength of the adhesive (Zhang et al., 2013; 2014). In the study by Zhang et al. (2013), the nano-scale MMT platelets most likely blocked some active groups of degraded soy protein (DSP) via intercalation/exfoliation, which effectively retarded the crosslinking reactions between Methylene Diphenyl Diisocyanate (MDI) and DSP, resulting in a longer pot life but with slightly reduced bond strength. The retarding effects of MMT on the crosslinking rate of the polyisocyanate-modified soy protein adhesive were also observed by Zhang et al. (2014). The reason is that the exfoliated nanoscale MMT platelets reduced the number and increased the steric hindrance of some active groups of soybean proteins via hydrogen and electrostatic bonds. MgO was speculated to react with carboxylic acid groups in the soy proteins to form salt bridges between Mg^{2+} and carboxylic acid side chains in the soy protein molecules, thus cross-linking the proteins and resulting in adhesive networks with improved water resistance. Figure 5 illustrates the adhesion mechanism between soy flour and MgO. Blocking carboxylic acid groups in the soy protein is an essential step in the conversion of the soy-based adhesive into a water-resistant adhesive for bonding wood (Jang and Li, 2015).

Enzymatic Treatment

Protein and carbohydrates are the two major components in soy meal. Partial hydrolysis of proteins or carbohydrates has been used to try to decrease the viscosity and improve the adhesion performance of soy-based adhesives. In the study by Chen et al. (2015), viscozyme L (from Aspergillus aculeatus; activity was 100 fungal beta-glucanase units

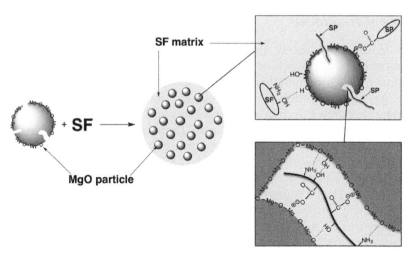

Figure 5. Proposed adhesion mechanisms between soy flour (SF) and soy protein (SP). Adapted from Jang and Li (2015).

per gram) was used to hydrolyze the carbohydrate in soy meal into reducing sugar. The hydrolyzed meal was treated separately with acid-salt and sodium hydroxide solutions to make the final solution with pH 11 and was used as adhesive. After hydrolization, the soy meal showed improved bonding strength, which was attributed to the Millard reaction between soy protein and the reducing sugars during the curing steps. Kumar et al. (2003) compared the effects of various proteolytic enzymes on the adhesion behavior of soy protein-based adhesive and concluded that papain-modified and urease-modified soy protein showed better adhesive strength compared to unmodified soy protein adhesive on rubber-wood, possibly due to the presence of carbohydrate in the protein which led to an increased number of functional groups for adhesion. The chymotrypsin-modified sample showed no adhesive strength. All enzymatically modified adhesives showed lower viscosity compared to unmodified protein adhesive (Kumar et al., 2003). Schmitz (2009) stated that the degree of hydrolysis (DH) is an integral factor for the bonding strength and durability of protein adhesives. Hydrolysis of soy flour was shown to lead to lower viscosity resins, but DH > 10% decreased shear strength of plywood specimens. Hydrolysis led to more reactive possibilities because it exposed more peptide side chains and increased the amount of amine groups. However, an additional cross-linker, such as PAE, is required to create finished adhesives that meet bonding strength of current industry standards.

Thermal Treatment

Heat treatment has been shown to potentially affect protein's structural characteristics, such as tertiary structure, rate of heat denaturation and refolding, and functional properties, such as gelation, emulsification, foaming, and hydration (Yamauchi et al., 1991; Sorgentini et al., 1995). Heating was shown to change the structures of glycinin and β-conglycinin, the two major components of soy protein. Therefore, adhesion performance of protein may be alternated by a combination of a chemical modification and preheating treatment (Wang et al., 2007b). 5% w/w of soy protein aqueous was heated at 60, 80, 110, and 130°C for 20 min before being used as adhesives (Wang et al., 2007b). The highest wet shear strength was observed with 80°C heat treatment, and wet shear strength decreased when soy protein adhesives were treated at 110°C and 130°C. After being heated, soy protein was denatured by disrupting the weak bonds, thereby furthering stabilizing the native confirmation. Moderate denaturation due to low-temperature treatments favored the binding strengths, whereas extensive denaturation and complete loss of the native structure from higher-temperature treatments adversely affected the binding strength of soy protein-based adhesives (Wang et al., 2007b). Similar results were reported by Vnucec et al. (2015), in which thermal modification at 50°C resulted in improved wet shear strength, but wet strength decreased with a heat treatment temperature of 100°C.

Summary

This review described adhesion properties of soy protein subunits, soy protein latex adhesives, and current technologies to improve adhesion performance of soy protein adhesives. Soy protein has high potential to produce adhesives with high bonding strength and water resistance. To develop better protein-based adhesives, more of the fundamental structure-property relationships related to adhesion should be understood. The objective of this review was to provide basic background information to help researchers design and modify soy protein-based adhesives that are comparable to synthetic resins in terms of bonding strength, rheology, and storage life. Future research should focus on making high solid content of soy protein adhesives with acceptable viscosities besides the improved adhesion properties. In addition, soy meal contains about half carbohydrates, which are likely to reduce the adhesiveness of soy protein due to their hydrophilic nature. Seeking the cross-linkers for both protein and polysaccharide could improve the adhesion properties of soy meal, thus exerting the role of carbohydrates on the adhesion performance of soy protein instead of removing them. Full utilization, both of the carbohydrate and protein, to make durable adhesives, would make soybean more competitive with fossil fuel-based adhesives and of greater economic value.

Keywords: Soy protein subunits, Wood adhesive, Modification, Cross-linking, Blending

References

Aracri, E., D. Blanco and T. Tzanov. 2014. An enzymatic approach to develop a lignin-based adhesive for wool floor coverings. Green Chem. 16: 2597–2603.

Babajimopoulos, M., S. Damodaran. Syed S.H. Rizvi and J.E. Kinsella. 1983. Effects of various anions on the rheological and gelling behavior of soy proteins: Thermodynamic Observations. 31: 1270–1275.

Badley, R.A., D. Atkinso, H. Hauser, D. Oldani, J.P. Green and J.M. Stubbs. 1975. The structure, physical and chemical properties of the soybean protein glycinin. Biochim. Biophys. Acta. 412: 214–228.

Bogracheva, T. Ya., N.Yu. Bespalova and A.L. Leont'ev. 1996. Isolation of 11S and 7S globulins from seeds of glycine max. Appl. Biochem. Biotech. 32 (4): 473–477.

Catsimpoolas, N., J.A. Kenny, E.W. Meyer and B.F. Szuhaj. 1971. Molecular weight and amino acid composition of glycinin subunits. J. Sci. Food Agric. 22: 448–450.

Chen, H., Z. Sun and L. Xue. 2004. Properties of nano SiO2 modified PVF adhesive. J. Wuhan Uni. Technol. Mat. Sci. Ed. 19: 73–75.

Chen, N., Q. Zeng., Q. Lin and J. Rao. 2015. Effect of enzymatic pretreatment on the preparation and properties of soy-based adhesive for plywood. Bioresources 10(3): 5071–5082.

Danielson, B and R. Simonson. 1998. Kraft lignin in phenol formaldehyde resin. Part 1. Partial replacement of phenol by kraft lignin in phenol formaldehyde adhesives for plywood. J. Adhes. Sci. Technol. 12: 923–939.

Deak, N.A., P.A. Murphy and L.A. Johnson. 2006. Effects of reducing agent concentration on soy protein fractionation and functionality. J. Food Sci. 71: 200–208.

Feeney, R.E. and J.R. Whitaker. 1985. Chemical and enzymatic modification of plant proteins. *In*: A.M. Altschul and H.L. Wilcke (eds.). New Food Proteins. New York: Academic Press. 5: 181–219.

Feldman, D. 2002. In Lignin and Its Polyblends—A Review. T. Hu (ed.). Springer, U.S. pp. 81–99.

Fennema, O.R. 1996. Amino acids, peptides and proteins. pp. 321–432. *In*: Food Chemistry. New York, N.Y.: Marcel Dekker.

Frihart, C. and M. Birkeland. 2014. Soy properties and soy wood adhesives. pp. 167–192. *In*: R. Brentin (ed.). Soy-Based Chemicals and Materials. Chapter 8. Omni Tech International, Midland, Michigan. Chapter DOI:10.1021/bk-2014-1178.ch008.

Frihart, C and H. Satori. 2013. Soy flour dispersibility and performance as wood adhesive. J. Adhe. Sci. Tech. 27: 2043–2052.

Gao, Q., S. Shi., S. Zhang, J. Li., X. Wang, W. Ding, K. Liang and J. Wang. 2012. Soybean meal-based adhesive enhanced by MUF resin. J. Appl. Ploym. Sci. 125: 3676–3681.

Gao, Z., G. Yu, Y. Bao and M. Guo. 2011. Whey-protein based environmentally friendly wood adhesives. Pigm. Resin. Technol. 40: 42–48.

Gilbert, E.N., B.S. Hayes and J.C. Seferis. 2003. Nano-alumina modified epoxy based film adhesives. Polym. Eng. Sci. 43: 1096–1104.

Gui, C., G. Wang, D. Wu, J. Zhu and X. Liu. 2013. Synthesis of a bio-based polyamidoamine-epichlorohydrin resin and its application for soy-based adhesives. Int. J. Adhe. Adhe. 44: 237–242.

Hatakeyama, H. 2002. pp. 41–56. *In*: T. Hu (ed.). Polyurethanes Containing Lignin, Springer, US. 2002.

Heitner, C., D. Dimmel and J. Schmidt. 2010. Lignin and Lignans: Advances in Chemistry. CRC Press Boca Raton LF, USA.

Hettiarachchy, N and U. Kalapathy. 1999. pp. 379–411. *In*: K. Liu (ed.). Soybeans, Chemistry, Technology and Utilization. Aspen Publishers, Inc.: Gaithersburg, MD.

Hemmila, V., J. Trischler and D. Sandberg. 2013. Lingin- an adhesive raw material of the future or waste of research energy? pp. 98–103. *In*: C. Brischke and L. Meyer (eds.). Proc. 9th Meeting of the Northern European Network for Wood Science and Engineering (WSE), Hannover, Germany, September 11–12, 2013.

Hou, H.J. and K.C. Chang. 2004. Structural characteristics of purified beta-conglycinin from soybeans stored under four conditions. J. Agric. Food Chem. 52: 7931–7936.

Hse, C., F. Fu and B. Bryant. 2001. Development of formaldehydebased wood adhesives with co-reacted phenol/soybean flour. pp. 13–19. *In*: Proceedings of the wood adhesives 2000 conference, 22–23 June. South Lake Tahoe, NV.

Huang, J. 2007. Development and characterization of new formaldehyde-free soy flourbased adhesives for marking interior plywood. PhD diss. Corvallis, OR.: Oregon State University, Department of Wood Science.

Huang, W.N. and X.S. Sun. 2000a. Adhesive properties of soy proteins modified by urea and guanidine hydrochloride. J. Am. Oil Chem. Soc. 77: 101–104.

Huang, W.N. and X.S. Sun. 2000b. Adhesive properties of soy proteins modified by sodium dodecyl sulfate and sodium dodecylbenzene sulfonate. J. Am. Oil Chem. Soc. 77: 705–708.

Hussain, M., A. Nakahira and K. Niihara. 1996. Mechanical property improvement of carbon fiber reinforced epoxy composites by Al2O3 filler dispersion. Mater Lett. 26: 185–91.

Iwabuchi, S. and F. Yamauchi. 1987. Electrophoretic analysis of whey proteins present in soybean globulin fractions. J. Agric. Food Chem. 35: 205–209.

Jang, Y. and K. Li. 2015. An all-natural adhesive for bonding wood. J. Am. Oil Chem. Soc. 92: 431–438.

Kalapathy, U., N. Hettiarachchy, D. Myers and K. Rhee. 1996. Alkali-modified soy proteins: Effect of salts and disulfide bond cleavage on adhesion and viscosity. J. Am. Oil Chem. Soc. 73: 1063–1066.

Kella, N.K.D., W.E. Barbeau and J.E. Kinsella. 1986. Effect of disulfide bond cleavage on the structure and conformation of glycinin. Int. J. Peptide Protein Res. 27: 421–432.

Kim, M and X.S. Sun. 2014. Adhesion Properties of Soy Protein Crosslinked with Organic Calcium Silicate Hydrate Hybrids. J. Appl. Polym. Sci. 131(17). DOI: 10.1002/app.40693.

Koshiyama, I. 1972. Comparison of acid-induced conformation changes between 7S and 11 S globulins in soybean seeds. J. Sci. Food Agric. 34: 853–859.

Kumar, R., V., Choudhary, S. Mishra, I. Varma and B. Mattiason. 2002. Adhesives and plastics based on soy protein products. Ind. Crops Prod. 16: 155–172.

Kumar, R., V. Choudhary, S. Mishra and I. Varma. 2003. Enzymatically-modified soy protein part 2: adhesion behavior. J. Adhe. Sci. Technol. 18(2): 261–273.

Lambuth, A.L. 1989. Adhesives from renewable resources: Historical Perspective and Wood Industry Needs. pp. 1–10. *In*: R.W. Hemingway, A.H. Conner and S.J. Branham (eds.). Adhesives from Renewable Resources. Washington, DC: American Chemical Society.

Lei, H., G. Du, Z. Wu., X. Xi and Z. Dong. 2014. Cross-linked soy-based wood adhesives for plywood. Int. J., Adhe. Adhe. 50: 199–203.

Li, J., J. Luo, X. Li, Z. Yi, Q. Gao and J. Li. 2015. Soybean meal-based wood adhesive enhanced by ethylene glycol diglycidyl ether and diethylenetriamine. Ind. Crop. Prod. 74: 613–618.

Li, K., S. Peshkova and X. Geng. 2004. Investigation of soy protein-kymene® adhesive systems for wood composites. J. Am. Oil Chem. Soc. 81(5): 487–491.

Liu, H., C. Li and X.S. Sun. 2015. Improved water resistance in undecylenic acid (UA)-modified soy protein isolate (SPI)-based adhesives. Ind. Crop Prod. 74: 577–584.

Liu, Y. and K. Li. 2007. Development and characterization of adhesives from soy protein for bonding wood. Int. J. Adhes Adhes. 27: 59–67.

Liu, D., H. Chen, P.R. Chang, Q. Wu, K. Li and L. Guan. 2010. Biomimetic soy protein nanocomposites with calcium carbonate crystalline arrays for use as wood adhesive. Bioresource Technol. 101: 6235–6241.

Luo, J., C. Li, X. Li, J. Luo, Q. Gao and J. Li. 2015a. A new soybean meal-based bioadhesive enhanced with 5,5-dimethyl hydantoin polyepoxide for the improved water resistance of plywood. RSC Adv. 5: 62957–62965.

Luo, J., J. Luo, C. Yuan, W. Zhang, J. Li, Q. Gao and H. Chen. 2015b. An eco-friendly wood adhesive from soy protein and lignin: performance properties. RSC Adv. 5: 100849–100855.

Mo, X., X.S. Sun and D. Wang. 2004. Thermal properties and adhesion strength of modified soybean storage proteins. J. Am. Oil Chem. Soc. 81: 395–400.

Mo, X., D. Wang and X.S. Sun. 2011. Physicochemical Properties of β and α′α Subunits Isolated from Soybean β-Conglycinin. J. Agric. Food Chem. 59: 1217–1222.

Mo, X., Z. Zhong, D. Wang and X.S. Sun. 2006. Soybean glycinin subunits: Characterization of physicochemical and adhesion properties. J. Agric. Food Chem. 54: 7589–7593.

Nagano, T., M. Hirotsuka, H. Mori, K. Kohyama and K. Nishinari. 1992. Dynamic viscoelastic study on the gelation of 7S globulin from soybeans, Ibid. 40: 941–944.

Nielsen, N.C. 1974. Structure of soy proteins. pp. 27–64 *In*: Aaron M. Altschul and Harold Ludwig Wilcke (eds.). New protein foods. Academic Press: New York, NY.

O'Keefe, S.F., L.A. Wilson, A.P. Resurreccion and P.A. Murphy. 1991. Determination of the binding of hexanal to soy glycinin and β-conglycinin in an aqueous model system using a head-space Technique. Ibid. 39: 200–205.

Peng, I.C., D.W. Quass, W.R. Dayton and C.E. Allen. 1984. The physicochemical and functional properties of soybean 11s globulin—A review. Cereal Chem. 61: 480–490.

Qi, G. and X.S. Sun. 2011. Soy protein adhesive blends with synthetic latex on wood veneer. J. Am. Oil Chem. Soc. 88: 271–281.

Qi, G., N. Li, D. Wang and X.S. Sun. 2012. Physicochemical properties of soy protein adhesives obtained by *in situ* sodium bisulfite modification during acid precipitation. J. Am. Oil Chem. Soc. 89(2): 301–312.

Qi, G., N. Li, D. Wang and X.S. Sun. 2013a. Adhesion and physicochemical properties of soy protein modified by sodium bisulfite. Journal of the American Oil Chemists' Society 90(12): 1917–1926.

Qi, G., N. Li, D. Wang and X.S. Sun. 2013b. Physicochemical properties of soy protein adhesives modified by 2-octen-1-ylsuccinic anhydride. Ind. Crop. Prod. 46: 165–172.

Qi, G., K. Venkateshan, X. Mo, L. Zhang and X.S. Sun. 2011. Physicochemical properties of soy protein: Effects of subunits composition. J. Agric. Food Chem. 59: 9958–9964.

Qu, P., H. Huang, G. Wu, E. Sun and Z. Chang. 2015. The effect of hydrolyzed soy protein isolate on the structure and biodegradability of urea–formaldehyde adhesives. J. Adhe. Sci. Tech. 29(6): 502–517.

Ren, X and M. Soucek. 2014. Soya-Based Coatings and Adhesives. pp. 207–254. *In*: R. Brentin (ed.). Soy-Based Chemicals and Materials. Chapter 10. Omni Tech International, Midland, Michigan. Chapter DOI:10.1021/bk-2014-1178.ch010.

Rickert, D., L.A. Johhson and P.A. Murphy. 2004. Improved fractionation of glycinin and β-conglycinin and partitioning of phytochemicals. J. Agric. Food Chem. 52: 1726–1734.

Renkema, M.S., H. Gruppen and T. Vliet. 2002. Influence of pH and ionic strength on heat induced formation and rheological properties of soy protein gels in relation to denaturation and their protein compositions. J. Agric. Food Chem. 50(21): 606–6071.

Rogers, J., X. Geng and K. Li. 2004. Soy-based adhesives with 1, 3-Dichloro-2-propanol as a curing agent. Wood Fiber Sci. 36: 186–194.

Soystats. 2015. http://soystats.com/planting-data-crop-area-planted/. Accessed by 01/11/2016.

Schmitz, J. 2009. Enzyme modified soy flour adhesives. Dissertation. Iowa State University. Ames, Iowa.

Sorgentini, D., J. Wagner and M. Añón. 1995. Effects of thermal treatment of soy protein isolate on the characteristics and structure-function relationship of soluble and insoluble fractions. J. Agric. Food Chem. 43: 2471–2479.

Steele, P., R. Kreibich, P. Steynberg and R. Hemingway. 1998. Finger jointing green southern yellow pine with a soy-based adhesive. Adhesive Age. 10: 49–54.

Sun, X.S., L. Zhu and D. Wang. 2008. Latex based adhesives derived from soybeans. U.S. Patent 0287635 A1.

Tejado, A., G. Kortaberria, C. Peña, J. Labidi, J.M. Echeverria and I. Mondragon. 2008. Isocyanate curing of novolac-type ligno-phenol–formaldehyde resins. Indu. Crops Prod. 27: 208–213.

Thanh, V.H. and K. Shibasaki. 1976. Heterogeneity of β-conglycinin. Biochim. Biophys. Acta. 439: 326–338.

Thanh, V.H. and K. Shibasaki. 1977. Beta-conglycinin from soybean proteins. Isolation and immunological and physicochemical properties of the monomeric forms. Biochim. Biophys. Acta. 490: 370–384.

Thanh, V.H. and K. Shibasaki. 1978. Major proteins of soybean seeds. subunit structure of beta-conglycinin. J. Agric. Food Chem. 26: 692–695.

Thiering, R., G. Hofland, N. Foster, G.J. Witkamp and L. Van De Wielen. 2001. Fractionation of soybean protein with pressurized carbon dioxide as a volatile electrolyte. Biotechno Bioeng. 73: 1–11.

Utsumi, S., H. Inaba and T. Mori. 1981. Heterogeneity of soybeans glycinin. Phytochemistry. 20: 585–589.

Vnucec, D., A. Gorsek, A. Kutnar and M. Mikuljan. 2015. Thermal modification of soy proteins in the vacuum chamber and wood adhesion. Wood Sci. Technol. 49: 225–239.

Wang, Y., X. Mo, X.S. Sun and D. Wang. 2007a. Soy protein adhesion enhanced by glutaraldehyde crosslink. J. Appl Polymer. Sci. 104: 130–136.

Wang, Y., X.S. Sun and D. Wang. 2007b. Effects of preheating treatment on thermal property and adhesion performance of soy protein isolates. J. Adhes. Sci. Technol. 21(15): 1467–1481.

Wang, D., X.S. Sun, G. Yang and Y. Wang. 2009. Improved water resistance of soy protein adhesion at isoelectric point. ASABE. 52: 173–177.

Wescott, J. and C. Frihart. 2004. Competitive soybean flour/phenol–formaldehyde adhesives for oriented strandboard. In: Proceedings of 38th international wood composites symposium, 6–9 April. Pullman, Washington, pp. 199–206.

Wolf, W.J. 1970. Soybean proteins their functional, chemical and physical properties. J. Agric. Food Chem. 18: 969–976.

Wolf, W.J. 1993. Sulfhydryl content of glycinin: effect of reducing agents. J. Agri Food. Chem. 41: 168–176.

Wu, S., P.A. Murphy, L.A. Johnson, A.R. Fratzke and M.A. Reuber. 1999. Pilot-plant fractionation of soybean glycinin and β-conglycinin. J. Am. Oil Chem. Soc. 76: 285–293.

Xiao, Z., Y. Li, X. Wu, G. Qi, N. Li, K. Zhang, D. Wang and X.S. Sun. 2013. Utilization of sorghum lignin to improve adhesion strength of soy protein adhesives on wood veneer. Ind. Crop Prod. 50: 501–509.

Xu, H., J. Luo, Q. Gao, S. Zhang and J. Li. 2014. Improved water resistance of soybean meal-based adhesive with SDS and PAM. Bioresources. 9(3): 4667–4678.

Xu, H., S. Ma., W. Lv and Z. Wang. 2011. Soy protein adhesives improved by SiO2 nanoparticles for plywoods. Pigm. Resin Technol. 40(3): 191–195.

Xu, Y., C. Wang, F. Chu, C. Frihart, L. Lorenz and M. Stark. 2012. Chemical modification of soy flour protein and its properties. Adv. Mat. Res. 343–344: 875–881.

Yamauchi, F., T. Yamagishi and S. Iwabuchi. 1991. Molecular understanding of heat induced phenomena of soybean protein. Food Rev. Int. 7: 283–322.

Yang, S., T. Yuan, M. Li and R. Sun. 2015. Hydrothermal degradation of lignin: Products analysis for phenol formaldehyde adhesive synthesis. Int. J. Biol. Macromol. 72: 54–62.

Zhang, L. and X.S. Sun. 2008. Effect of sodium bisulfite on properties of soybean glycinin. J. Agric. Food Chem. 56: 11192–11197.

Zhang, L. and X.S. Sun. 2010. Sodium bisufite induced changes in the physicochemical, surface and adhesive properties of soy β-conglycinin. J. Am. Oil Chem. Soc. 87: 583–590.

Zhang, Y., W. Zhu, Y. Lu, Z. Gao and J. Gu. 2013. Water-resistant soybean adhesive for wood binder employing combinations of caustic degradation, nano-modification and chemical crosslinking. Bioresources. 8(1): 1283–1291.

Zhang, Y., W. Zhu, Y. Lu, Z. Gao and J. Gu. 2014. Nano-scale blocking mechanism of MMT and its effects on the properties of polyisocyanate-modified soybean protein adhesive. Ind. Crops Prod. 57: 35–42.

Zhang, Z. and Y. Hua. 2007. Urea-modified soy globulin proteins (7S and 11S): Effect of wettability and secondary structure on adhesion. J. Am. Oil Chem. Soc. 84: 853–857.

Zhong, Z., X.S. Sun, X. Fang and J.A. Ratto. 2001. Adhesion strength of sodium dodecyl sulfate modified soy protein to fiberboard. J. Adhesion Sci. Tech. 15(12): 1417–1427.

Zhong, Z., X.S. Sun and D. Wang. 2007. Isoelectric pH of Polyamide-Epichlorohydrin modified soy protein improved water resistance and adhesion properties. J. Appl. Polymer. Sci. 103: 2261–2270.

Zhong, Z. and X.S. Sun. 2007. Plywood adhesives by blending soy protein polymer with phenol–formaldehyde resin. J. Biobased Mater. Bio. 1: 380–387.

Zhong, Z., X.S. Sun, X. Fang and J.A. Ratto. 2002. Adhesive strength of guanidine hydrocholoride-modified soy protein for fiberboard application. Int. J. Adhes Adhes. 22: 267–272.

Zhu, D. and S. Damodaran. 2014. Chemical phosphorylation improves the moisture resistance of soy flour-based wood adhesive. J. Appl. Polymer. Sci. DOI: 10.1002/APP.40451.

3

Modification of Soy-based Adhesives to Enhance the Bonding Performance

Sheldon Q. Shi,* Changlei Xia and *Liping Cai*

ABSTRACT

Soy protein (SP) which exists in soy flour, soybean meal, soy protein concentrate, and soy protein isolate is one of the vegetable proteins. It represents a very practical and inexpensive material for bio-based adhesives. The positive public perception of SP makes it very competitive as a wood adhesive. While SP is weak in adhesion strength and water resistance compared to the traditional formaldehyde-based resins, the performance of the SP adhesive bonded composites would be significantly improved if a strong chemical bonding between the SP adhesive with the cellulosic materials can be achieved. In order to enhance the bonding performance, this chapter presents the current modification methods of soy-based adhesives, which describes: (1) Chemical modifications including alkali denaturation, surfactant denaturation, organic solvent denaturation, enzymatic treatment, epoxy groups crosslinking, formaldehyde compounds crosslinking, acylation and silanation; (2) Physical modification including compression, high temperature treatment, ultrasound treatment and blending with hydrophobic additives; (3) Nanoparticle modification using cellulosic nanofibers, inorganic nanoparticles and nano-clays. In addition, it also illustrates the use of the soy adhesive as a substitute for other adhesives in this chapter.

Department of Mechanical and Energy Engineering, University of North Texas, Denton, TX 76203, USA.
* Corresponding author: Sheldon.Shi@unt.edu

Introduction

Animal and vegetable proteins are practical for making food packaging films because of their high availability as byproducts from the food processing industry and their relatively low cost (Janjarasskul and Krochta, 2010; Mangavel et al., 2004; Paetau et al., 1994; Reddy and Yang, 2013). As one of the vegetable proteins, soy protein (SP) exists in soy flour (SF), soybean meal (SM), soy protein concentrate (SPC), and soy protein isolate (SPI). With the advantages of renewability, processability, biocompatibility, film-forming capacity, and biodegradability, SP has great potential to be used in the food industry, agriculture, bioscience, and biotechnology (Khan et al., 2012; Monedero et al., 2009; Song et al., 2011). SF is an interesting alternative for producing environment-friendly materials. As an abundant and renewable resource with high biodegradability, SF has aroused further interest in many researchers (Cui et al., 2014; Nishinari et al., 2014). SP adhesive is produced from SM, a processing residue of soybean oil and other soy-based food products. The SP isolated from soybean has many unique properties, such as low cost, ease of handling, and the ability to bond biomass materials at relatively high moisture content (MC).

Wood based composites, such as plywood, medium-density fiberboard (MDF), and particleboard, have been used extensively in housing, furniture, construction, and transportation. The production of structural wood panel products in the U.S. was approximately 26.6 million m^3 in 2013, 63.5% of which was oriented strand board (OSB) (http://www.fx168.com/). Currently, most adhesives used in wood composites are formaldehyde-based, such as phenol formaldehyde (PF) resin, urea formaldehyde (UF) resin, melamine urea formaldehyde (MUF) resin, etc. These adhesives are not sustainable or renewable resources, which has caused considerable human health and environmental problems. In 2007, the California Air Resources Board (CARB) approved regulations to set a limitation on formaldehyde emissions for wood-based composites. The U.S. Department of Housing and Urban Development (HUD) standards limit formaldehyde emissions from wood products, and voluntary standards which are even more restrictive have been set by industry. In 2010, Congress enacted legislation mandating a national emissions standard for composite wood products. In the past few years, the addition of metal and mineral nano-particles to composite matrix has helped in improving thermal conductivity of composite mats (Taghiyari et al., 2013a; Taghiyari et al., 2013c), eventually decreasing the amount of formaldehyde-based adhesives to be used in wood composite panels (Taghiyari et al., 2013b). However, the development of sustainable and environment-friendly adhesives to replace the formaldehyde-based adhesives is still required to completely eliminate formaldehyde emission (Huang and Netravali, 2009).

As an adhesive, SP represents a very practical and inexpensive material for bio-based adhesives (Li et al., 2004). SP has been used as a substitute for MUF adhesive in the plywood-production industry (Guezguez et al., 2013) for reducing formaldehyde release. Recently, significant progress on the research of SP adhesives has been realized and new technologies have been developed or are under development (Mo and Sun, 2013). The positive public perception of SP makes it very competitive as a wood adhesive. However, the two major drawbacks in the application of SF are: (1) the overall performance of these modified SP, in terms of adhesive strength and water resistance, is not comparable to formaldehyde-based resins; (2) their fragility in the wet state and poor properties of moisture barrier (Andreuccetti et al., 2011). The performance of the SP adhesive bonded composites will be significantly improved if a strong chemical bonding between the SP adhesive and the cellulosic materials can be achieved. Hence, physical, chemical and enzymatic modifications have been used to improve the functional properties of SF, including changing drying conditions and processing methods (Ciannamea et al., 2014; Denavi et al., 2009b), enzymatic treatment with horseradish peroxidase (Stuchell and Krochta, 1994), ultrasound treatment (Hu et al., 2013; Jambrak et al., 2009; Wang et al., 2014), heat curing (Gennadios et al., 1996), and blending with hydrophobic additives such as neutral lipids, fatty acids, waxes or polyvinyl alcohol (PVA) (Huang et al., 2014; Rhim, 2004).

Chemical Modifications of Soy-based Adhesive

Chemical Denaturation

The methods of chemical denaturation for protein include alkali, surfactants, and organic solvents (Wu and Inglett, 1974). The major storage proteins of soybean which are referred to as glycinin (11S) and β-conglycinin (7S) are globulins and the functional properties of soy-based products (such as flour, concentrates and isolates) are reflected in their composition and structure (Barać et al., 2004). The targets of chemical denaturation of SP are varied for different purposes. However, the main objective of this treatment is to increase the bonding property of the SP as an adhesive. After chemical denaturation, the chains of these globular proteins become linear and interactive with others. The SP are modified using sodium hydroxide (NaOH) (Kalapathy et al., 1995; Mackay, 1998) and certain reagents such as guanidine hydrochloride, urea, sodium dodecyl benzene sulphonate (SDBS) and sodium dodecyl sulphate (SDS), which could denature the soy protein and improve the gluing strength and water resistance (Bian and Sun, 1998; Huang and Sun, 2000a; Huang and Sun, 2000b).

Alkali Denaturation

The tensile strength of beech substrates bonded with dispersions of alkali-denatured (such as NaOH) SPI and wheat gluten was measured (Nordqvist et al., 2010) for the comparisons of bond strength and resistance to cold water. The adhesive properties of SPI were superior, particularly with regard to water resistance. The water resistance of wheat gluten was improved to some extent. Similar TS values were obtained for the dispersions of alkali-denatured SPI regardless of pH or salt concentration. The response surface methodology was applied to optimize the preparation conditions of SF adhesive (Chen et al., 2013). The effects and interactions of treatment time, temperature and additive amount of acid-salt solution on wet strength were examined. The optimal preparation conditions were: 3.0 g acid-salt solution, 33 min treatment time and 29°C treatment temperature. Under these conditions, the wet strength of tensile strength was 1.18 ± 0.08 MPa, which was in agreement with the predicted value (1.11 MPa).

Surfactant Denaturation

The effects of using SDS and a modified polyacrylic acid (MPA) solution on a SM adhesive were investigated (Gao et al., 2012b). The results showed that using 1% SDS improved the water resistance of the SM adhesive by 30%. After incorporating 20% MPA, the water resistance of the SM/SDS/MPA adhesive was further improved by 60%, the solid content of the adhesive increased by 15%, and the viscosity of the adhesive was reduced by 81%. The plywood bonded with the SM/SDS/MPA adhesive met the interior plywood requirements of China National Standard (GB/T 9846.3-2004) (Gao et al., 2012b). A study was conducted on SP adhesive and water-resistance by introducing SDS or SDBS by Huang and Sun (2000a). After using 1% SDS or SDBS, the shear strengths of the glued pine wood were improved about 1 time after the incubation (from 21 to 43 or 45 kg/cm²), and approximately 5 times after water soaking (from 6 to 41 or 45 kg/cm²).

Organic Solvent Denaturation

Protein-disrupting chaotropic agents (guanidine hydrochloride, urea, and dicyandiamide), and the cosolvent propylene glycol were able to provide increased protein/protein and protein/polyamidoamine-epichlorohydrin interactions (Kumar et al., 2002). Urea-treated SP showed greater shear strengths compared with unmodified SPI (Huang and Sun, 2000b). The 3 M urea modification gave SP the highest shear strength of 42 kg/cm². Additionally, Huang and Sun (Huang and Sun, 2000b) reported that the guanidine hydrochloride-treated SPI had greater shear strength than unmodified SPI after using 0.5 and 1 M guanidine hydrochloride. Frihart and Lorenz (2013) reported that the utilization of dicyandiamide in the system of soy flour adhesives with polyamidoamine-epichlorohydrin polymeric coreactant could significantly increase the mechanical and water-resistance properties.

Enzymatic Treatment

Enzymatic treatment causes the SP degradation in addition to crosslinking. In order to improve the wet shear strength and decrease the viscosity of SM adhesive, the SM adhesive, enzyme-treated SM (ESM) adhesive, and the glycidyl methacrylate (GMA) ESM (ESM-g-GMA) adhesive were prepared by Qin et al. (2013). The apparent viscosity of the SM adhesives decreased due to the presence of GMA and the enzymatic treatment. The wet shear strength of the plywood panels bonded with the ESM-g-GMA adhesive was significantly improved from 0.45 MPa to 1.05 MPa. The change in wet shear strength was not obvious within 24 h water immersion. The optimum hot-press temperature and remaining time were 150°C and 6 min, respectively. The results from the Fourier transform infrared spectroscopy (FTIR) testing showed that GMA molecules were successfully grafted onto the ESM adhesives. The results of the thermogravimetric analysis indicated that the peak degradation temperature of the ESM-g-GMA adhesives were higher than that of the SM adhesive due to the cross-linking reaction creating a dense structure of the macromolecular between the ESM adhesive and the GMA monomer.

In the research of Chen et al. (2014), the Viscozyme® L enzymatic pretreatment of the carbohydrates (e.g., polysaccharides) in defatted soy flour (DSF) improved the water resistance of the soy-based adhesives as a result of the Maillard reaction that occurred between proteins and monosaccharides. In their recent research (Chen et al., 2015), it was reported that the optimal enzymatic pretreatment conditions of SBA were the pretreatment time of 20.0 min, pretreatment temperature of 54°C, and pretreatment pH value of 5.1. Under these conditions, the reduced sugar content and bonding strength (boiling-water test) of SBAs were 2.9% and 0.62 MPa, which were increased by 113.9% and 30.6% compared to that of the control ones, respectively.

Chemical Crosslinking

Epoxy Groups

Soy-based films are considered as an alternate to plastic packing materials. Also, the properties of the soy-based films could be utilized as an indicator for the bonding performance of soy-based adhesives. More than 90%

proteins were contained in SPI, counting 6.8% lysine, 7.7% arginine, and 2.5% cysteine that has abundance of –NH$_2$ groups (Mateos-Aparicio et al., 2008). The total percentage of that type of amino acids containing –NH$_2$ groups can be 17.0% in the protein. These amino acids are able to react with the epoxy groups. In the studies on the SPI-based films (Xia et al., 2015d; Xia et al., 2016c), two cross-linking agents with multiple epoxy groups were investigated, including epoxidized soybean oil (ESO) (Fig. 1) and TriSilanolPhenyl polyhedral oligomeric silsesquioxanes (POSS) (Fig. 2). After being treated with ESO and POSS, the tensile strengths of the SPI-based films were increased by 139.8% and 34.0%, respectively, and the 24-h water absorptions decreased by 78.1% and 54.7%, respectively. Li et al. (2015) reported that SM-based wood adhesive was enhanced by ethylene glycol diglycidyl ether (EGDE) and diethylenetriamine (DETA). In terms of EGDE/DETA (Fig. 3), the wet shear strength of the plywood bonded with the resultant adhesive reached 1.15 MPa, increased by 30.7%, meeting the requirements of interior use plywood ($\geq$ 0.7 MPa) according to the China

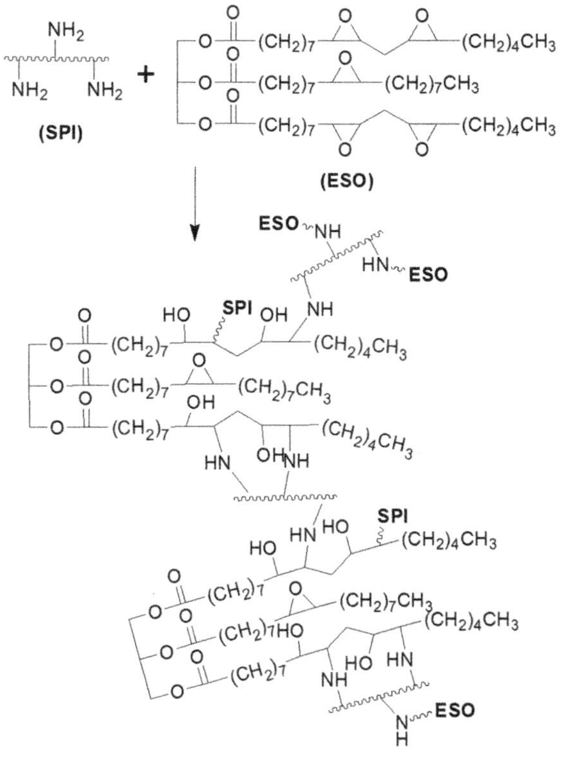

Figure 1. Cross-linking reaction between soy protein isolate (SPI) and epoxidized soybean oil (ESO).

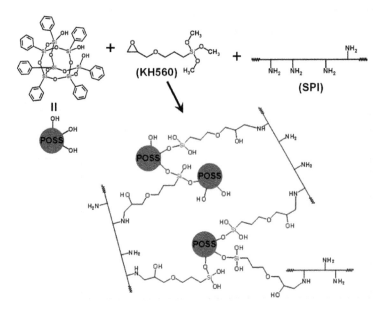

Figure 2. Illustration of the cross-linking reactions between SPI, γ-glycidoxy-propyltrimethoxysilane (KH560) and TriSilanolPhenyl polyhedral oligomeric silsesquioxanes (POSS).

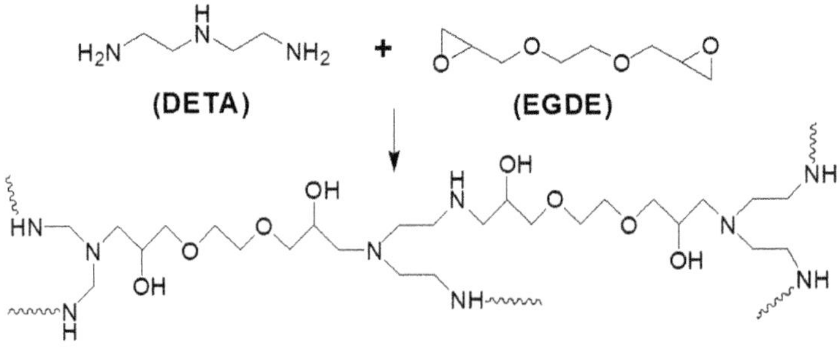

Figure 3. Cross-linking reactions between diethylenetriamine (DETA) and ethylene glycol diglycidyl ether (EGDE).

National Standard (GB/T 9846.3-2004). Additionally, compared with the un-treated SM-based adhesives, the viscosity of the cured adhesive was decreased by 39.4–50.3%, which made the resultant adhesive practical for plywood industrial application.

Formaldehyde Compounds

It was investigated that the crosslinking and denaturation of proteins by adding formaldehyde donors or adducts could enhance the water resistance, granular consistency, longer pot life, improved assembly time tolerance and curing behavior, and increased water holding capacity (Lambuth, 2003). Examples of compounds of this group included: urea-formaldehyde, tris-hydroxymethyl nitromethane, aldehydic starch, dimethylol urea, glyceraldehyde, sodium formaldehyde bisulphite, hexamethylene tetramine, methylolated phenols, etc. Since the formaldehyde and paraformaldehyde are very active crosslinking agents, pre-mature gelation of the SP might be easy to occur. Therefore, very low concentration, ranging from 0.1% to 1%, of aldehyde-active compounds was sufficient and was added during the final stage of the glue mixing.

In order to reduce the formaldehyde emissions and improve the wet shear strength of the SM/MUF adhesive-bonded plywood, the hot pressing parameters in plywood manufacturing were optimized (Gao et al., 2011). The optimum hot pressing parameters were: hot-press temperature of 150°C, hot-press time of 70 s/mm, hot-press pressure of 1.2 MPa, and adhesive spread rate of 165 g/m². In accordance with these optimum processing conditions, the wet shear strength of the plywood was improved by 24.2% from 0.95 to 1.18 MPa, while the formaldehyde emissions was decreased by 21.4% from 0.28 to 0.22 mg/L.

The SM flour, polyethylene glycol (PEG), NaOH, and a MUF resin were used to formulate SM/MUF resin adhesive (Gao et al., 2011). It was reported that the wet shear strength of plywood bonded by SM/NaOH adhesive increased by 33% to 0.61 MPa after the addition of NaOH into the SM adhesive. After adding PEG, the viscosity of the SM/NaOH/PEG adhesive was reduced by 91% to 34,489 cP. Compared to the MUF resin, the solid content of the SM/MUF resin adhesive was reduced to 39.2%, the viscosity of the adhesive was further decreased by 37% to 21,727 cP, and the wet shear strength of plywood bonded with the adhesive was increased to 0.95 MPa. The formaldehyde emission of plywood bonded by the SM/MUF resin adhesive was obtained at the level of 0.28 mg/L.

Other Compounds

As strong crosslinkers of dispersed SP, sulphur compounds, such as CS_2, ethylene di- or tri-thiocarbonate, thiourea and potassium xanthate, were used to improve the water resistance, pot life and consistency of composite resins (Bjorksten, 1951). An adhesive system prepared by mixing SP and

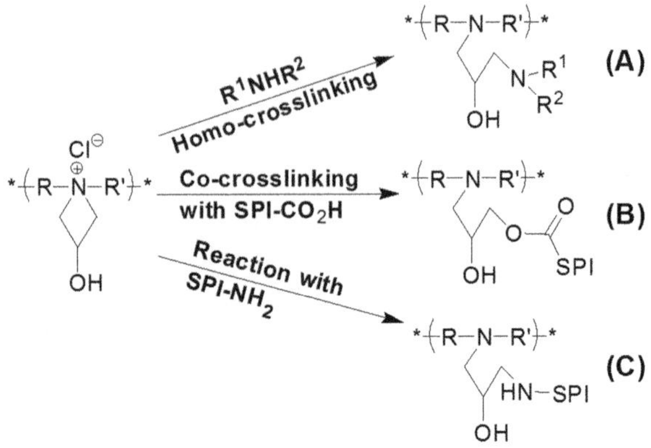

Figure 4. Scheme of interactions between Kymene® 557H and soy protein.

Kymene® 557H was developed (Li et al., 2004). The interaction between Kymene® 557H and SP is shown in Fig. 4. Wood composites bonded with this adhesive showed a shear strength comparable to or higher than those bonded with commercial PF resins. In addition, the water resistance of the composites was also increased. To improve the water resistance of soy-based adhesive for wood panels, three kinds of cross-linking agents, including epoxy resin (EPR), melamine-formaldehyde (MF) and their mixture EPR+MF, were investigated by Lei et al. (2014). The results indicated that all three cross-linking agents improved the water resistance of soy-based adhesive, from which the hybrid cross-linking agent EPR+MF was the best.

Functional Group Modification

Acylation

Reported by Liu and Li (2007), SPI was first modified with maleic anhydride (MA) to form MA-grafted SPI (MSPI). In terms of the amide linkages and ester linkages, MA was grafted onto SPI molecules. The formation of the amide linkages, i.e., the reaction between amino groups of SPI and MA, was faster than the formation of the ester linkages, i.e., the reaction between MA and hydroxyl groups of SPI. A combination of MSPI and polyethylenimine (PEI) dramatically increased the water resistance and strength of the resulting wood composites.

Through the treatment of SP with succinic anhydride or acetic anhydride, the ease of protein acylation could be demonstrated (Franzen and Kinsella, 1976; Lawal and Adebowale, 2004). During the reaction, the pH value was maintained between 7 and 8 with NaOH. Both succinylation and acetylation were utilized to improve the functional properties of the adhesive, such as solubility and surface hydrophobicity. Additionally, the phthalic anhydride was reported as another acylation agent (Guo et al., 2015; Jong, 2009; Jong, 2011), which was able to improve the water-resistance property.

Silanation

It was reported that, when silane compounds at nano-scale were mixed with urea-formaldehyde resins, the water-repellent property and liquid permeability in medium-density fiberboard (MDF) were significantly improved (Taghiyari, 2013; Taghiyari et al., 2015). Silane coupling agents, silicon-based chemicals, can also be employed for the silanation of SP molecules. Silane coupling agents contain two types of reactive groups, where one is Si-O-H group, and the other one is various. Silane coupling agents are able to create a chemical bridge due to their bifunctional structures, for the purpose of bonding with various materials (Fang et al., 2014). Li et al. (2014a) used γ-amino, γ-glycidyl, and γ-methacryloyloxy-propyltrimethoxysilane (KH550, KH560, and KH570, the structures are shown in Fig. 5) silane coupling agents as enhancers in the soybean flour adhesive. It was found that KH560 was the best on improving the water-

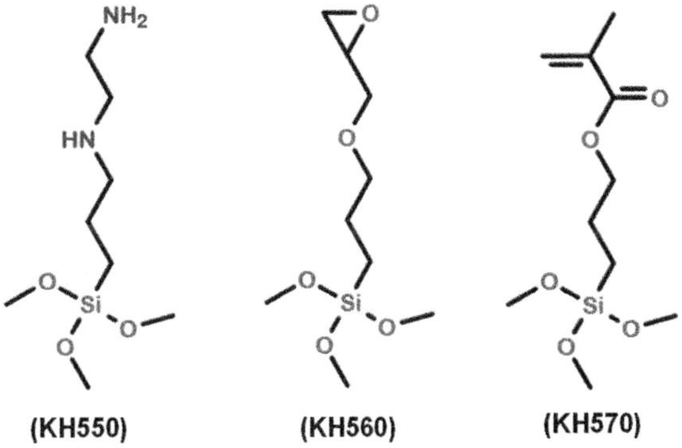

(KH550)　　　　　**(KH560)**　　　　　**(KH570)**

Figure 5. Chemical structures of γ-amino, γ-glycidyl, and γ-methacryloyloxy-propyltrimethoxysilane (KH550, KH560, and KH570).

resistant bonding strength. The wet shear strength of the plywood bonded was significantly increased from 0.41 MPa to 0.98 MPa, if 3% KH560 was added into the SF adhesive. Another literature (Liang and Wang, 1999) reported that the 3-(2-aminoethyl)-aminopropyltrimethoxy silane modification could enhance the interfacial bonding between the SP matrix and glass fibers in the production of fiber-reinforced composites.

Other Compounds

Mixed with various synthetic resins, such as, PF, diphenylmethane diisocyanate (MDI), epoxy, MF and polymeric PF (PMF), and an aminepolymer-epichlorohydrin adduct, non-hydrolyzed or hydrolyzed soy flour could be potentially used for exterior plywood (He et al., 2012). A plywood adhesive was produced using total proteins extracted from Spirulina platensis and Chlamydomonas reinhardtii by Roy et al. (2014). With NaOH and chemical cross-linking, the TS and water resistance of the adhesive were improved. Among the three aldehydes tested, glyoxal was found to be the best cross-linking agent. The optimum concentration of NaOH was found to be approximately 50 mM and of glyoxal was 2 wt%.

It was reported that the soy-based adhesive interacted with polysaccharides for improving water stability (Shih, 1994). Mixing SPI with sodium alginate/propylene glycol alginate under alkaline conditions would format electrostatic/covalent complexes. As a result, the films formed from covalent complexes had greater stability in water as compared to that obtained from protein-alginate complexes.

Physical Modification

The physical modifications of the soy-based adhesive that have been used to improve the functional properties include compression, ultrasound treatment, changing drying conditions and blending with hydrophobic additives.

Compression

Using a thermal compaction technique, SPI, as a biopolymer, and glycerol, as a plasticizer, were utilized to make light-yellow, transparent protein films successfully (Cunningham et al., 2000). At a processing temperature of 150°C, compaction pressure of 10 MPa and pressing time of 2 min, the flexible films, approximately 180 micrometers, were produced. It was found that the dispersion of glycerol among the SPI macromolecules was a diffusion-limited process that could be accelerated by intensive mixing. Made from the intensively mixed material, the films containing 30 wt% glycerol had higher averaged tensile strength and EB values, counting 6% and 300% higher than those of films made from unaged, manually mixed material, respectively. SPI was mixed with four polyol-based plasticizers and molded into plastics using a hot press by Mo and Sun (2002). The denaturation temperatures and enthalpies of the plasticized SPI powder was examined. Thermal properties of the SPI plastics with propylene glycol were depressed to the largest degree. The plastics with glycerol showed the largest strain at break, whereas the plastics with 1, 3-butanediol had the highest tensile strength. Compared to the ones processed by casting, the SPI based films processed by compression molding have demonstrated much improved mechanical properties, higher tensile strength and EB (Guerrero et al., 2010).

Two methods, intensive mixing followed by compression molding and solution-casting, were used to glycerol-plasticize soybean protein concentrate-based films by Ciannamea et al. (2014). It was found that, compared to the solution-casting films at the same plasticizer level, the thermo-pressed SPC films were much more stretchable and less soluble, lower water vapor permeability but greater oxygen permeability coefficient. It could be caused by the intermolecular forces involved in the film formation. As revealed by solubility of obtained films in denaturing solutions and infrared spectroscopy, hydrophobic interactions and hydrogen bonding dominated the formation solution-casting films, whereas disulphide bonding played a more important role in the formation of compression molded films. The forming process played a major role in determining the final properties of SPC-based films. It revealed the possibility of SPC-glycerol mixtures to be transformed through thermo-mechanical processing into biodegradable films with the potential application in food packaging.

Heating/Drying

SP film properties were modified by heat-curing (Gennadios et al., 1996). Glycerin-plasticized SPI films cast from alkaline aqueous solutions were heated at 80 or 95°C for 2, 6, 14, or 24 h. The tensile strength (TS) and -\-b (yellowness) Hunter color values of heated films were increased and EB, MC, water solubility, water vapor permeability (WVP) values reduced. These effects were enhanced as the heating time and temperature increased.

Denavi et al. (2009a) prepared two types of SPI films: a commercial one (CSPI) and lab-made one (LSPI). The influence of drying conditions (air temperature and relative humidity, RH) on mechanical properties, solubility in water, and color of the films was investigated. The films were dried in a chamber with air circulation under conditions of RH (24, 30, 45, 60 and 66%) and air temperatures (34, 40, 55, 70 and 76°C). It was revealed that the mechanical properties of films made from LSPI and CSPI were influenced in a very different way by the drying conditions due to a diverse initial protein conformation in both materials. The solubility of the LSPI film was the lowest (about 50%) when the films obtained at high RH and temperatures ranging from 45 to 76°C. For the CSPI films, in contrast, the solubility did not depend on the drying process and it remained relatively constant (about 40%). It was found that the optimal drying conditions were

70°C and 30% RH for CSPI films and 60°C and 60% RH for LSPI films. Dried under these conditions, the CSPI films presented a higher tensile strength, lower EB, lower solubility and better water and oxygen permeability than that of LSPI ones.

The effects of temperature and vacuum pressure on the MC, WVP, color (L, a, b, and ΔE), TS, EB, and total soluble matter (TSM) of SPI films were examined by Kim et al. (2002b). SPI films were cured at 60, 72.5, or 85°C and at 101.3, 81.32, or 61.32 kPa for 24 h. As a result of heat-curing, MC, WVP, EB, and TSM decreased, and total color difference and TS increased. Pressure, individually and interactively with temperature, significantly affected film MC, TS, and TSM. Kim et al. (2002a) also investigated the effect of heat curing at atmospheric or sub-atmospheric conditions on selected properties (MC, WVP, color, TS, EB, and TSM content) of the cast SPI films. Films were heat-cured at 85°C for 6, 12, 18, or 24 h at absolute pressures of 101.3, 81.32, or 61.32 kPa. Heat-cured films increased the TS and decreased the WVP and EB compared to the unheated films. Compared to the heat-curing at atmospheric pressure, the heat treatment under vacuum reduced the film WVP faster. High TS values, low EB values, and low TSM values were also obtained within shorter heating time under vacuum. However, the size and number of cavities in cured films were increased by the vacuum treatment.

Ultrasound Treatment

SPI and SPC were treated with ultrasound 20 kHz probe and ultrasound baths (40 and 500 kHz) system by Jambrak et al. (2009). The results indicated that the ultrasound treatment significantly affected the texture of model systems prepared with SPC that gelled during the ultrasound treatment with probe 20 and 40 kHz bath for 15 min. The model system prepared with SPI creamed during the ultrasound treatment with probe 20 kHz for 15 min. The treatment with 20 kHz probe ultrasound significantly changed the conductivity, increased the solubility for SPC, enlarged the specific surface area and raised values of the emulsion activity index. The weight mean diameter and volume-surface average diameter were reduced considerably for all samples and all treatments. There was no improvement in foaming and emulsifying properties of SP model systems after 500 kHz bath treatment.

At varying powers (200, 400 or 600 W) and times (15 or 30 min), the effects of low-frequency (20 kHz) ultrasonication on functional and structural properties of reconstituted SPI dispersions were examined by Hu et al. (2013). Using the ultrasonic treatments, both the storage modulus and loss modulus of SPI dispersions were reduced and formed more viscous SPI dispersions (fluid character). Moreover, the ultrasound treatment significantly decreased the consistency coefficients and increased the flow behavior index of SPI dispersions. Shown in the Scanning

Electron Microscopy (SEM), lyophilized ultrasonicated SPI had different microstructure with larger aggregates compared to the non-treated SPI. With the ultrasonic treatment, the free sulfhydryl content, surface hydrophobicity and protein solubility of SPI dispersions were all increased. The differences in solubility profiles in the presence versus absence of denaturing and reducing (mercaptoethanol) agents after the ultrasonic treatment suggested a decrease in non-covalent interactions of SPI in dispersion. Using the secondary structure analysis by circular dichroism, it was found that the lower α-helix and random coil were obtained in SPI treated at lower power, in contrast to higher α-helix and lower β-sheet in SPI treated with higher power (600 W). The ultrasonic treatment lead to partial unfolding and reduction of intermolecular interactions as demonstrated by increases in free sulfhydryl groups and surface hydrophobicity, resulting in improved solubility and fluid character of SPI dispersions, while larger aggregates of ultrasonic-treated SPI in the dry state were formed after the lyophilization.

The effect of high intensity ultrasound (HIU, 105–110 W/cm^2 for 5 or 40 min) pre-treatment of SPI on the physicochemical properties of ensuing transglutaminase-catalyzed SPI cold set gel (TSCG) was investigated by Zhang et al. (2016b). After 40 min HIU pre-treatment applied to SPI, the gel strength of TSCG increased from 34.5 to 207.1 g and the gel yield and water holding capacity also increased. A more uniform and denser microstructure of TSCG was observed by the Scanning electron microscopy. The content of free sulfhydryl (SH) groups was higher in HIU-treated TSCG than non-HIU treated TSG, even though greater decrease of the SH groups present in HIU treated SPI was observed when the TSCG was formed. It was due to the involvement of disulfide bonds in gel formation. After the HIU pretreatment, the protein solubilities of TSCG in both denaturing and non-denaturing solvents were increased. The changes in hydrophobic amino acid residues as well as in polypeptide backbone conformation and secondary structure of TSCG were validated by Raman spectroscopy. It was suggested that the increased inter-molecular ε-(γ-glutamyl) lysine isopeptide bonds, disulfide bonds and hydrophobic interactions probably contributed to the HIU TSCG gel network. HIU treatment changed physicochemical and structural properties of SPI, producing improved substrates for TGase. The resulting TSCG network structure was formed with greater involvement of covalent and non-covalent interactions between SPI molecules and aggregates than in the TSCG from non-HIU SPI.

Hu et al. (2015) reported that both the HIU-treated and non-HIU-treated SPIs were cross-linked by transglutaminase to form hydrogels. HIU treatment increased the amount of high molecular weight aggregates due to the formation of ε-(γ-glutamyl) lysine bonds, improved the hydrophobic nature of transglutaminase gels and changed the 3D-network structure of transglutaminase induced SPI gel with riboflavin (TSGR). The 40-min

HIU treatment also increased gel yield, riboflavin encapsulation efficiency and gel strength of TSGR. HIU treatment decreased the swelling and protein erosion of TSGR in simulated gastrointestinal fluids. In addition, it led to reduced riboflavin release rate and altered the release mechanism in simulated gastrointestinal fluids both in the absence and presence of digestive enzymes. The HIU treatment could facilitated covalent cross-linking, increased hydrophobicity and changed the 3D network of TSGR, resulting in differences in hydrogel stability, as well as riboflavin encapsulation and release profiles.

With the assistance of ultrasonic/microwave treatment, edible films were prepared using SPI (4 g/100 g), oleic acid (0–2 g/100 g) and stearic acid (0–2 g/100 g) by Wang et al. (2014). The effect of the oleic acid to stearic acid ratio and ultrasonic/microwave assisted treatment on WVP and contact angle of the prepared films were evaluated. The WVP and contact angle were affected significantly by the changes in the oleic acid to stearic acid ratios. When the treatment temperature, time and power were 20°C, 15 min, and 500 W, respectively, it was discovered that the prepared films (oleic acid:stearic acid = 2:3) possessed the lowest WVP value $(0.1 \times 10^{-12} \text{ g cm}^{-1} \text{ s}^{-1} \text{ Pa}^{-1})$ and highest contact angle (135°).

Blending with Hydrophobic Additives

The effect of dialdehyde starch (DAS) on the physical properties of cast-SPI films was explored by Rhim et al. (1998). Films were casted from heated (70°C for 20 min) alkaline (pH 10) aqueous solutions of SPI at 5 g/100 ml water, glycerin (50 wt% of SPI), and DAS at 0, 5, 10, 15, or 20 wt% of SPI. For all film types used, after the SPI films were conditioned at 50% RH and 25°C for 48 h, the Hunter color values (L, a, and b), TS, EB, WVP, MC were measured. After the immersion in water at 25°C for 24 h, the total soluble matter (TSM) was examined. The addition of DAS increased the film yellowness (+b values), implying the occurrence of cross-linking between SPI and DAS. Compared to the control films, TS of the films with 5 or 10% DAS increased, whereas film EB was not significantly affected by DAS. A small increase in WVP or MC was found for the DAS-containing films, which was most likely due to the water absorption by hydrophilic groups along the DAS polymer chains. The TSM of SPI films was significantly reduced by DAS (approximately 50%, i.e., from 28.6% for control films to 14.4% for films with 20% DAS). It was shown that DAS had a potential to increase resistance to breakdown in water for the SPI film. The tensile strength of the film was increased from 8.2 to 15.8 or 14.7 MPa. However, the EB of film decreased from 30 to 6% due to the heat curing. The DAS-containing, heat-cured, and UV-irradiated films showed darker, because their L values were lower than that of the control films. It was demonstrated that PS of SPI films can be remarkably modified through chemical or physical treatments prior to or after the casting. The water barrier properties of biopolymer-based edible films were improved using lipid materials such as wax, fatty acids, neutral lipids and resins by Rhim (2004). Depending on the nature of lipid materials used, water vapor barrier properties of lipid-based biopolymer films and coatings were affected by the film structure and thermodynamic factors, such as temperature, vapor pressure or physical state of water in contact with films.

Mechanical, swelling, and optic properties of composite films made from SPI and gelatin were investigated by Cao et al. (2007). TS, EB, elastic modulus and swelling property of the SPI/gelatin composite films increased with the increase of the gelatin ratio in composite films. The films became more transparent and easier to handle. When the ratio of SPI/gelatin was 4:6/2:8, the TS, EB, and other properties of composite film approached those of gelatin film and were better than those of SPI film. Because the composite film is more economic than the gelatin film, it is possible to be used as edible films instead of gelatin films for food packaging.

Nanoparticle Modification of Polymers and Soy-based Adhesives

In an abundance of literature, nanophase reinforcement in polymers and adhesives has been shown to be very effective in improving their performance (Gao et al., 2012a; Liang and Shi, 2010; Liang and Shi, 2011; Xia et al., 2015a; 2015b; 2015c; 2016a; 2016b; 2016d; Yan et al., 2015). The multi-walled carbon nanotube (MWNT) was used to reinforce soy-based polyurethane foam (Liang and Shi, 2010). Compared to the neat PU foam, the compressive, flexural, and tensile properties of MWNTs-PU foams were improved by 24, 30 and 30%, respectively. Compared to the unaligned ones, the higher tensile strength and more effective load transfer from the polymer matrix to carbon nanotubes were found to be because of the aligned carbon nanotubes. The aligned MWNTs successfully prevented the crack development, leading to an increased tensile strength, and thus creating a new energy-dissipation path for foam composites. The polyurethane foam was made from soy-based polyol and the polymeric diphenylmethane diisocyanate (PMDI). The foam systems were incorporated with the nanoclay Cloisite 30B to improve the thermal stabilities and mechanical properties (Liang and Shi, 2011). The densities of nanoclay soy-based polyurethane foams were higher than that of the pure soy-based polyurethane foam. The mechanical properties of the soy-based polyurethane foam were increased by 22–98%, after loading 0.5 parts per hundred of polyols nanoclay by weight. The increase in the compressive strength and modulus of nanoclay soy-based polyurethane foams was due to the higher density and smaller cell size of foam composites. To prepare nano-scale to macro-scale cellulosic fibers from kenaf bast fibers for polymer composite reinforcement, various chemical processes were conducted by Shi et al. (2011b), including alkaline retting for obtaining the single cellulosic retted fiber, bleaching treatment for obtaining delignified bleached fiber, and acidic hydrolysis for obtaining both pure-cellulose microfiber and cellulose nanowhisker. The incorporation of 9 wt% cellulosic nano-whiskers (CNW) in PVA composites increased the tensile strength by 46%. The polymer composites were fabricated using the inorganic nanoparticle impregnation (INI)-treated fiber as reinforcement and polypropylene (PP) as the matrix (Shi et al., 2011a). The results showed that the INI treatments improved the compatibility between kenaf fibers and PP matrix. Reinforced with INI-treated fibers, the tensile modulus and tensile strength of the composites increased by 25.9% and 10.4%, respectively, compared to those reinforced with untreated kenaf fibers. Phenol resorcinol formaldehyde (PRF) resins were also modified with two Cloisite (montmorillonite) clay types, Cloisite 10A and Cloisite 30B, both at 3 wt% (Peng and Shi, 2009).

Cellulosic Nanofibers

Cellulosic nanofibers (CNF) can be obtained from natural fibers, such as wood, cotton and kenaf, by reducing the amorphous regions of the fiber, removing the lignin and hemicellulose, and breaking down the chains of the cellulose to the nanoscale units. CNF has a special rod-like fiber structure and a great amount of hydroxyl groups on its surface, which shows the great potential application in improving the performance of the matrix. In a recent study about SPI-based films (Zhang et al., 2016a), the tensile strength and 24-h water up-take of the CNF-modified films significantly increased by 53.0% and decreased by 29.1%, respectively. Cellulose nano-fibrils could also be used as a filler to enhance the fracture mechanical properties of adhesives (Arboleda et al., 2013; Wang et al., 2008).

SM-based adhesive reinforced with CNW was investigated by Gao et al. (2012a). The SM flour, CNW, NaOH, and PEG were used as the adhesive. The results showed that the formulation improved the water-resistance property of plywood by 20%. The soybean-meal/CNW/NaOH/PEG bonded plywood met the interior plywood requirement (200 (ANSI/HPVA HP-1)). There were fewer holes and cracks, as well as a smooth surface, observed on the cross section of the cured adhesive after the incorporation of CNW.

The modification of the SP adhesive was developed by increasing the physical bonding performance between the adhesives and wood fibers. Organic aerogels based on the two important and widely abundant renewable resources, SP and nano-fibrillar cellulose (NFC), were developed from precursor aqueous dispersions and a facile method conducive of channel- and defect-free systems yielded apparent densities on the order of 0.1 g/cm^3, after cooling and freeze-drying cycles (Arboleda et al., 2013).

The internal morphology of the composite aerogels was driven to transition from network- to fibrillar-like, with high density of interconnected cells by loading NFC. A compression modulus of 4.4 MPa of the composite aerogels with SP loadings of ca. 70% was obtained, which was very close to that obtained from the pure NFC aerogels. Based on soy-flour mass, the best combination was found to be 10% lime, 2% milk and 20 wt% NaOH (Wang et al., 2008). The viscosity was increased and the mold growth was restrained after early stage when 0.5 wt% sodium benzoate or 25 wt% PF was added. The above modifications are limited within the physical bonding between the adhesive and the cellulosic material.

Inorganic Nanoparticles

By investigating the structure and morphology of the adhesive and its fracture bonding interface and adhesion strength, the biomimetic SP/CaCO$_3$ hybrid wood glue was developed by (Liu et al., 2010). The results showed that the water resistance and bonding strength of SP adhesives were significantly improved by the compact rivets or interlocking links, and ion crosslinking of calcium carbonate/hydroxyl in the adhesive. Glue strength of SP hybrid adhesive was higher than 6 MPa even after three water-immersion cycles.

Three curing agents (SF1, SF2 and SF3) were synthesized from epichlorohydrin and ammonium hydroxide (Huang et al., 2013). The combinations of SF with one of the curing agents were investigated as adhesives for making interior plywood. The results from the water resistance tests showed that the plywood panels bonded with SF1 and SF3 adhesives could be used as interior plywood, whereas those bonded with SF2 could not.

Nano-clays

Using the renewable glycerol polyglycidyl ether as a cross-linking agent and low-cost attapulgite (ATP) as an enhancer, an excellent SP adhesive was developed for the plywood industry (Li et al., 2014a). The results indicated that the shear strength and water resistance of the AMSF adhesives were improved significantly. After ATP was introduced, the viscosity and solid content of the MSF adhesive increased and met the requirements of ANSI/HPVA HP-1-2000. The characterization of bio-based composites comprising jute fabric and SPC modified with glycerol and/or halloysite nanotubes (HNT) was reported by Nakamura et al. (2013). The results revealed that SPC had lower flammability (heat release capacity) than that of the petrochemical-based resins. Furthermore, it was found that incorporating 5% mass fraction of HNT would reduce the composite flammability without compromising the mechanical properties.

Soy Adhesive as a Substitute of Other Adhesives

The use of a soy adhesive system as a substitute for melamine urea formaldehyde (MUF) adhesive in plywood production was explored by Guezguez et al. (2013). The results demonstrated that the addition of soy adhesive system significantly reduced the formaldehyde release from the plywood. It was recommended that the 25% substitution was the highest level for the panel production with very low formaldehyde emission. Using SF and a renewable curing agent, a new formaldehyde-free wood adhesive was prepared for the fabrication of particleboards (Gu et al., 2013). The new curing agent was derived from ammonia and epichlorohydrin that was originated from the renewable glycerol. The composition of the adhesive was SF/NaOH/curing agent at a dry weight ratio of 9/0.3/1.

Nonporous-composite membranes owing good physical properties, were developed using SPC and cellulose (Zhu et al., 2013). Utilizing the glutaraldehyde as a crosslinking agent, the molecular network structure of the blended membranes was stabilized. The results showed that the membranes were strong, flexible, and the exposure to the crosslinking agent demonstrated the structural and thermal improvement of the network membranes. The developed blend of biopolymer membranes with improved physical abilities could be promisingly applied for food packaging, filtration systems, or even medical applications.

To examine the water resistance of the adhesive (three-cycle soak test), 3-ply plywood specimens were fabricated (Gao et al., 2012b). The results showed that the using of SDS improved the water resistance of the SM adhesive by 30%. After incorporating the MPA, the viscosity of the adhesive was decreased by 81%, the water resistance of the SM/SDS/MPA adhesive was further improved by 60%, and the solid content of the adhesive was increased by 15%. It could cause by the reaction of the ammonia group in SP and the carboxyl group to form a chemical structure network within the soy adhesive. The plywood bonded by the SM/SDS/MPA adhesive met the interior plywood requirements. From the FTIR analysis, it was found that more peptide linkages were formed, if MPA was incorporated into the cured adhesive. The polyethyleneglycol diacrylate (PEGDA) was used as a crosslinker and viscosity reducer to add into the SM-based adhesive system (Gao et al., 2013). After using 4% PEGDA-modified SM adhesive, the viscosity of modified SM adhesives were decreased by 35% and the wet shear strength of plywood was increased by 114.2%. SF-based adhesives were prepared with either γ-amino, γ-glycidoxy, or γ-methacryloyloxy-propyltrimethoxysilane (KH550, KH560, and KH570) silane coupling agents (SCAs) (Li et al., 2014b). The results showed that KH560 was the most efficient SCA for improving the wet bonding strength. After adding 3 wt% KH560 into the SM adhesive, the wet bonding strength of the sample was maximized at 0.98 MPa, meeting the requirements of interior plywood. KH560 can be a crosslinking agent to enhance the water-resistant bonding strength of plywood.

Conclusions

Soy protein (SP) existing in soy flour (SF), soy meal (SM), soy protein concentrate (SPC), and soy protein isolate (SPI), is one of the vegetable proteins. With the advantages of renewability, processability, biocompatibility, film-forming capacity, and biodegradability, SP has a great potential to be used in food industry, wood industry, as well as in the fields of agriculture, bioscience, and biotechnology. The positive public perception of SP makes it very competitive as a wood adhesive so that it represents a very practical and inexpensive material for bio-based adhesives. While it is weak in adhesion strength and water resistance compared to the traditional formaldehyde-based resins, the performance of the SP adhesive bonded composites can be significantly improved after proper physical, chemical and enzymatic modifications. Future studies should be done on understanding the bonding mechanism of the modified soy-based resin with the cellulosic fibers.

Keywords: Soy, Adhesive, Modification, Bonding performance

References

Andreuccetti, C., R.A. Carvalho, T. Galicia-Garcia, F. Martinez-Bustos and C.R.F. Grosso. 2011. Effect of surfactants on the functional properties of gelatin-based edible films. J. Food Eng. 103: 129–136. doi:10.1016/j.jfoodeng.2010.10.007.

Arboleda, J.C., M. Hughes, L.A. Lucia, J. Laine, K. Ekman and O.J. Rojas. 2013. Soy protein-nanocellulose composite aerogels. Cellulose 20: 2417–2426. doi:10.1007/s10570-013-9993-4.

Barać, M.B., S.P. Stanojević, S.T. Jovanović and M.B. Pešić. 2004. Soy protein modification: A review. Acta Periodica Technologica 3–16.

Bian, K. and S. Sun. 1998. Adhesive performance of modified soy protein polymers. Abstracts of Papers of the American Chemical Society 216, U1-U1.

Bjorksten, J. 1951. Cross Linkages in Protein Chemistry. *In*: Anonymous Advances in Protein Chemistry. Academic Press, New York, pp. 353–358.

Cao, N., Y. Fu and J. He. 2007. Preparation and physical properties of soy protein isolate and gelatin composite films. Food Hydrocoll. 21: 1153–1162. doi:10.1016/j.foodhyd.2006.09.001.

Chen, N., Q. Lin, Q. Zeng and J. Rao. 2013. Optimization of preparation conditions of soy flour adhesive for plywood by response surface methodology. Ind. Crop. Prod. 51: 267–273. doi:10.1016/j.indcrop.2013.09.031.

Chen, N., Q. Zeng, Q. Lin and J. Rao. 2015. Effect of enzymatic pretreatment on the preparation and properties of soy-based adhesive for plywood. Bioresources 10: 5071–5082.

Chen, N., Q. Zeng, J. Rao and Q. Lin. 2014. Effect of preparation conditions on bonding strength of soy-based adhesives via viscozyme L action on soy flour slurry. Bioresources 9: 7444–7453.

Ciannamea, E.M., P.M. Stefani and R.A. Ruseckaite. 2014. Physical and mechanical properties of compression molded and solution casting soybean protein concentrate based films. Food Hydrocoll. 38: 193–204. doi:10.1016/j.foodhyd.2013.12.013.

Cui, Z., X. Kong, Y. Chen, C. Zhang and Y. Hua. 2014. Effects of rutin incorporation on the physical and oxidative stability of soy protein-stabilized emulsions. Food Hydrocoll. 41: 1–9. doi:10.1016/j.foodhyd.2014.03.006.

Cunningham, P., A.A. Ogale, P.L. Dawson and J.C. Acton. 2000. Tensile properties of soy protein isolate films produced by a thermal compaction technique. J. Food Sci. 65: 668–671. doi:10.1111/j.1365-2621.2000.tb16070.x.

Denavi, G., D.R. Tapia-Blacido, M.C. Anon, P.J.A. Sobral, A.N. Mauri and F.C. Menegalli. 2009a. Effects of drying conditions on some physical properties of soy protein films. J. Food Eng. 90: 341–349. doi:10.1016/j.jfoodeng.2008.07.001.

Denavi, G.A., M. Perez-Mateos, M.C. Anon, P. Montero, A.N. Mauri and M. Carmen Gomez-Guillen. 2009b. Structural and functional properties of soy protein isolate and cod gelatin blend films. Food Hydrocolloid. 23: 2094–2101. doi:10.1016/j.foodhyd.2009.03.007.

Fang, L., L. Chang, W. Guo, Y. Chen and Z. Wang. 2014. Influence of silane surface modification of veneer on interfacial adhesion of wood-plastic plywood. Appl. Surf. Sci. 288: 682–689. doi:10.1016/j.apsusc.2013.10.098.

Franzen, K.L. and J.E. Kinsella. 1976. Functional properties of succinylated and acetylated soy protein. J. Agric. Food Chem. 24: 788–795.

Frihart, C.R. and L. Lorenz. 2013. Protein modifiers generally provide limited improvement in wood bond strength of soy flour adhesives. For. Prod. J. 63: 138–142.

Gao, Q., J. Li, S.Q. Shi, K. Liang and X. Zhang. 2012a. Soybean meal-based adhesive reinforced with cellulose nano-whiskers. Bioresources 7: 5622–5633.

Gao, Q., Z. Qin, C. Li, S. Zhang and J. Li. 2013. Preparation of wood adhesives based on soybean meal modified with PEGDA as a crosslinker and viscosity reducer. Bioresources 8: 5380–5391.

Gao, Q., S.Q. Shi, J. Li, K. Liang and X. Zhang. 2012b. Soybean meal-based wood adhesives enhanced by modified polyacrylic acid solution. Bioresources 7: 946–956.

Gao, Q., S. Shi, S. Zhang, J. Li and K. Liang. 2011. Improved plywood strength and lowered emissions from soybean meal/melamine urea-formaldehyde adhesives. For. Prod. J. 61: 688–693.

Gennadios, A., V. M. Ghorpade, C.L. Weller and M.A. Hanna. 1996. Heat curing of soy protein films. Trans. ASAE 39: 575–579.

Gu, K., J. Huang and K. Li. 2013. Preparation and evaluation of particleboard bonded with a soy flour-based adhesive with a new curing agent. J. Adhes. Sci. Technol. 27: 2053–2064. doi:10.1080/01694243.2012.696950.

Guerrero, P., A. Retegi, N. Gabilondo and K. de la Caba. 2010. Mechanical and thermal properties of soy protein films processed by casting and compression. J. Food Eng. 100: 145–151. doi:10.1016/j.jfoodeng.2010.03.039.

Guezguez, B., M. Irle and C. Belloncle. 2013. Substitution of formaldehyde based adhesives with soy based adhesives in production of low formaldehyde emission wood based panels. Part 1-Plywood. International Wood Products Journal 4: 30–32.

Guo, G., C. Zhang, Z. Du, W. Zou, A. Xiang and H. Li. 2015. Processing and properties of phthalic anhydride modified soy protein/glycerol plasticized soy protein composite films. J. Appl. Polym. Sci. 132: 42221. doi:10.1002/app.42221.

He, G., M. Feng and C. Dai. 2012. Development of soy-based adhesives for the manufacture of wood composite products. Holzforschung 66: 857–862. doi:10.1515/hf-2011-0196.

Hu, H., J. Wu, E.C. Y. Li-Chan, L. Zhu, F. Zhang, X. Xu, G. Fan, L. Wang, X. Huang and S. Pan. 2013. Effects of ultrasound on structural and physical properties of soy protein isolate (SPI) dispersions. Food Hydrocoll. 30: 647–655. doi:10.1016/j.foodhyd.2012.08.001.

Hu, H., X. Zhu, T. Hu, I.W.Y. Cheung, S. Pan and E.C.Y. Li-Chan. 2015. Effect of ultrasound pre-treatment on formation of transglutaminase-catalysed soy protein hydrogel as a riboflavin vehicle for functional foods. J. Funct. Foods 19: 182–193. doi:10.1016/j.jff.2015.09.023.

Huang, J., K. Gu and K. Li. 2013. Development and evaluation of new curing agents derived from glycerol for formaldehyde-free soy-based adhesives in wood composites. Holzforschung 67: 659–665. doi:10.1515/hf-2012-0102.

Huang, W.N. and X.Z. Sun. 2000a. Adhesive properties of soy proteins modified by sodium dodecyl sulfate and sodium dodecylbenzene sulfonate. J. Am. Oil Chem. Soc. 77: 705–708. doi:10.1007/s11746-000-0113-6.

Huang, W.N. and X.Z. Sun. 2000b. Adhesive properties of soy proteins modified by urea and guanidine hydrochloride. J. Am. Oil Chem. Soc. 77: 101–104. doi:10.1007/s11746-000-0016-6.

Huang, X. and A. Netravali. 2009. Biodegradable green composites made using bamboo micro/nano-fibrils and chemically modified soy protein resin. Composites Sci. Technol. 69: 1009–1015. doi:10.1016/j.compscitech.2009.01.014.

Huang, Y., P. Wu, M. Zhang, W. Ruan and E.P. Giannelis. 2014. Boron cross-linked graphene oxide/polyvinyl alcohol nanocomposite gel electrolyte for flexible solid-state electric double layer capacitor with high performance. Electrochim. Acta 132: 103–111. doi:10.1016/j.electacta.2014.03.151.

Jambrak, A.R., V. Lelas, T.J. Mason, G. Kresic and M. Badanjak. 2009. Physical properties of ultrasound treated soy proteins. J. Food Eng. 93: 386–393. doi:10.1016/j.jfoodeng.2009.02.001.

Janjarasskul, T. and J.M. Krochta. 2010. Edible packaging materials. Annual Review of Food Science and Technology 1: 415–448. doi:10.1146/annurev.food.080708.100836.

Jong, L. 2011. Aggregate structure and effect of phthalic anhydride-modified soy protein on the mechanical properties of styrene-butadiene copolymer. J. Appl. Polym. Sci. 119: 1992–2001. doi:10.1002/app.32870.

Jong, L. 2009. POLY 593-Effect of phthalic anhydride modified soy protein on viscoelastic properties of polymer composites. Abstracts of Papers of the American Chemical Society 238.

Kalapathy, U., N. Hettiarachchy, D. Myers and M.A. Hanna. 1995. Modification of soy proteins and their adhesive properties on woods. J. Am. Oil Chem. Soc. 72: 507–510. doi:10.1007/BF02638849.

Khan, M.K.I., M.A.I. Schutyser, K. Schroen and R. Boom. 2012. The potential of electrospraying for hydrophobic film coating on foods. J. Food Eng. 108: 410–416. doi:10.1016/j.jfoodeng.2011.09.005.

Kim, K.M., C. Weller, M. Hanna and A. Gennadios. 2002a. Heat curing of soy protein films at atmospheric and subatmospheric conditions. J. Food Sci. 67: 708–713.

Kim, K.M., C.L. Weller, M.A. Hanna and A. Gennadios. 2002b. Heat curing of soy protein films at selected temperatures and pressures. LWT-Food Sci. Technol. 35: 140–145. doi:10.1006/fstl.2001.0825.

Kumar, R., V. Choudhary, S. Mishra, I.K. Varma and B. Mattiason. 2002. Adhesives and plastics based on soy protein products. Ind. Crop. Prod. 16: 155–172. doi:10.1016/S0926-6690(02)00007-9.

Lambuth, A.L. 2003. Protein adhesives for wood. pp. 172–180. *In*: I. Skeist (ed.). Handbook of Adhesive Technology. Marcel Dekker, Inc., New York, NY.

Lawal, O.S. and K.O. Adebowale. 2004. Effect of acetylation and succinylation on solubility profile, water absorption capacity, oil absorption capacity and emulsifying properties of mucuna bean (*Mucuna pruriens*) protein concentrate. Nahrung-Food 48: 129–136. doi:10.1002/food.200300384.

Lei, H., G. Du, Z. Wu, X. Xi and Z. Dong. 2014. Cross-linked soy-based wood adhesives for plywood. Int. J. Adhes. Adhes. 50: 199–203. doi:10.1016/j.ijadhadh.2014.01.026.

Li, C., H. Li, S. Zhang and J. Li. 2014a. Preparation of reinforced soy protein adhesive using silane coupling agent as an enhancer. Bioresources 9: 5448–5460.

Li, H., C. Li, Q. Gao, S. Zhang and J. Li. 2014b. Properties of soybean-flour-based adhesives enhanced by attapulgite and glycerol polyglycidyl ether. Industrial Crops and Products 59: 35–40. doi:10.1016/j.indcrop.2014.04.041.

Li, J., J. Luo, X. Li, Z. Yi, Q. Gao and J. Li. 2015. Soybean meal-based wood adhesive enhanced by ethylene glycol diglycidyl ether and diethylenetriamine. Ind. Crop. Prod. 74: 613–618. doi:10.1016/j.indcrop.2015.05.066.

Li, K.C., S. Peshkova and X.L. Geng. 2004. Investigation of soy protein-Kymene((R)) adhesive systems for wood composites. J. Am. Oil Chem. Soc. 81: 487–491. doi:10.1007/s11746-004-0928-1.

Liang, F. and Y. Wang. 1999. Effects of silane coupling agents on the interface of soy protein and glass fiber. Proceedings of American Society for Composites 511–520.

Liang, K.W. and S.Q. Shi. 2010. Soy-based polyurethane foam reinforced with carbon nanotubes. Key Engineering Materials 419: 477–480.

Liang, K. and S.Q. Shi. 2011. Nanoclay filled soy-based polyurethane foam. J. Appl. Polym. Sci. 119: 1857–1863. doi:10.1002/app.32901.

Liu, D., H. Chen, P.R. Chang, Q. Wu, K. Li and L. Guan. 2010. Biomimetic soy protein nanocomposites with calcium carbonate crystalline arrays for use as wood adhesive. Bioresour. Technol. 101: 6235–6241. doi:10.1016/j.biortech.2010.02.107.

Liu, Y. and K. Li. 2007. Development and characterization of adhesives from soy protein for bonding wood. Int. J. Adhes. Adhes. 27: 59–67. doi:10.1016/j.ijadhadh.2005.12.004.

Mackay, C.D. 1998. Good adhesive bonding starts with surface preparation. Adhes. Age 41: 30–32.

Mangavel, C., N. Rossignol, A. Perronnet, J. Barbot, Y. Popineau and J. Gueguen. 2004. Properties and microstructure of thermo-pressed wheat gluten films: A comparison with cast films. Biomacromolecules 5: 1596–1601. doi:10.1021/bm049855k.

Mateos-Aparicio, I., A.R. Cuenca, M.J. Villanueva-Suarez and M.A. Zapata-Revilla. 2008. Soybean, a promising health source. Nutr. Hosp. 23: 305–312.

Mo, X.Q. and X.Z. Sun. 2002. Plasticization of soy protein polymer by polyol-based plasticizers. J. Am. Oil Chem. Soc. 79: 197–202. doi:10.1007/s11746-002-0458-x.

Mo, X. and X. S. Sun. 2013. Soy proteins as plywood adhesives: formulation and characterization. J. Adhes. Sci. Technol. 27: 2014–2026. doi:10.1080/01694243.2012.696916.

Monedero, F.M., M.J. Fabra, P. Talens and A. Chiralt. 2009. Effect of oleic acid–beeswax mixtures on mechanical, optical and water barrier properties of soy protein isolate based films. J. Food Eng. 91: 509–515.

Nakamura, R., A.N. Netravali, A.B. Morgan, M.R. Nyden and J.W. Gilman. 2013. Effect of halloysite nanotubes on mechanical properties and flammability of soy protein based green composites. Fire Mater. 37: 75–90. doi:10.1002/fam.2113.

Nishinari, K., Y. Fang, S. Guo and G. Phillips. 2014. Soy proteins: A review on composition, aggregation and emulsification. Food Hydrocoll. 39: 301–318.

Nordqvist, P., F. Khabbaz and E. Malmstrom. 2010. Comparing bond strength and water resistance of alkali-modified soy protein isolate and wheat gluten adhesives. Int. J. Adhes. Adhes. 30: 72–79. doi:10.1016/j.ijadhadh.2009.09.002.

Paetau, I., C. Chen and J. Jane. 1994. Biodegradable plastic made from soybean products. 1. Effect of preparation and processing on mechanical properties and water absorption. Ind. Eng. Chem. Res. 33: 1821–1827.

Peng, Y. and S.Q. Shi. 2009. Thermal property improvement of phenol resorcinol formaldehyde resin by incorporation of cloisite clay particles. Wood Adhesives 2009: Session 2B-Composites 152–156.

Qin, Z., Q. Gao, S. Zhang and J. Li. 2013. Glycidyl methacrylate grafted onto enzyme-treated soybean meal adhesive with improved wet shear strength. Bioresources 8: 5369–5379.

Reddy, N. and Y. Yang. 2013. Thermoplastic films from plant proteins. J. Appl. Polym. Sci. 130: 729–738. doi:10.1002/app.39481.

Rhim, J.W. 2004. Increase in water vapor barrier property of biopolymer-based edible films and coatings by compositing with lipid materials. Food Sci. Biotechnol. 13: 528–535.

Rhim, J.W., A. Gennadios, C.L. Weller, C. Cezeirat and M.A. Hanna. 1998. Soy protein isolate dialdehyde starch films. Ind. Crop. Prod. 8: 195–203. doi:10.1016/S0926-6690(98)00003-X.

Roy, J.J., L. Sun and L. Ji. 2014. Microalgal proteins: a new source of raw material for production of plywood adhesive. J. Appl. Phycol. 26: 1415–1422. doi:10.1007/s10811-013-0169-2.

Shi, J., S.Q. Shi, H.M. Barnes, M.F. Horstemeyer and G. Wang. 2011a. Kenaf bast fibers-Part II: Inorganic nanoparticle impregnation for polymer composites. Int. J. Polym. Sci. 2011: 736474. doi:10.1155/2011/736474.

Shi, J., S.Q. Shi, H.M. Barnes, C.U. Pittman, Jr. 2011b. A chemical process for preparing cellulosic fibers hierarchically from kenaf bast fibers. Bioresources 6: 879–890.

Shih, F.F. 1994. Interaction of soy isolate with polysaccharide and its effect on film properties. J. Am. Oil Chem. Soc. 71: 1281–1285. doi:10.1007/BF02540552.

Song, F., D. Tang, X. Wang and Y. Wang. 2011. Biodegradable soy protein isolate-based materials: A review. Biomacromolecules 12: 3369–3380. doi:10.1021/bm200904x.

Stuchell, Y.M. and J.M. Krochta. 1994. Enzymatic treatments and thermal effects on edible soy protein films. J. Food Sci. 59: 1332–1337.

Taghiyari, H.R., K. Mobini, Y.S. Samadi, Z. Doosti, F. Karimi, M. Asghari, A. Jahangiri and P. Nouri. 2013a. Effects of nano-wollastonite on thermal conductivity coefficient of medium-density fiberboard. J. Nanomater. Mol. Nanotechnol. 2: 1. doi:10.4172/2324-8777.1000106.

Taghiyari, H.R., A. Karimi and P.M. Tahir. 2013b. Nano-wollastonite in particleboard: physical and mechanical properties. BioResources 8: 5721–5732.

Taghiyari, H.R., A. Moradiyan and A. Farazi. 2013c. Effect of nanosilver on the rate of heat transfer to the core of the medium density fiberboard mat. International Journal of Bio-Inorganic Hybrid Nanomaterials 2: 303–308.

Taghiyari, H.R. 2013. Nano-zycosil in MDF: gas and liquid permeability. Eur. J. Wood Wood Prod. 71: 353–360. doi:10.1007/s00107-013-0691-6.

Taghiyari, H.R., A. Karimi and P.M. Tahir. 2015. Organo-silane compounds in medium density fiberboard: physical and mechanical properties. J. For. Res. 26: 495–500. doi:10.1007/s11676-015-0033-0.

Wang, W.H., X.P. Li and X.Q. Zhang. 2008. A soy-based adhesive from basic modification. Pigm. Resin Technol. 37: 93–97. doi:10.1108/03699420810860446.

Wang, Z., J. Zhou, X. Wang, N. Zhang, X. Sun and Z. Ma. 2014. The effects of ultrasonic/microwave assisted treatment on the water vapor barrier properties of soybean protein isolate-based oleic acid/stearic acid blend edible films. Food Hydrocoll. 35: 51–58. doi:10.1016/j.foodhyd.2013.07.006.

Wu, Y.V. and G.E. Inglett. 1974. Denaturation of plant proteins related to functionality and food applications. A review. J. Food Sci. 39: 218–225.

Xia, C., S.Q. Shi, L. Cai and J. Hua. 2015a. Property enhancement of kenaf fiber composites by means of vacuum-assisted resin transfer molding (VARTM). Holzforschung 69: 307–312. doi:10.1515/hf-2014-0054.

Xia, C., S.Q. Shi, L. Cai and S. Nasrazadani. 2015b. Increasing inorganic nanoparticle impregnation efficiency by external pressure for natural fibers. Ind. Crop. Prod. 69: 395–399. doi:10.1016/j.indcrop.2015.02.054.

Xia, C., S.Q. Shi and L. Cai. 2015c. Vacuum-assisted resin infusion (VARI) and hot pressing for CaCO3 nanoparticle treated kenaf fiber reinforced composites. Composites Part B 78: 138–143. doi:10.1016/j.compositesb.2015.03.039.

Xia, C., S.Q. Shi, Y. Wu and L. Cai. 2016a. High pressure-assisted magnesium carbonate impregnated natural fiber-reinforced composites. Ind. Crop. Prod. 86: 16–22. doi:10.1016/j.indcrop.2016.03.023.

Xia, C., L. Wang, Y. Dong, S. Zhang, S.Q. Shi, L. Cai and J. Li. 2015d. Soy protein isolate-based films cross-linked by epoxidized soybean oil. RSC Adv. 5: 82765–82771. doi:10.1039/C5RA15590H.

Xia, C., S. Zhang, H. Ren, S.Q. Shi, H. Zhang, L. Cai and J. Li. 2016b. Scalable fabrication of natural-fiber reinforced composites with electromagnetic interference shielding properties by incorporating powdered activated carbon. Materials 9: 10. doi:10.3390/ma9010010.

Xia, C., S. Zhang, S.Q. Shi, L. Cai, A.C. Garcia, H.R. Rizvi and N.A. D'Souza. 2016c. Property enhancement of soy protein isolate-based films by introducing POSS. Int. J. Biol. Macromol. 82: 168–173. doi:10.1016/j.ijbiomac.2015.11.024.

Xia, C., S. Zhang, S.Q. Shi, L. Cai and J. Huang. 2016d. Property enhancement of kenaf fiber reinforced composites by *in situ* aluminum hydroxide impregnation. Ind. Crop. Prod. 79: 131–136. doi:10.1016/j.indcrop.2015.11.037.

Yan, Y., Y. Dong, J. Li, S. Zhang, C. Xia, S.Q. Shi and L. Cai. 2015. Enhancement of mechanical and thermal properties of Poplar through the treatment of glyoxal-urea/nano-SiO2. RSC Adv. 5: 54148–54155. doi:10.1039/C5RA07294H.

Zhang, S., C. Xia, Y. Dong, Y. Yan, J. Li, S.Q. Shi and L. Cai. 2016a. Soy protein isolate-based films reinforced by surface modified cellulose nanocrystal. Ind. Crop. Prod. 80: 207–213. doi:10.1016/j.indcrop.2015.11.070.

Zhang, P., T. Hu, S. Feng, Q. Xu, T. Zheng, M. Zhou, X. Chu, X. Huang, X. Lu, S. Pan, E.C.Y. Li-Chan and H. Hu. 2016b. Effect of high intensity ultrasound on transglutaminase-catalyzed soy protein isolate cold set gel. Ultrason. Sonochem. 29: 380–387. doi:10.1016/j.ultsonch.2015.10.014.

Zhu, Y., E. Douglass, T. Theyson, R. Hogan and R. Kotek. 2013. Cellulose and soy proteins based membrane networks. pp. 70–86. *In*: Anonymous Macromolecular Symposia. Wiley Online Library.

4

Canola Protein and Oil-based Wood Adhesives

Ningbo Li,[1] Guangyan Qi,[2] Xiuzhi Susan Sun[2] and Donghai Wang[3],*

ABSTRACT

This chapter reviews the application of canola protein and oil for wood adhesives. The most prevalent canola products include oil for human consumption and meal for livestock feed. Canola seeds consist of 19.5%–23.5% protein and 37.5%–50.9% oil, depending on the breeds and growing environments. Canola meal, which contains 30%–50% protein and is a by-product of canola oil extraction, is usually not used in human food applications due to the presence of glucosinolates, erucic acid, phytates, and phenolics. Development of value-added product from canola meal will provide new market opportunities for the canola industry and US farmers, as well as expand capabilities for an "eco green" biorenewable economy and reduce reliance on heavily utilized fossil fuel feedstocks. From previous researches, canola protein-based adhesives have shown good wood bonding performance in terms of dry, wet, and soaked shear strength. Drawbacks of canola protein-based adhesives, such as high viscosity and relatively low water resistance, can be overcame by using

---

[1] Sunhai Bioadhesive Technologies LLC, 2005 Research Park Cir, Manhattan, KS 66502.
[2] Biomaterials and Technology Laboratory, Department of Grain Science and Industry, Kansas State University, Manhattan, KS 66506.
[3] Department of Biological and Agricultural Engineering, Kansas State University, Manhattan, KS 66506.
* Corresponding author: dwang@ksu.edu

reducing agents to unfold protein's structure and grafting functional groups to the amino acid chains of canola protein. Modification through cross-linking could significantly increase the bonding strength and water resistance of canola protein-based adhesives. This chapter also describes blending canola meal with synthetic resins for wood adhesives. Canola meal and synthetic adhesives showed potential for use as bonders in particle board and wood flooring. Canola oil can also be used to make wood adhesives by synthesizing canola oil into polyurethanes.

CANOLA PROTEIN-BASED ADHESIVES

Introduction

Canola Seed

Canola, a crop developed from rapeseed in Canada in 1976, produces high-quality oil that is low in saturated fat (Bell, 1993; Canola Council of Canada, 1990). It originated from the Brassica family and is also known as rape, oilseed rape, rapa, or rapeseed. It contains 40% oil and 19.5%–23.5% protein (Barthet, 2014). As an important edible and industrial oil source, worldwide canola production has been increasing in recent years. Canola currently ranks as the second-largest oilseed crop (12.6% in year 2015/16) produced after soy (60.5% in year 2015/2016), with increased global production from 61.46 million metric tons (MMT) in 2011–2012 to approximately 67.09 MMT in 2015–2016 (Table 1) (USDA, 2015a). Most canola production in the United States occurs in northern states adjacent to Canada, such as North Dakota, Idaho, and Montana. North Dakota alone produces over 80% of the nation's canola crop (USDA, 2015b). Canola production in the United States accounts for approximately 2% of global production, and the production of canola increased to 1.26 MMT in 2014 from 1.20 MMT in

Table 1. Annual global oilseed production from 2011 to 2016. Adapted from USDA (2015a).

Oilseeds	Production (10^6 Mg)				
	2011/12	2012/13	2013/14	2014/15	2015/16
Soybean	240.43	268.82	283.15	318.68	321.02
Canola/Rapeseed	61.46	63.62	71.96	71.9	67.09
Peanut	38.46	40.45	41.15	39.48	40.79
Sunflower seed	39.21	35.52	42.35	39.99	39.69
Cottonseed	48.02	46.15	45.68	44.32	39.77
Palm	13.86	14.88	15.74	16.29	17.09
Copra	5.59	5.79	5.43	5.43	5.51

2012 although the growing area decreased to 1.55 million acres in 2014 from 1.71 million acres in 2012. The production increase was attributed to yield per acre increase from 1,392 lbs. in 2012 to 1,614 lbs. in 2014 (USDA, 2015a).

Because rapeseed contains high levels of erucic acid and glucosinolates, it is unsuitable for human and animal consumption. Therefore, rapeseed was bred to contain less than 2% erucic acid and less than 30 micromoles of glucosinolates and was classified as canola (Bell, 1993; Canola Council of Canada, 1990). The most commonly used canola products include oil for human consumption and meal for livestock feed. Depending on the breeds and environment, oil content in canola seeds can range from 37.5% to 50.9%. For any known canola variety, cool growing conditions result in higher oil content compared to warm growing conditions. In addition, oil and protein contents follow an inverse relationship: the lower the oil, the higher the protein content in canola seed (Barthet, 2014).

Canola Proteins

Canola seeds consist of 19.5%–23.5% protein, depending on the breeds and growing environments (Barthet, 2014). Canola meal, a by-product of canola oil extraction, is comprised of an average of 38% protein and 40% carbohydrate (Bell, 1993). Canola meal is usually not used in human food applications due to the presence of glucosinolates, erucic acid, phytates, and phenolics (Bell, 1993; Canola Council of Canada, 1990), which are typically associated with toxia effects, dark color, bitter taste, and astringency for human consumption (Hale, 2013). However, glucosinolate levels in canola are low enough to allow the meal to be used in animal feed applications (Bell, 1993). Two major storage protein fractions, napin (2S) and cruciferin (11S), constitute 20% and 60% of the total protein, respectively, in mature canola seeds (Hoglund et al., 1992; Li et al., 2012a; Schwenke et al., 1983). Napin, which belongs to albumin storage proteins, exhibits molecular weights from 12.5 to 14.5 KDa (Monsalve and Rodriguez, 1990). Disulfide bonds primarily stabilize napin protein structure by generating two disulphide-linked polypeptide chains (Krzyzaniak et al., 1998). Napin has a high content of α-helical structure (40%–46%) and a low content of β-sheet conformation (12%) in the secondary structure (Schwenke, 1994). Cruciferin, however, has a low content of α-helical structure (10%) and a high content of β-sheet conformation (50%) (Zirwer et al., 1985). Cruciferin, belongs to globulin storage protein, has a hexamer structure similar to soy glycinin protein (Berot et al., 2005) and is a neutral protein with a molecular weight of approximately 300 KDa (Schwenke et al., 1983). Covalent (disulfide bonds) and non-covalent bonds dominate cruciferin protein structure (Wu and Muir, 2008).

Using alkaline solution to solubilize canola protein and then using HCl for protein precipitation at the isoelectric point (pI) is a popular method to isolate canola protein (Ghodsvali et al., 2005; Kalapathy et al., 1996; Wäsche et al., 1998; Pedroche et al., 2004; Tzeng et al., 1988). High pH (> 11) is necessary to maximally hydrate the protein in canola meal, and pH 3.5–5.5 is primarily used to precipitate canola proteins (Ghodsvali et al., 2005; Tzeng et al., 1990; Pedroche et al., 2004; Tzeng et al., 1988). In the protein extraction process in previous research, canola meal was initially added to the alkaline solution at pH 12 and stirred for a period of time (0.5–2 h) to solubilize the protein in the canola meal. The mixture was then centrifuged to separate the solubilized protein and pellet. The pH of the supernatant was adjusted with dilute acids to precipitate the protein (pH 3.5–5.5), and the protein was separated using centrifuging (Manamperi et al., 2010a; Li et al., 2012a). More than one pI of canola protein was reported, and protein fractions precipitated at various pHs showed unique physicochemical properties. Manamperi et al. (2010a) stated that canola protein fractions precipitated sequentially from pH 11 to 3 demonstrated different thermal, rheological, and mechanical properties. Lönnerdal and Janson (1972) found that the basic napin protein fraction had a pI close to pH 11, and they suggested that the pI of the other canola protein fractions was ranged between pH 4 and 8. Predroche et al. (2004) extracted two canola protein fractions at pH 3.5 and pH 5.5. Membrane technology was also applied for canola protein preparation. Canola protein was initially solubilized at pH 12 and then harvested with membrane (molecular weight cutoff of 10 kDa). Almost 90% of the proteins were recovered using the membrane method (Ghodsvali et al., 2005).

Industrial Application of Canola Protein

Canola meal is primarily used as feedstock protein supplements for animals. However, many researches have been conducted to explore value-added products from canola meal. Manamperi and Pryor (2011) studied the properties of canola protein-based plastics and concluded that denaturation of canola protein with sodium dodecyl sulfate (SDS) and sodium dodecyl benzene sulfonate (SDBS) improved mechanical properties such as tensile strength and toughness of protein-based plastics. Manamperi et al. (2011) attempted to determine the effect of canola protein isolation conditions on the properties of protein-based plastics, stating that tensile strength and plastic toughness were affected by solubilization and precipitation pH of protein isolates, whereas elongation and water absorption were impacted only by solubilization pH. Studies on canola protein-based plastics or polymers were also conducted by Baganz et al. (1999), Wasche et al. (1998), and Manamperi et al. (2010b). Peptides from canola protein showed potential for use in cosmetics and medicines. Rivera et al. (2015) enzymatically hydrolyzed canola proteins into short-chain peptides, resulting in biocompounds that can function as active ingredients in skin care applications. Pancreatin hydrolysate from canola protein showed blood pressure lowering effects (Alashi et al., 2014). Other applications of canola protein-based industrial products included protein-based films for package application (Chang and Nickerson, 2015; Shi and Dumont, 2014), antioxidant activities (Cumby et al., 2008; He et al., 2013; Pan et al., 2011; Zhang et al., 2008), and emulsifiers (Tan et al., 2014).

Protein-Based adhesives

Adhesion is the tendency of dissimilar particles and/or surfaces to cling to one another with the aid of adhesive by chemical or physical forces (Schultz and Nardin, 1999). In order to achieve bonding, adhesives spread and wet the surface of the substrate, penetrating into the fibers cells through the capillary path and acting as a mechanical anchor (Schultz and Nardin, 1999). Adhesives for wood bonding range from natural starch, protein, and lignin to very durable synthetic resins created from petrochemicals (Eckelman, 2010). Before World War II, essentially all adhesives used for wood bonding were originated from natural resources such as mud, dung, clay, or a mixture of these substances. However, better-performing synthetic resins with higher bonding strength and water resistance from petrochemicals, such as urea-formaldehyde (UF), melamine-formaldehyde (MF), and phenol-formaldehyde (PF), quickly dominated the wood composite market and usage surpassed a majority of the previous natural glues for wood bonding (Koch et al., 1987). Recent environmental concerns and awareness of finite natural resources, however, have led to a resurgence in the development of biodegradable, durable, and sustainable bio-based adhesives such as protein-based adhesives, animal glue, blood-based adhesives, casein-based adhesives, lignin adhesives, and vegetable protein-based adhesives (Lin and Gunasekaran, 2010; Pizzi, 2006; Yang et al., 2006).

Extensive study of protein-based adhesives, such as soy protein-based binders, has led to the conclusion that adhesion between proteins and wood surfaces occurs as the protein spreads, wetting and penetrating the porous wood structure to achieve mechanical interlocking. Physical attraction and chemical bonding occurs between wood and protein during the setting period, followed by entanglements and cross-links caused by physical attraction and chemical bonding between protein-protein and protein-wood surfaces during thermal setting (Seller, 1994; Wool and Sun, 2005).

Soy-based adhesive is one of the most popular biodegradable adhesives currently under development because of its high gluing strength, biodegradability, and renewability. Soy protein fraction glycinin has proven to be the main contributor to adhesive bonding strength, especially wet strength of soy protein isolate (SPI) adhesive. Adhesives with high amounts of glycinin have shown higher dry and wet adhesion strengths due to increased content of hydrophobic amino acids compared to other soy protein fractions (Mo et al., 2004).

Similar to soy protein, canola proteins have abundant reactive groups, such as carboxyl, hydroxyl, amino, disulfide, imidazole, indole, phenolic and sulfhydryl groups, on the molecule chains (Feeney and Whitaker, 1985). Therefore, canola protein has great potential for use in wood adhesives (Li et al., 2012a; Li et al., 2012b). Current US canola production is approximately 1.26 MMT (USDA, 2015a), potentially generating 0.3 MMT of canola protein. Canola meal is fed primarily to cattle and pigs as part of a protein feed ration; however, feeding rates are limited due to the presence of erucic acid and glucosinolates in canola seeds. Utilization of coproducts from canola would increase the number of canola growing areas and canola crop production, consequently benefitting US farmers and enhancing rural economic development.

Preparation and Characterization of Canola Protein and Oil for Wood Adhesives

Canola Protein-based Adhesive Preparation

Both canola protein isolates and canola meal can be used as wood adhesives, and canola protein-based adhesives have shown higher bonding strength and water resistance than canola meal because protein isolates have more functional groups that can react with wood (Li et al., 2012a; Li et al., 2012b; Wang et al., 2014; Yang et al., 2006; Yang et al., 2014). When canola meal is used in adhesives, pretreatment is necessary to unfold the protein structures and improve bonding strength. Yang et al. (2006) prepared rapeseed flour-based adhesives by partially hydrolyzing the proteins using acid and alkali solutions with various concentrations (3%, 5%, and 7%) at 70°C for 60 min. Hydrolysis degree (HD) significantly impacts the adhesion performance; excessive hydrolysis breaks down protein's molecules, resulting in decreased bonding strength.

Canola protein isolates have shown improved adhesion performance compared to canola meal and are extractable with methods adopted by previous studies. Using alkaline to solubilize the protein and then precipitating the protein with acid solvents is the most popular method to isolate canola protein due to resulting high protein yield, easy operation, and low cost (Pedroche et al., 2004; Tzeng et al., 1988; Wäsche et al., 1998). Li et al. (2012a) and Hale (2013) demonstrated the preparation method of canola protein adhesives by mixing defatted canola meal with particle size

< 0.25 mm with distilled water at a solid/liquid ratio of 1:12 (w/v). The slurry was presoaked for 1 h with stirring, and then the pH value of the slurry was adjusted to 12 with 6 mol/L NaOH solution. The slurry was stirred for another 2 h at room temperature to solubilize the protein in the canola meal. The slurry mixture was centrifuged at 12,000 × g for 15 min, and the supernatant was decanted through a six-layer cheesecloth to remove impurities on top of the supernatant. The pH value of the supernatant was then slowly adjusted to 3.5 with 2 mol/L HCl solution in order to precipitate the protein. The mixture was centrifuged again at 12,000 × g for 15 min to isolate the wet canola protein. Modification step of canola protein can be conducted either during protein isolation or after the protein is isolated.

The albumin fraction, which is water soluble and highly hydrophilic, is unfavored for protein-based adhesives. Therefore, bonding strength of canola protein-based adhesives can be improved if the albumin fractions are removed. Wang et al. (2014) demonstrated a method to eliminate the albumin fraction by mixing the canola meal (100 mesh) with 10 folds of water, stirring for 30 min, adjusted to pH 4 with 2 N HCl, and then centrifugation in order to remove the water-soluble albumin fraction. After albumin was removed, the remaining canola protein was solubilized with 2.0 N NaOH at pH 12 and then precipitated at pH 4.0 with 2.0 N HCl. The protein was harvested through centrifugation.

Physicochemical Properties of Canola Protein Adhesives

Physicochemical properties such as molecular weight (MW), amino acids profiles, and rheological and thermal properties determine adhesion performances of protein-based adhesives. An in-depth study of these properties increases understanding of the bonding property and provides clues to further improve the bonding performance of canola protein-based adhesives.

Rheology: Viscosity is an important physical property that affects behavior of wood adhesives. Low viscosity of protein-based adhesives allow for good flowing and wetting effect on wood surfaces. Li et al. (2012a) observed that apparent viscosity decreased as shear rate increased (Fig. 1A), indicating that canola protein isolates showed shear thinning properties. Unmodified canola protein adhesive exhibited maximum viscosity around 9,000 Pa.s., and apparent viscosities of modified proteins decreased considerably as sodium bisulfide ($NaHSO_3$) concentration increased, thus improving handling and flowability of canola protein. Minimum apparent viscosity of approximately 1,200 Pa.s. was observed at an $NaHSO_3$ concentration of 15 g/L (Fig. 1A). The apparent viscosity of protein was manipulated by intermolecular interaction, such as electrostatic interactions and disulfide bonds (Cheng, 2004). As a reducing agent, $NaHSO_3$ broke disulfide bonds among polypeptides, dissociating canola protein into smaller subunits and resulting in weakened protein-protein interactions that reduced the apparent viscosity of $NaHSO_3$-canola protein adhesives (Zhang and Sun, 2008).

Morphology: Transmission electron microscopy (TEM) images of $NaHSO_3$-treated canola proteins are displayed in Fig. 1B (Li et al., 2012a). In unmodified canola protein, irregular, highly dense protein clusters were observed as a mixture of spherical and rod-shaped clusters with diameters ranging from 30 to 250 nm. The protein clusters were dissociated into smaller protein aggregates when canola protein was treated with $NaHSO_3$. The highly dense protein clusters in the control canola protein indicated that strong protein-protein aggregation occurred. Most of the protein aggregate surfaces were considered to be hydrophilic because the hydrophobic portion was buried through protein aggregation (Sun et al., 2008). Reduced protein aggregation was achieved in $NaHSO_3$-modified canola protein by breaking disulfide bonds in proteins and introducing negatively charged groups (SO_3^-) to protein molecules (Sun et al., 2008). As a result, partial hydrophobic groups were exposed on the canola protein surface, followed by the aggregates' dissociation process. Exposure of more hydrophobic groups is considered to be beneficial for adhesion performance of canola protein.

Amino acid profiles: Functional groups on protein chains, such as carboxyl, hydroxyl, amino, disulfide, imidazole, indole, phenolic, and sulfhydryl groups, can chemically bond with abundant polar hydroxyl groups on wood, thereby achieving bonding. Therefore, amino acid profiles comprise a key factor that governs adhesion performance of protein-based adhesives (Hettiarachchy et al., 1995). Amino acid profiles of canola protein fractions extracted at different pH values varied (Li et al., 2012a,b). Protein fractions at pH 3.5 had higher glutamate content than fractions at pH 5.5 and 7.0, but the fraction extracted at pH 7.0 had higher amounts of asparagine, threonine, alanine, arginine, tyrosine, valine, phenylalanine, isoleucine, and lysine than the other two fractions. Amino acid profiles of canola proteins at pH 5.5 were characterized as an intermediate position between pH 7.0 and 3.5. Based on hydrophobicity, amino acids typically are grouped into hydrophobic (non-polar amino acid) and hydrophilic (polar amino acid) types. Alanine, methionine, phenylalanine, isoleucine, leucine, and proline belong to hydrophobic amino acids, and they account for 25.69%–28.17% of canola protein fractions (Table 2) (Li et al., 2012a,b). Hydrophobic proteins repelled water, ensuring the intactness of the bonded boundaries between protein adhesives and wood surfaces. On the other hand, hydrophilic protein absorb water, which could destroy the cohesion between the adhesives and the wood surface. Thus, proteins-based adhesives with more hydrophobic amino acids are projected to show higher water resistance

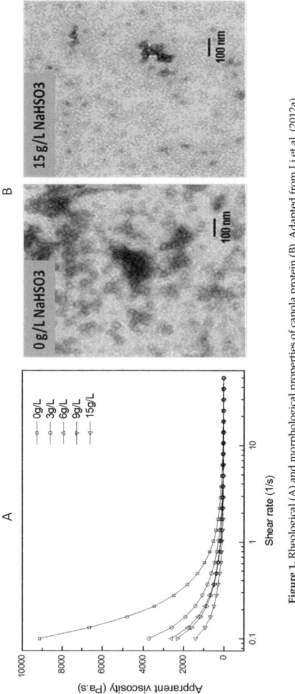

Figure 1. Rheological (A) and morphological properties of canola protein (B). Adapted from Li et al. (2012a).

Table 2. Amino acid compositions of native and NaHSO$_3$-modified canola protein (CP) adhesives extracted at various pH. Adapted from Li et al. (2012a,b).

Amino acid (% of total)	Native CP[1] fraction extracted at various pH			CP[1] fraction modified with NaHSO$_3$ (6 g/L) at various pH			CP[2]
	pH 7.0	pH 5.5	pH 3.5	pH 7.0	pH 5.5	pH 3.5	
Aspartate	9.21	8.78	8.15	9.88	9.79	8.24	10.54
Glutamate	22.14	24.01	26.97	20.02	21.6	25.73	20.79
Serine	5.35	5.27	5.32	5.56	5.25	5.06	5.42
Histidine	3.3	3.35	3.48	3.24	3.15	3.41	3.03
Glycine	6.05	5.99	5.95	6.04	5.99	6.33	6.25
Threonine	4.79	4.5	4.37	5.18	4.76	4.19	5.0
Alanine	5.27	5.08	4.86	5.46	5.16	4.78	5.26
Arginine	6.95	6.73	6.67	7.19	6.97	7.3	7.01
Tyrosine	3.77	3.41	2.75	4.1	3.69	2.89	3.81
Valine	5.73	5.67	5.52	5.85	5.94	5.58	5.97
Methionine	2.07	2.12	2.14	2.23	2.21	2.19	1.97
Phenylalanine	5.48	5.34	4.94	5.45	5.53	5.04	5.62
Isoleucine	5.62	5.54	5.27	5.75	5.82	5.37	5.77
Leucine	9.15	8.92	8.58	9.28	8.97	8.31	9.11
Lysine	5.13	5.3	5.04	4.76	5.19	5.59	4.46
T-AA (%)	100	100	100	100	100	100	100
T-protein (%)	65.19	70.09	69.57	64.16	66.92	60.32	65.76
Hydrophobic[3]	27.59	26.99	25.78	28.17	27.69	25.69	27.72
Hydrophilic[4]	72.41	73.01	74.22	71.83	72.31	74.31	72.28

[1] Canola protein fractions isolated at different pH values.
[2] Total canola protein.
[3] Alanine, methinine, phenylalanine, isoleucine, leucine, and proline.
[4] Lysine, tyrosine, arginine, threonine, glycine, histidine, serine, glutamine, and asparagine.

(He et al., 2015). However, as a low hydrophobic protein-based adhesive (25.69–28.17%), canola showed better water resistance (3.97 MPa) compared with sorghum protein (58%, 3.15 MPa) and soy protein (37%, 1.63 MPa) (Li et al., 2011). This is probably due to the fact that the hydrophobic property of canola protein may have been changed during curing process; however, the mechanism causing this phenomenon should be investigated (Li et al., 2011).

Molecular weight: Previous research used sodium dodecyl sulfate polyacrylamide gel electrophoresis (SDS-PAGE) to characterize MW distribution of canola protein-based adhesives (Fig. 2) (Li et al., 2012b). In the absence of 2-mercaptol ethanol (ME), two major components found in native canola protein, cruciferin (12S) and napin (2S), were observed with polypeptides' MWs ranging from 17 to 55 kDa. This finding was in good agreement with previous reports (Aluko and McIntosh, 2001; Krzyzaniak et al., 1998; Wu and Muir, 2008) in which canola protein demonstrated MW from 14 to 59 kDa. For $NaHSO_3$-modified protein, intensity of polypeptide bands at 17 kDa and 55 kDa was shown to fade, indicating the reducing effects of $NaHSO_3$ for breaking disulfide linkages and dissociating the protein into smaller polypeptide chains (Li et al., 2012b; Wu and Muir, 2008). The dissociating effect of protein molecules is the main reason for lower viscosity and less aggregation of $NaHSO_3$-modified canola proteins (Fig. 1B).

Thermal property: Thermal properties of proteins, such as denaturation and degradation temperatures, are important factors for assessing their utility for industrial applications (Manamperi et al., 2010b); use of a reducing agent to unfold the structure could decrease the denaturation temperature of protein adhesive (Li et al., 2012a). Canola protein-based adhesive prepared by Li et al. (2012a) exhibited a broad endothermic peak around 115°C, which was attributed to thermal denaturation of two major canola proteins, cruciferin and napin (Manamperi et al., 2010a). Protein denaturation temperature (T_d) was affected by $NaHSO_3$. The T_d of unmodified protein decreased gradually from 124°C to 115°C (Fig. 3).

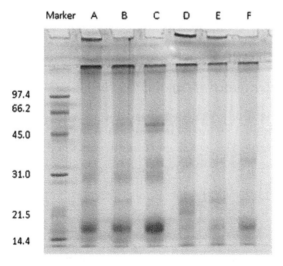

Figure 2. Non-reducing SDS-PAGE pattern of native and $NaHSO_3$-modified canola protein extracted at various pH. Native canola protein: pH 7.0 (lane A); pH 5.5 (lane B); pH 3.5 (lane C); 6 g/L $NaHSO_3$-modified canola protein: pH 7.0 (lane D); pH 5.5 (lane E); pH 3.5 (lane F). Adapted from Li et al. (2012b).

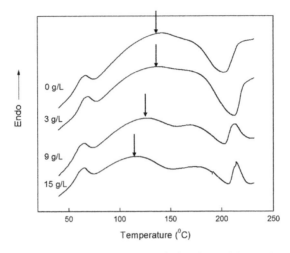

Figure 3. DSC thermograms of canola protein modified with NaHSO₃ at various concentrations. Adapted from Li et al. (2012a).

Another report indicated that T_d of canola protein decreased significantly after being treated with β-ME as a reducing agent. However, cross-linking of protein has been shown to increase protein's thermal stability. A study by Wang et al. (2014) showed that T_d increased after grafting poly (glydidyl methacrylate) to protein chains because additional energy was required to unfold the protein structure. Similar results have been demonstrated in other research (Fitzpatrick et al., 2010; Liu et al., 2008).

Canola Protein's Adhesion Performance and Modification

Adhesion Performance of Canola Protein

Adhesive strength of a protein adhesive, especially wet shear strength, has been determined by interactions of functional groups of protein with the wood substrate. The wet shear strength of unmodified canola protein adhesives have been shown to be 1.95, 2.23, and 3.97 MPa at curing temperature of 150°C, 170°C, and 190°C, respectively (Table 3) (Li et al., 2012a; Hale, 2013). Curing temperature positively affects the shear strength of canola protein-based adhesives because heat enhances immobilization of the protein adhesive and increases the possibility of chemical reactions at the interface between the protein adhesive and the wood surface (Mo et al., 1999). Compared to fractions isolated at pH 5.5 and pH 7.0, canola protein fractions isolated at pH 3.5 showed the highest wet shear strength because pH 3.5 is close to the pI (pH 4.0) of canola protein. Protein fractions showed the strongest protein-protein interaction at pI, resulting in increased

Table 3. Effect of NaHSO$_3$ on adhesion properties of canola protein (CP) adhesives. Adapted from Li et al. (2012a,b).

Canola adhesives	Wet strength (MPa)			Dry strength (MPa)		
	150°C	170°C	190°C	150°C	170°C	190°C
CP-0[1]	1.95±1.07	2.23±0.45	3.97±0.53	4.63±0.57	5.15±0.33	5.44±0.12
CP-3	2.34±0.14	2.86±0.1	3.92±0.05	5.78±0.32	5.08±0.35	5.40±0.37
CP-6	2.37±0.06	2.78±0.12	3.33±0.11	5.06±0.21	5.15±0.21	5.15±0.26
CP-9	2.39±0.19	2.85±0.1	3.39±0.15	5.17±0.13	5.40±0.12	4.98±0.39
CP-15	2.76±0.16	2.77±0.11	3.79±0.08	5.40±0.51	5.21±0.41	5.02±0.08
CP-0-pH7	2.15±0.14	2.87±0.15	3.83±0.17	5.23±0.33	5.75±0.23	5.40±0.16
CP-0-pH5.5	2.22±0.20	3.11±0.09	4.04±0.15	5.38±0.54	5.87±0.14	5.70±0.07
CP-0-pH3.5	2.39±0.16	3.41±0.21	4.07±0.16	5.60±0.18	5.19±0.35	5.28±0.47
CP-6-pH7	2.25±0.06	3.52±0.18	3.54±0.14	5.33±0.25	5.22±0.43	5.18±0.28
CP-6-pH5.5	2.42±0.22	3.74±0.13	3.89±0.08	5.66±0.30	5.76±0.09	5.56±0.20
CP-6-pH3.5	2.55±0.15	3.03±0.12	3.87±0.09	5.34±0.13	5.38±0.29	5.54±0.33

[1] Values 0–13 represent concentration of NaHSO$_3$ in the adhesives.

adhesion strength (Li et al., 2012b). Wang et al. (2014) determined that the wet shear strength of unmodified canola protein was around 1 MPa; low shear strength may be attributed to the low curing temperature (110°C). Canola protein-based adhesives showed lower water resistance than UF adhesive (4.66 MPa at curing temperature 170°C) (Qi and Sun, 2011).

Canola protein-based adhesive has demonstrated reasonable water resistance but is not comparable to UF adhesives (Qi and Sun, 2011). One of the reasons is the natural aggregated structure of proteins. If aggregated, the hydrophobic functional group is trapped inside the proteins and has less chance to react with the wood surface, leading to low water resistance (Li et al., 2012a; 2012b).

Canola Protein Unfolding with Reducing Agents

When proteins and subsequent aggregation of individual proteins fold, many functional groups that can chemically react with wood to achieve bonding are buried, resulting in decreased bonding strength and water resistance of protein-based adhesives (Frihart and Birkeland, 2014). One way to improve the wet bonding strength of protein adhesives is to open the tertiary structures using unfolding agents. Various unfolding agents have been investigated to improve adhesion strength of protein adhesives. Alkali, such as sodium hydroxide, has been the most common chemical used to unfold protein molecules and expose functional groups. Hettiarachchy et al. (1995) found that alkali treatment of soy protein improved the water resistance of soy protein under moderate alkaline conditions (pH 10.0 and 50°C). Urea and guanidine hydrochloride are also denaturation chemicals that can interact actively with hydroxyl groups of soy protein and then break down the hydrogen bonding, resulting in the unfolded protein structure (Sun and Bian, 1999; Zhang and Hua, 2007; Zhong et al., 2002).

Li et al. (2012a) investigated the effect of $NaHSO_3$ on adhesion performance of canola protein-based adhesive. Canola protein was modified with $NaHSO_3$ solutions at 0, 3, 6, 9, or 15 g/L. Results showed that $NaHSO_3$ concentration significantly affected wet adhesion properties of canola protein adhesives (Table 3). In general, 100% wood cohesive failure (WCF) was observed with all dry and soaked plywood specimens, indicating that bonding strength between wood and canola protein adhesives was stronger than the mechanical strength of the wood. Only partial WCF occurred for the wet assembled plywood samples (Table 3); the percentage of WCF for samples varied greatly with $NaHSO_3$ concentration and curing temperature. As shown in Table 3, optimum wet strength occurred at 190°C for all unmodified and $NaHSO_3$-modified canola protein adhesives. $NaHSO_3$ had a negative effect on adhesion performance of canola protein adhesives under a curing temperature of 190°C and almost no effect when cured at 150 or 170°C. Wet shear strength decreased slightly as $NaHSO_3$ concentration increased from 0 g/L to 6 g/L, then increased slightly as $NaHSO_3$ concentration rose from 9 g/L to 15 g/L. Wet shear strength remained lower than that of unmodified canola protein adhesives. Furthermore, at curing temperatures of 150°C and 170°C, adhesion improvement occurred only at levels of $NaHSO_3$ from 0 and 3% g/L, and all other levels of modification showed little differences in mean values obtained due to reduced viscosity and improved flowability of $NaHSO_3$-modified canola proteins.

Canola protein fractions isolated at various pH values also exhibited varied adhesion performance (Li et al., 2012b). Under a curing temperature of 190°C, native canola protein fractions isolated at pH 7.0 that were not treated by $NaHSO_3$ had insignificantly reduced wet strength of 3.83 MPa compared to protein fractions isolated at pH 5.5 and pH 3.5. $NaHSO_3$-modified canola protein fraction adhesives at other pH values displayed similar trends. The pI of canola protein proved to have an approximate pH of 5.5, at which pH point protein fractions had the strongest protein-protein interaction and highest hydrophobicity (Wang et al., 2008), resulting in increased adhesion strength. At pH 7.0, the protein had redundant charges on its surface because the pH point was far from the pI (pH 4.0); therefore, the protein-water interaction was preferred over the protein-protein interactions during the water-soaking process, and the adhesive had decreased wet strength.

The weakening effect of $NaHSO_3$ on protein adhesion performance was also observed for soy protein adhesives (Zhang and Sun, 2008). $NaHSO_3$

can cleave disulfide bonds in proteins to form sulfhydryls (R-SH) groups, and during the reducing reaction, some R-SH resulting from deoxidization were blocked as a sulfonate group (RS-SO$_3^-$). Therefore, negative effects of NaHSO$_3$ were attributed to the extra negative RS-SO$_3^-$ group that bonded with water via formation of a chemical bond in the protein adhesive, stimulating hydrophilic behavior and causing the RS-SO$_3^-$ group to absorb more water and disrupt the continuous adhesive matrix, an action that is detrimental to the wet shear strength of protein. Cruciferinciferin and napin contain disulfide bonds, especially in the polypeptide chains of napin fraction that are mainly held together by disulfide bridges (Schwenke et al., 1983); consequently, the induced extra negative RS-SO$_3^-$ group is one of the reasons that NaHSO$_3$ decreases adhesion strength of canola protein adhesives.

Hale (2013) studied the effect of denaturation agent of SDS (loading level: 0.5%, 1.0%, 3%, and 5% based on protein) on physicochemical and adhesion properties of canola protein adhesive with a curing temperature of 170°C. In general, SDS loading levels of 3% and 5% can sufficiently unfold the protein. Dry shear strength increased as SDS concentration increased, and SDS loading levels higher than 3% resulted in 100% WCF. Wet shear strength decreased as SDS concentration increased because SDS created negative surface charges on the adhesive and made the adhesive hydrophilic. Hydrophilic adhesive absorbs water, therefore disrupting the bond between adhesive and wood. Overall, the SDS increased dry and soak shear strength but decreased the wet shear strength of canola protein-based adhesives.

Cross-linking

As mentioned above, functional groups such as carboxyl, hydroxyl, amino, disulfide, imidazole, indole, phenolic, and sulfhydryl groups on amino acid side chains of protein are available for chemical/cross-linking reaction with cross-linkers (Feeney and Whitaker, 1985). Cross-linked protein adhesives have proven to have a more compact complex than uncross-linked proteins during thermal setting and are able to maintain complex structure better than unmodified protein adhesives under wet condition by reducing penetration of water molecules into the bonding interface of protein and wood, thus improving the water resistance of protein adhesives. Such cross-linking agents studied include 1,3-dichloro-2-propanol (Rogers et al., 2004), polyamide-epichlorohydrin (PAE) (Zhong and Sun, 2007), epoxies (Huang, 2007; Lambuth, 1989), maleic anhydride (Liu and Li, 2007), glutaraldehyde (Wang et al., 2007), and MgO (Jang and Li, 2015).

Wang et al. (2014) developed canola protein-poly (glycidyl methacrylate) (GMA) conjugates adhesives for wood, which demonstrated significant improvement of dry and wet bonding strength compared to unmodified proteins. In the preparation process 2 g of canola protein were mixed with 50 mL water at 40°C in a water bath. The protein solution was purged with nitrogen gas for 10 min. Ammonium persulfate (0.108 g) was added to the solution, and stirred for 10 min. Then GMA monomer (0.9, 1.8, or 2.7 g) were added to the flask. Ammonium persulfate functioned as a radical polymerization initiator to initiate graft copolymerization, and ethanol was added as the reaction terminator and protein precipitation agent. Proposed grafting of GMA onto canola protein is summarized in Fig. 4. Grafting degrees of 23.32% (CG1), 52.65% (CG2), 60.80% (CG3), and 81.97% (CG4) were obtained by controlling the grafting time and GMA loading levels, and the grafting degree could be identified by solid-state NMR. Natural birch wood veneers were bonded with CG adhesives at the designated curing time, pressure, and temperature. Results showed that CG4 improved dry strength to 8.25 ± 0.12 MPa, wet strength to 3.68 ± 0.29 MPa, and soaked strength to 7.1 ± 0.10 MPa compared to unmodified canola protein at 3.72 ± 0.21 MPa, 1.02 ± 0.14 MPa, and 1.25 ± 0.10 MPa, respectively. Several factors contributed to the improved adhesive strength for the conjugates. The secondary structure of the protein was partially disrupted due to exposure of the hydrophobic groups, thereby allowing an increase of intermolecular interactions among proteins that resulted in good distribution of interactions between the protein and the wood. Structural and thermal analysis indicated strong interactions (i.e., hydrogen bonding, hydrophobic interactions, or covalent bonding) in the conjugate; therefore, intrinsic strength of the

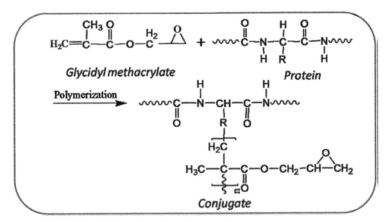

Figure 4. Grafting scheme of glycidyl methacrylate onto canola protein. Adapted from Wang et al. (2014).

adhesive improved in the bulk conjugate. In addition, exposure of the inner hydrophobic amino acids also increased the water resistance.

Blending with Synthetic Resins

Blending proteins with synthetic resins is a short-term solution for reducing dependence on petrochemicals and lowering emissions of volatile organic compounds. Previous studies have proven that blends of soy-based proteins with polyvinyl alcohol, polyvinyl acetate resin, urea formaldehyde resins, and phenolic resins improve water resistance for wood applications (Kumar et al., 2002; Qi and Sun, 2011; Qu et al., 2015; Steele et al., 1998; Frihart and Wescott, 2004; Hse et al., 2001; Yang et al., 2006; Zhong and Sun, 2007).

Yang et al. (2011) conducted a preliminary study using defatted rapeseed flour (RF)-based wood adhesives for wood flooring. RF was hydrolyzed with 1% NaOH at 70°C for 60 min and then mixed with PF resins that were synthesized at molar ratios of formaldehyde to phenol at 1.8, 2.1, and 2.4. RF/PF resin was brushed on the plywood and fancy-veneer plywood, respectively, at loading a level of 300 g/m² and cured at 140°C for 5 min under 10 kg/cm² (approximate 1 MPa). Average density, moisture content, formaldehyde emissions, and critical stress met or exceeded the standards of fiberboards (Korean Standard Association, 2006). The critical stress increased as the formaldehyde/phenol (PF) ratio increased.

Yang et al. (2014) investigated adhesive properties of medium-density fiberboards bonded with RF-based adhesive resins. The resin was a mixture of PF (30%), sulfuric acid hydrolyzed RF (35%), and sodium hydroxide hydrolyzed RF (35%) (w/w/w). 1% (weight basis) of lignosulfonate-based wax emulsion was added to increase the water repellency of the RF/PF adhesive. Hydrolysis treatment broke down the protein molecule and achieved improved wetting and penetration to wood substrates. PF was synthesized by mixing formaldehyde with phenol at a molar ratio of 1.5, 1.8 and 2.1.5% of RF/PF adhesive (based on larch fiber weight) was sprayed onto the larch fiber with sufficient mixing. The blends of fiber were pressed at 2.95 MPa at 180°C for 8 or 10 min with a hot press. Results showed that the concentration of acid or alkali negatively affected the bonding performance due to decomposition of the rapeseed protein. The PF ratio also significantly influenced the adhesion performance of the resin: The bonding strength increased when the ratio increased from 1.5 to 1.8 and then decreased at a ratio of 2.1. Ratio 1.8 exhibited optimum adhesion performance because it had more cross-linked network than that of ratio 1.5. However, further increasing of the ratio to 2.1 decreased the adhesion performance due to insufficient curing of the RF/UF adhesive.

CANOLA OIL-BASED ADHESIVES

Canola Oil

Canola oil, which has a homogeneous fatty acid composition of more than 90% of 18 carbon fatty acids, is usually hydrogenated to produce shortenings and margarines because the *trans* isomers have higher melting points than *cis* fatty acids (Table 4) (Milchert and Smagowicz, 2009). Triglycerides are the most abundant lipid class found in canola oil (Przybylski et al., 2005). Commercial production of canola oil includes mechanical pressing and solvent extraction incorporation with degumming, refining, bleaching, interesterification, and dewaxing processes. Canola oil is an essential oil crop to humans, valued for its culinary versatility and nutritional qualities in the human diet. In addition, oil is the most important component of the canola seed in terms of market value. Many researches have been conducted to discover the application of canola oil in industrial application. Baroi and Dalai (2015) evaluated the sustainability of homogeneous and heterogeneous acid-catalyzed biodiesel production process from canola oil and concluded that this process is more energy efficient and environmentally friendly than the homogeneous process. Other canola oil-based value-added products include wood coatings (Kong et al., 2013; Philipp and Eschig, 2012), plasticizers (Dabbagh, 2011), and biodegradable lubricants from epoxidation of rapeseed oil (Arumugam et al., 2014; Wu et al., 2000).

Table 4. Fatty acid profiles of rapeseed oil. Adapted from Milchert et al. (2009).

Number of carbon atoms: Number of unsaturated bonds		Fatty acids composition in rapeseed oil	wt (%)
14:00	Myristic	$CH3(CH2)_{12}COOH$	0.1
16:00	Palmitic	$CH3(CH2)_{14}COOH$	4.5
16:01	Oleopalmitic	$CH3(CH2)_5CH=CH(CH2)_7COOH$	0.4
18:00	Stearic	$CH3(CH2)_{16}COOH$	2.1
18:01	Oleic	$CH3(CH2)_7CH=CH(CH2)_7COOH$	64.5
18:02	Linoleic	$CH3(CH2)_3(CH2CH=CH)_2(CH2)_7COOH$	18.3
18:03	Linolenic	$CH3(CH2CH=CH)_3(CH2)_7COOH$	6.8
20:00	Arachidic	$CH3(CH2)_{18}COOH$	0.8
20:01	Gadoleic	$CH3(CH2)_9CH=CH(CH2)_7COOH$	1.3
22:00	Behenic	$CH3(CH2)_{20}COOH$	0.4
22:01	Erucic	$CH3(CH2)_7CH=CH(CH2)_{11}COOH$	0.8

Canola Oil-based Polyurethane Adhesives for Wood

Contrary to canola protein, canola oil cannot be used directly as wood adhesives. However, vegetable oils have fatty acid chains with unsaturated double bonds, and the ester group enables them to be used in production of a wide variety of valuable products such as polyols, glycol, carbonyl compound, plasticizer for polymer, etc. (Saurabh et al., 2011). Intensive studies have proved that vegetable oil-based polyurethane form a versatile class of polymers used in broad range of applications, such as elastomers, sealants, fibers, foams, coatings, lubricants, foams, adhesives, and biomedical materials (Guo et al., 2000; Hojabri et al., 2010b; Jia and Liang, 1994; Monono et al., 2015; Tan and Chow, 2010; Vallat and Bessaha, 2000).

Canola oil can be synthesized into polyurethanes to be used as adhesives to bond porous or non-porous substrates, such as wood to wood, wood to metal, or metal to metal (Patel et al., 2009). However, bonding mechanisms of polyurethane adhesives differ on wood and metal surfaces. Wood bonding involves complex chemistry such as celluloid hydrogen bonding (i.e., Vander-walls forcers), while metal bonding happens through the polarity of metal substrate (i.e., orientation of the polar group) (Patel et al., 2009). The process to synthesize canola oil into polyurethane is as follows: (1) double bonds on fatty acid chains of canola oil are epoxidized into oxirane rings, (2) the epoxidized canola oil is converted to polyols through opening the oxirane rings, and (3) canola oil-based polyurethane is synthesized with polyols and multi-isocyanates (Miao et al., 2014; Mosiewicki et al., 2014; Zieleniewska et al., 2015).

Kong et al. (2011) characterized canola oil-based polyurethane wood adhesives and compared canola oil-based polyurethane adhesives to commercial polyurethane adhesives. The polyurethane adhesive was synthesized from canola oil-derived polyols and commercially available polymeric aromatic diphenymethane diisocyanate (PMDI) at ratios calculated by NCO group from PMDI to OH group from polyols (i.e., NCO/OH ratios were 1.2/1.0, 1.5/1.0, 1.8/1.0). Polyols were synthesized via epoxidation of canola oil and the subsequent reaction due to ring-opening with 1,3-propane diol according to the method described by Curtis and Liu (2010). Birch wood was bonded with the canola oil-based polyurethane adhesive at curing conditions of 0.1 MPa press for 48 h at room temperature or 100°C. Results indicated that canola oil-based polyurethane adhesive showed similar dry shear strength and significantly higher wet shear strength than commercial polyurethane adhesives. In addition, the adhesion performance of canola oil-based polyurethane adhesive synthesized at NCO/OH molar ratio of 1.8/1.0 was similar to the adhesion strength of 1.5/1.0 ratio but higher the ratio of 1.2/1.0. A lower NCO/OH ratio may result in most isocyanate groups being consumed when urethane bonds are formed, thereby failing to leave enough free isocyanate to react with the wood and resulting in decreased shear strength. Increased NCO/OH molar ratio increased cross-linking, leading to a higher rigidity and strength of the adhesive bond (Kong et al., 2011). Similar results were also reported for argemone oil and castor oil-based polyurethane adhesives (Desai et al., 2003; Somani et al., 2003).

Production of Canola Oil-based Polyurethane Adhesives

Polyurethanes are synthesized primarily by reacting polyols with petroleum-derived multi-isocyanate. However, polyols and multi-isocyanate can be derived from triglycerides and their derivatives in vegetable oils (Lligadas et al., 2010; Kong et al., 2011; More et al., 2013; Miao et al., 2014; Szycher, 2012).

Vegetable Oil-based Polyols

Vegetable oils, which are triglycerides of predominantly unsaturated fatty acids, are relatively chemically unreactive and must be transformed into polyols that contain at least two hydroxyl groups capable of forming polyurethane in reaction with diisocyanates (Rojek and Prociak, 2011). Production of polyols from vegetable oil occurs via epoxidation and oxirane ring-opening pathway. Epoxidation of vegetable oil, the first step in making oil-based adhesives, is achieved by hydroformylation and hydrogenation, ozonolysis, thiol-ene coupling, or transesterification/amidation strategies (Li et al., 2015).

Figure 5 describes one pathway to synthesize polyols from canola oil (Mosiewicki et al., 2014; Zieleniewska et al., 2015). Rapeseed polyol was synthesized using a two-step method: epoxidation of the double bonds of rapeseed oil and the reaction from opening oxirane rings using diethylene glycol. Epoxidation of rapeseed oil required peracetic acid generated *in situ* as a result of the reaction of H_2O_2 and glacial acetic acid at 60°C with concentrated sulfuric acid as a catalyst. The resulting mixture was washed with water, and the separated organic phase was dried under vacuum. Epoxidized oils can be converted into polyols by reacting epoxidized oils with diethylene glycol. Stirna et al. (2013) synthesized and characterized polyols from rapeseed oil with diethanolamine, triethanolamine, and glycerol at various ratios. Rapeseed oil was heated to 140–230°C, and then diethanolamides, traethanolamine, or monoglyceride were added to synthesize the polyols with zinc acetate as a catalyst. The polyurethane was then synthesized by mixing polyols and NCO/OH at 70°C for 2 h. In order to reduce the reaction time and energy consumption, epoxidation of rapeseed oil was conducted using peracetic acid generated *in situ*, originating from the reaction of hydrogen peroxide and glacial acetic acid in acidic medium.

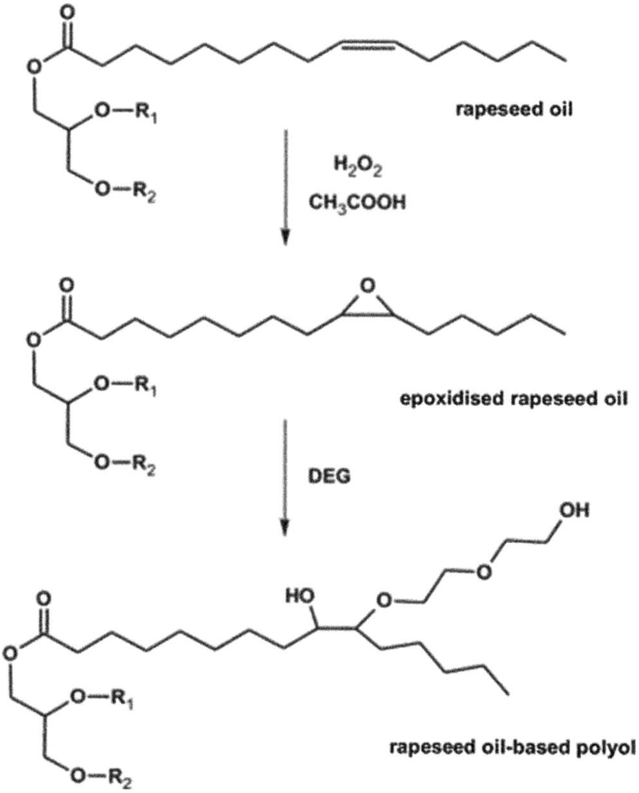

Figure 5. Chemical reaction steps to produce rapeseed oil-based polyol. Adapted from Mosiewicki et al. (2014) and Zieleniewska et al. (2015).

The epoxidized oil was converted into the polyol using diethylene glycol (DEG) in the presence of sulfuric acid with the aid of microwave irradiation (Dworakowska et al., 2012).

Monono et al. (2015) studied the scale-up production of canola oil epoxidation. H_2O_2 adding rates, water bath temperature, and reaction times were characterized to investigate the effect of those factors on the properties of epoxidated canola oil epoxidation. At reaction time 2.5 h, the epoxidation of canola oil was positively affected by the H_2O_2 addition rate and water bath temperature. The oxirane conversion rate reached 95% at the H_2O_2 addition rate of 4.6 g/min/mole and temperature of 75°C. Increasing temperature increased the reaction rate of epoxidation and the reaction rates of side products. Therefore, a balance must be maintained between maximizing oxirane formation and minimizing side reactions when selecting optimal temperature. A high temperature (75°C) results in increased viscosity of samples potentially due to the formation of hydroxyl, carboxylic, and ketone groups from oxirane cleavage. Hydrogen bonds formed as the amount of hydroxyl groups increased, thereby increasing viscosity levels of the product (Gamage et al., 2009; Monono et al., 2015).

Vegetable Oil-based Multi-isocyanates

Isocyanates, the primary building block of polyurethanes, are usually petroleum-derived (Miao et al., 2014). Although isocyanates can be prepared via several routes, reaction of gaseous phosgene with amines or their corresponding salts is the only method for practiced multi-isocyanate production (Hojabri et al., 2010b). Health and safety concerns associated with isocyanate chemistry motivate the search for non-phosgene methods for diisocyanate preparation. Using triglycerides or fatty acid from vegetable oil as starting materials, various chemical approaches have been explored to synthesize multi-isocyanates (Miao et al., 2014). Hojabri et al. (2009; 2010a,b) synthesized 1,7-heptamethylene diisocyanate (HPMDI) and 1,16-diisocyanatohexadec-8-ene (HDEDI) from oleic acid. Fatty acid-based diisocyanate was also synthesized from diesters. Hydrazinolysis of dimethyl sebacate was conducted in absolute ethanol to form diacyl hydrazide, which was then converted into diacyl azide. Diisocyanate was synthesized by the Curtius rearrangement of diacyl azide (More et al., 2013). Synthesis of diacylazide from diacyl hydrazide was the determining step because the side reaction, or formation of secondary amide, was observed in this process (Miao et al., 2014).

Vegetable Oil-based Polyurethane

Although polyurethane can be synthesized using a variety of methods, the most widely used method starts from di-functional or poly-functional hydroxyl-compounds (polyol) with di-functional or poly-functional isocyanates (Szycher, 2012). Figure 6 illustrates the general route to synthesis polyurethane (Wang, 2006). Polyurethane can be synthesized by the reaction between polyols and diisocyanate at desired molar ratios of OH group from polyols and NCO group from diisocyanates. The desired OH/NCO molar ratio satisfies the equation

$$MR = \frac{Wpolyol/EWpolyol}{(Wpolyurethane - Wpolyol)/EWisocyanate} \tag{1}$$

Where W_{polyol} is the weight of the polyol, EW_{polyol} is the equivalent weight of polyol, $W_{polyurethane}$ is the total weight polyurethane to produce, and $EW_{isocyanate}$ is the equivalent weight of the isocyanate (Hojabri et al., 2010a,b).

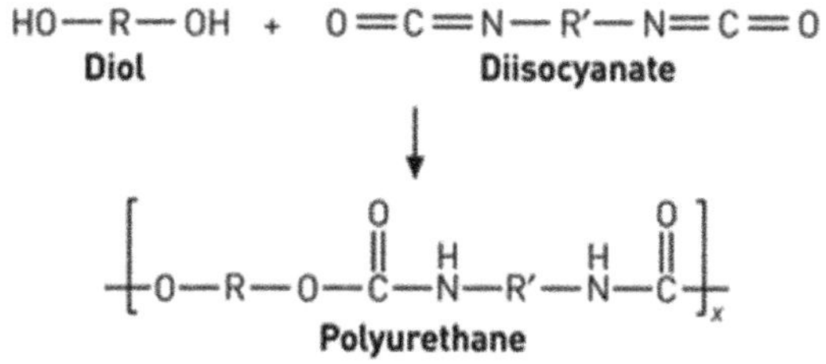

Figure 6. Polyurethane synthesis from diol and diisocyanate. Adapted from Wang (2006).

In order to synthesize polyurethanes, polyols and diisocyanates (i.e., HPMDI and HDEDI) were mixed at a designated OH/NCO molar ratio with dibutyltindilaurate (DBTDL). The mixture was poured into another container and placed in a vacuum oven at 40°C for 5–8 min to degas CO_2 released during the side reaction of diisocyanate with moisture or carboxylic acid and air trapped during mixing. Air was then introduced to the oven to avoid deformation of the sample under vacuum, and the sample was postcured for approximately 24 h at 70°C and 10 h at 110°C (Hojabri et al., 2010a,b).

The physical properties and crystalline structure of polyurethanes synthesized from HPMDI and HDEDI differ. The HDEDI-based polyurethane exhibited a triclinic crystal form, whereas the HPMDI-based polyurethane showed a hexagonal crystal lattice. In addition, HDEDI-based polyurethane demonstrated a higher tensile strength at break than HPMDI-based polyurethane, attributed to the higher degree of hydrogen bonding associated with HDEDI-based polyurethane. However, a lower Young's modulus and higher elongation in HDEDI-based polyurethane were obtained due to the flexibility of the long chain introduced by the HDEDI diisocyanate (Hojabri et al., 2010a,b).

Summary

Canola protein and canola oil exhibited adhesive potential for the wood industry. Canola protein-based adhesives can be prepared from protein isolates or defatted canola meal. Canola meal-based adhesive is suitable for making particle-board or veneer flooring used in indoor environments since it shows relatively low water resistance compared to pure canola protein-based adhesive. Canola protein-based adhesive, however, has high protein content and is more hydrophobic, making it suitable for use in indoor and outdoor conditions. Physiochemical properties and adhesion performance of canola protein-based adhesives can be enhanced through modifications such as cross-linking, hydrolization, unfolding, or blending with commercial adhesives. A low viscosity of protein-based adhesive is favored because low viscosity benefits the wetting and penetrating ability to wood surfaces. Water resistance, high curing temperature, and feasibility of large-scale production of canola protein-based adhesives are important concerns for future research.

Polyurethane resins synthesized from canola oil can be used to bond wood substrates. However, canola oil-based adhesive is underexploited compared to intensive studies conducted on protein-based adhesives. Although canola oil showed potential for use as wood adhesives, it requires 24 h to cure. Therefore, future work should focus on using canola oil to develop fast-curing polyurethane adhesive with high water resistance. Additional concerns regarding vegetable oil-based polyurethane adhesives may be limited shelf life, toxicity, and high sensitivity to humidity.

Keywords: Canola protein, Canola oil, Wood adhesive, Protein modification, Polyurethane

References

Alashi, A., C. Blanchard, R. Mailer, S. Agboola, A. Mawson, R. He, S. Malomo, A. Girgih and R. Aluko. 2014. Blood pressure lowering effects of Australian canola protein hydrolysates in spontaneously hypertensive rats. Food Res. Int. 55: 281–287.

Aluko, R. and T. McIntosh. 2001. Polypeptide profile and functional properties of defatted meals and protein isolates of canola seeds. J. Sci. Food Agric. 81(4): 391–396.

Arumugam, S., G. Sriram and R. Ellappan. 2014. Bio-lubricant-biodiesel combination of rapeseed oil: An experimental investigation on engine oil tribology, performance, and emissions of variable compression engine. Energy. 72: 618–627.

Baganz, K., H. Lang and G. Meißner. 1999. Industrial use of oilseed meal: a reasonable injection moulding compound. Eur. J. Lipid Sci. Tech. 101(8): 306–307.

Baroi, C. and A. Dalai. 2015. Process sustainability of biodiesel production process from green seed canola oil using homogeneous and heterogeneous acid catalysts. Fuel Process. Tech. 133: 105–119.

Barthet, V. 2014. Quality of western Canadian canola. Canadian Grain Commission. ISSN 1700–2222.

Bell, J. 1993. Factors affecting the nutritional value of canola meal: A review. Can. J. Anim. Sci. 73: 679–697.

Berot, S., J. Compoint, C. Larre, C. Malabat and J. Gueguen. 2005. Large scale purification of rapeseed proteins (*Brassica napus* L.). J. Chromatogr., B: Anal. Technol. Biomed. Life Sci. 818(1): 35–42.

Canola Council of Canada. 1990. Canola oil and meal: standards and regulations. Winnipeg. Manitoba: Canola Council of Canada Publication. 4 p.

Chang, C. and M. Nickerson. 2015. Effect of protein and glycerol concentration on the mechanical, optical, and water vapor barrier properties of canola protein isolate-based edible films. Food Sci. Technol. Int. 21(1): 33–44.

Cheng, E. 2004. Adhesion mechanism of soybean protein adhesives with cellulosic materials. Ph.D. dissertation. Kansas State University, Department of Grain Science and Industry. Manhattan, KS.

Cumby, N., Y. Zhong, M. Naczk and F. Shahidi. 2008. Antioxidant activity and water-holding capacity of canola protein hydrolysates. Food Chem. 1(1): 144–148.

Curtis, J. and G. Liu. 2010. Polyol synthesis from fatty acids and oils. US patent application, 61,366,416.

Dabbagh, S. 2011. Oxidative degradation of polyether in contact with minerals. Diploma Work.

Desai, S., J. Patel and V. Sinha. 2003. Polyurethane adhesive system from biomaterial-based polyol for bonding wood. Int. J. Adhes. Adhes. 23: 393–399.

Dworakowska, S., D. Bogdal and A. Prociak. 2012. Microwave-assisted synthesis of polyols from rapeseed oil and properties of flexible polyurethane foams. Polymers 4(3): 1462–1477.

Eckelman, C. 2010. Brief Survey of Wood Adhesives. Forestry and Natural Ressources. Annual Report.

Feeney, R.E. and J.R. Whitaker. 1985. Chemical and enzymatic modification of plant proteins. *In*: A.M. Altschul and H.L. Wilcke (eds.). New Food Proteins. New York: Academic Press. 5: 181–219.

Fitzpatrick, S.D., M.A.J. Mazumder, F. Lasowski, L.E. Fitzpatrick and H. Sheardown. 2010. PNIPAAm-Grafted-Collagen as an Injectable, *In Situ* Gelling, Bioactive Cell Delivery Scaffold. Biomacromolecules 11: 2261–2267.

Frihart, C.R. and J.M. Wescoott. 2004. Improved water resistance of bio-based adhesives for wood bonding. *In*: Proceedings of 1st International conference on Environmentally-Compatible Forest Products. 22–24 September. Oporto, Portugal. pp. 293–302.

Frihart, C. and M. Birkeland. 2014. Soy properties and soy wood adhesives. *In*: Soy-Based Chemicals and Materials; Brentin; ACS Symposium Series; American Chemical Society: Washington, DC.

Gamage, P., M. O'Brien and L. Karunanayake. 2009. Epoxidation of some vegetable oils and their hydrolysed products with peroxyformic acid–optimised to industrial scale. J. Natl. Sci. Foundation Srilanka 37(4): 229–240.

Ghodsvali, A., M. Khodaparast, M. Vosoughi and L. Diosady. 2005. Preparation of canola protein materials using membrane technology and evaluation of meals functional properties. Food Res. Int. 38(2): 223–231.

Guo, A., L. Javni and Z. Petrovic. 2000. Rigid polyurethane foams based on soybean oil. J. Appl. Polym. Sci. 77(2): 467–473.

Hale, K. 2013. The potential of canola protein for bio-based wood adhesives. A Thesis, Department of Biological and Agricultural Engineering. Kansas State University.

He, R., A. Alashi, S. Malomo, A. Girgih, D. Chao, X. Ju and R. Aluko. 2013. Antihypertensive and free radical scavenging properties of enzymatic rapeseed protein hydrolysates. Food Chem. 141(1): 153–159.

He, Z., H. Zhang and D.C. Olk. 2015. Chemical composition of defatted cottonseed and soy meal products. PLoS One 10(6): e0129933. DOI:10.1371/journal.pone.0129933.

Hettiarachchy, N.S., U. Kalapathy and D.J. Myers. 1995. Alkali-modified soy protein with improved adhesive and hydrophobic properties. J. Am. Oil Chem. Soc. 72: 1461–1464.

Hoglund, A.S., J. Rodin, E. Larsson and L. Rask. 1992. Distribution of napin and cruciferin in developing rape seed embryos. Plant Physiol. 98(2): 509–515.

Hojabri, L., X. Kong and S. Narine. 2009. Fatty acid-derived diisocyanate and biobased polyurethane produced from vegetable oil: synthesis, polymerization, and characterization. Biomacromolecules. 10(4): 884–91.

Hojabri, L., X. Kong and S. Narine. 2010a. Novel long chain unsaturated diisocyanate from fatty acid: synthesis, characterization, and application in bio-based polyurethane. J. Polym. Sci. Pol. Chem. 48(15): 3302–10.

Hojabri, L., X. Kong and S. Narine. 2010b. Functional thermoplastics from linear diols and diisocyanates produced entirely from renewable lipid sources. Biomacromolecules. 11(4): 911–8.

Hse, C.Y., F. Fu and B.S. Bryant. 2001. Development of formaldehyde-based wood adhesives with co-reacted phenol/soybean flour. *In*: Proceedings of the Wood Adhesives 2000 Conference. 22–23 June. South Lake Tahoe, NV, pp. 13–19.

Huang, J. 2007. Development and characterization of new formaldehyde-free soy flour-based adhesives for marking interior plywood. PhD diss. Corvallis, OR.: Oregon State University, Department of Wood Science.

Jang, Y. and K. Li. 2015. An all-natural adhesive for bonding wood. J. Am. Oil Soc. 92: 431–438.

Jia, D. and X. Liang. 1994. Mechanism of adhesion of polyurethane/polymethacrylate simultaneous interpenetrating networks adhesives to polymer substrates. J. Polym. Sci. Part B: Polym. Phys. 132: 817–823.

Kalapathy, U., N. Hettiarachchy, D. Myers and K. Rhee. 1996. Alkali-modified soy proteins: Effect of salts and disulfide bond cleavage on adhesion and viscosity. J. Am. Oil Chem. Soc. 73: 1063–1066.

Koch, G.S., F. Klareich and B. Exstrum. 1987. Introduction and summary. In: Adhesives for the Composite Wood Panel Industry, 1–11. Park Ridge, NJ: Noyes data corporation.

Kong, X., G. Liu and J. Curtis. 2011. Characterization of canola oil based polyurethane wood adhesives. Int. J. Adhes. Adhes. 31: 559–564.

Kong, X., G. liu, H. Qi and J. Curtis. 2013. Preparation and characterization of high-solid polyurethane coating systems based on vegetable oil derived polyols. Prog. Org. Coat. 76(9): 1151–1160.

Korean Standard Association. 2006. Fiberboards. KS F 3200.

Krzyzaniak, A., T. Burova and J. Barciszewski. 1998. The structure and properties of Napin-seed storage protein from rape (*Brassica napus* L.). Nahrung 42: 201–204.

Kumar, R., V. Choudhary, S. Mishra, I.K. Varma and B. Mattiason. 2002. Adhesives and plastics based on soy protein products. Ind. Crop Prod. 16: 155–172.

Lambuth, A.L. 1989. Adhesives from renewable resources: Historical Perspective and Wood Industry Needs. pp. 1–10. In: R.W. Hemingway, A.H. Conner and S.J. Branham (eds.). Adhesives from Renewable Resources. Washington, DC: American Chemical Society.

Li, N., Y. Wang, M. Tilley, S. Bean, X. Wu, X. Sun and D. Wang. 2011. Adhesive performance of sorghum protein extracted from sorghum DDGS and Flour. J. Polym. Environ. 19: 755–765.

Li, N., G. Qi, X. Sun, M. Stamm and D. Wang. 2012a. Physicochemical properties and adhesion performance of canola protein modified with sodium bisulfite. J. Am. Oil Chem. Soc. 89: 897–908.

Li, N., G. Qi, X. Sun and D. Wang. 2012b. Effects of sodium bisulfite on the physicochemical and adhesion properties of canola protein fractions. J. Polym. Environ. 20: 905–915.

Li, Y., X. Luo and S. Hu. 2015. Polyols and Polyurethanes from Vegetable Oils and Their Derivatives. In: Bio-based Polyols and Polyurethanes. DOI 10.1007/978-3-319-21539-6_2.

Lin, H. and S. Gunasekaran. 2010. Cow blood adhesive: Characterization of physicochemical and adhesion properties. Int. J. Adhes Adhes. 30: 139–144.

Liu, W., A. Mohanty, P. Askeland, L. Drzal and M. Misra. 2008. Modification of Soy Protein Plastic with Functional Monomer with Reactive Extrusion. J. Polym. Environ. 16(3): 177–182.

Liu, Y. and K. Li. 2007. Development and characterization of adhesives from soy protein for bonding wood. Int. J. Adhes. Adhes. 27: 59–67.

Lligadas, G., J. Ronda, M. Galia and V. Cadiz. 2010. Plant Oils as platform chemicals for polyurethane synthesis: current state-of-the-art. Biomacromolecules. 11(11): 2825–2835.

Lönnerdal, B. and J. Janson. 1972. Studies on Brassica seed proteins. I. The low molecular weight proteins in rapeseed. Isolation and characterization. Biochim. Biophys. Acta. 278(1): 175–183.

Manamperi, W., S. Chang, C. Ulven and S. Pryor. 2010b. Plastics from an improved canola protein isolate: preparation and properties. J. Am. Oil Chem. Soc. 87: 909–915.

Manamperi, W., M. Fuqua, C. Ulven, D. Wiesenborn and S. Pryor. 2011. Effect of canola protein extraction parameters on protein-based plastic properties. J. Biobased Mater. Bio. 5(4): 500–506.

Manamperi, W. and S. Pryor. 2011. Properties of canola protein-based plastics and protein isolates modified using SDS and SDBS. J. Am. Oil Chem. Soc. 89: 541–549.

Manamperi, W.A.R., S.K.C. Chang, C.A. Ulven and S.W. Pryor. 2010a. Impact of meal preparation method and extraction procedure on canola protein yield and properties. J. Am. Oil Chem. Soc. 87: 909–915.

Miao, S., P. Wang, Z. Su and S. Zhang. 2014. Vegetable-oil-based polymers as future polymeric biomaterials. Acta Biomater. 10: 1692–1704.

Milchert, E. and A. Smagowicz. 2009. The influence of reaction parameters on the epoxidation of rapeseed oil with peracetic acid. J. Am. Oil Chem. Soc. 86: 1227–1233.

Mo, X., X.S. Sun and D. Wang. 2004. Thermal properties and adhesion strength of modified soybean storage proteins. J. Am. Oil Chem. Soc. 81: 395–400.

Mo, X., X.S. Sun and Y. Wang. 1999. Effects of molding temperature and pressure on properties of soy protein polymers. J. Appl. Polym. Sci. 73: 2595–2602.

Monono, E., D. Haagenson and D. Wiesenborn. 2015. Characterizing the epoxidation process conditions of canola oil for reactor scale-up. Ind. Crop Prod. 67: 364–372.

Monsalve, R.I. and R. Rodriguez. 1990. Purification and characterization of proteins from the 2S fraction from seeds of the Brassicaceae family. J Expt. Bot. 41: 89–94.

More, A., T. Lebarbe, L. Maisonneuve, B. Gadenne, C. Alfos and H. Cramail. 2013. Novel fatty acid based di-isocyanates towards the synthesis of thermoplastic polyurethanes. Eur. Polym. J. 49: 823–33.

Mosiewicki, M., P. Rojek, S. Michalowski, M. Aranguren and A. Prociak. 2014. Rapeseed oil-based polyurethane foams modified with glycerol and cellulose micro/nanocrystals. J. Appl. Polym. 132(10). DOI: 10.1002/app.41602.

Pan, M., T. Jiang and J. Pan. 2011. Antioxidant Activities of Rapeseed Protein Hydrolysates. Food Bioprocess Tech. 4(7): 1144–1152.

Patel, M., J. Shukla, N. Patel and K. Patel. 2009. Biomaterial based novel polyurethane adhesives for wood to wood and metal to metal bonding. Mater. Res. 12(4): 385–393.

Pedroche, J., M. Yust, H. Lqari, J. Giron-Calle, M. Alaiz, J. Vioque and F. Millan. 2004. Brassicacarinata protein isolates: chemical composition, protein characterization and improvement of functional properties by protein hydrolysis. Food Chem. 88(3): 337–346.

Philipp, C. and S. Eschig. 2012. Waterborne polyurethane wood coatings based on rapeseed fatty acid methyl esters. Prog. Org. Coat. 74(4): 705–711.

Pizzi, A. 2006. Recent developments in eco-efficient bio-based adhesives for wood bonding: opportunities and issues. J. Adhesion Sci. Technol. 20: 829–846.

Przybylski, R., T. Mag, N. Eskin and B. McDonald. 2005. Canola oil. *In*: Bailey's Industrial Oil and Fat Products, 6th ed.; Shahidi, F., Ed.; Wiley Interscience, a John Wiley and Sons, Inc., Publication: Hoboken, NJ, 2005; pp. 61–121.

Qi, G. and X. Sun. 2011. Soy protein adhesives blends with synthetic latex on wood veneer. J. Am. Chem. Soc. 88: 271–281.

Qu, P., H. Huang, G. Wu, E. Sun and Z. Chang. 2015. Hydrolyzed soy protein isolates modified urea–formaldehyde resins as adhesives and its biodegradability. J. Adhe. Sci. Tech. 29(21): 2381–2398.

Rivera, D., K. Rommi, M. Fernandes, R. Lantto and T. Tzanow. 2015. Biocompounds from rapeseed oil industry co-stream as active ingredients for skin care applications. Int. J. Cosmetic Sci. 37(5): 496–505.

Rogers, J., X. Geng and K. Li. 2004. Soy-based adhesives with 1, 3-Dichloro-2-propanol as a curing agent. Wood Fiber Sci. 36: 186–194.

Rojek, P. and A. Prociak. 2011. Effect of different rapeseed-oil-based polyols on mechanical properties of flexible polyurethane foams. J. Appl. Polym. Sci. 125: 2936–2945.

Saurabh, T., M. Patnaik, S. Bhagt and V. Renge. 2011. Epoxidation of vegetable oils: A review. Int. J. Adv. Eng. Technol. 2(4): 491–501.

Schultz, J. and M. Nardin. 1999. Theories and mechanisms of adhesion. pp. 1–6 *In*: K.L. Mittal and A. Pizzi (eds.). Adhesion Promotion Techniques: Technological Applications. Marcel Dekker: New York.

Schwenke, K., B. Raab, P. Plietz and G. Damaschun. 1983. The structure of the 12 S globulin from rapeseed (*Brassica napus* L.). Nahrung. 27: 165–175.

Schwenke, K. 1994. Rapeseed protein. pp. 281–306. *In*: B.J.F. Hudson (ed.). New and Developing Sources of Food Proteins. London, U.K.: Chapman and Hall.

Seller, T.J. 1994. Adhesives in the wood industry. pp. 599–614. *In*: A. Pizzi and K.L. Mittal (eds.). Handbook of Adhesive Technology, Marcel Dekker, Inc., New York.

Shi, W. and M. Dumont. 2014. Process and physical properties of canola protein isolate-based films. Ind. Crop Prod. 52: 269–277.

Somani, K., S. Kansara, N. Patel and A. Pakshit. 2003. Castor oil based polyurethane adhesives for wood-to-wood bonding. Int. J. Adhes. Adhes. 23: 269–275.

Steele, P., R. Kreibich, P. Steynberg and R. Hemingway. 1998. Finger jointing green southern yellow pine with a soy-based adhesive. Adhes Age. 10: 49–54.

Stirna, U., A. Fridrihsone, B. Lazdina, M. Misane and D. Vilsone. 2013. Biobased polyurethanes from rapeseed oil polyols: Structure, mechanical and thermal properties. J. Polym. Environ. 21: 952–962.

Sun, X.S., D. Wang, L. Zhang, X. Mo, L. Zhu and D. Bolye. 2008. Morphology and phase separation of hydrophobic clusters of soy globular protein polymers. Macromol. Biosci. 8: 295–303.

Sun, X. and K. Bian. 1999. Shear strength and water resistance of modified soy protein adhesives. J. Am. Oil Chem. Soc. 76: 977–980.

Szycher, M. 2012. Szycher's Handbook of Polyurethanes. CRC Press: Boca Raton, FL, 1999.

Tan, S., R. Mailer, C. Blanchard and S. Agboola. 2014. Emulsifying properties of proteins extracted from Australian canola meal. LWT-Food Sci. Technol. 57(1): 376–382.

Tan, S. and W. Chow. 2010. Biobased epoxidized vegetable oils and its greener epoxy blends: a review. Polym. Plast. Technol. Eng. 49(15): 1581–1590.

Tzeng, Y., L. Diosady and L. Rubin. 1988. Preparation of rapeseed protein isolate by sodium hexametaphosphate extraction, ultrafiltration, diafiltration, and ion-exchange. J. Food Sci. 53(5): 1537–1542.

Tzeng, Y., L. Diosady and L. Rubin. 1990. Production of canola protein materials by alkaline extraction, precipitation, and membrane processing. J. Food Sci. 55: 1147–1156.

USDA. 2015a. Oilseeds: World markets and trade. December 2015. http://apps.fas.usda.gov/psdonline/circulars/oilseeds.pdf.

USDA. 2015b. Crop Production 2014 Summary. January 2015. ISSN: 1057–7823.

Vallat, M. and B. Bessaha. 2000. Adhesive behavior of polyurethane-based materials. J. Appl. Polym. Sci. 76(5): 665–671.

Wang, C., J. Wu and G. Bernard. 2014. Preparation and characterization of canola protein isolate–poly(glycidyl methacrylate) conjugates: A bio-based adhesive. Ind. Crop. Prod. 57: 124–131.

Wang, D., X. Sun, G. Yang and Y. Wang. 2008. Improved water resistance of soy protein adhesive at isoelectric point. Trans. ASASE. 52(1): 173–177.

Wang, L. 2006. Polyurethane Foam. Chem. Eng. News. 84(2): 48.

Wang, Y., X. Mo, X.S. Sun and D. Wang. 2007. Soy protein adhesion enhanced by glutaraldehyde crosslink. J. Appl. Polymer. Sci. 104: 130–136.

Wäsche, A., S. Wurst, A. Borcherding and T. Luck. 1998. Film forming properties of rape-seed protein after structural modification. Nahrung 42: 269–271.

Wool, R. and X.S. Sun. 2005. Soy protein adhesives. pp. 327–368. *In*: R.P. Wool and X.S. Sun (eds.). Bio-based Polymers and Composites, 1st edn., Elsevier Academic Press, Burlington.

Wu, J. and A. Muir. 2008. Comparative structural, emulsifying, and biological properties of 2 major canola proteins, cruciferin and napin. J. Food Sci. 73: 210–216.

Wu, X., X. Zhang, S. Yang, H. Chen and D. Wang. 2000. The study of epoxidized rapeseed oil used as a potential biodegradable lubricant. J. Am. Oil Chem. Soc. 77(5): 561–563.

Yang, I., S. Ahn, I. Choi, G. Han and S. Oh. 2011. Preliminary study of rapeseed flour-based wood adhesives for making wood flooring. J. Korean Wood Sci. & Tech. 39(5): 451–458.

Yang, I., G. Han, S. Ahn, I. Choi, Y. Kim and S. Oh. 2014. Adhesive properties of medium-density fiberboards fabricated with rapeseed flour-based adhesive resins. J. Adhesion. 90: 279–295.

Yang, I., M. Kuo, D.J. Myers and A. Pu. 2006. Comparision of protein-based adhesive resins for composites. J. Wood Sci. 52: 503–508.

Zhang, L. and X.S. Sun. 2008. Effect of sodium bisulfite on properties of soybean glycinin. J. Agric. Food Chem. 56: 11192–11197.

Canola Protein and Oil-based Wood Adhesives 139

Zhang, S., Z. Wang and S. Xu. 2008. Antioxidant and antithrombotic activities of rapeseed peptides. J. Am. Oil Chem. Soc. 85(6): 521–527.

Zhang, Z. and Y. Hua. 2007. Urea-modified soy globulin proteins (7S and 11S): Effect of wettability and secondary structure on adhesion. J. Am. Oil Chem. Soc. 84: 853–857.

Zhong, Z. and X.S. Sun. 2007. Plywood adhesives by blending soy protein polymer with phenolformaldehyde resin. J. Biobased Mater. Bio. 1: 380–387.

Zhong, Z., X.S. Sun, X. Fang and J.A. Ratto. 2002. Adhesive strength of guanidine hydrocholoride-modified soy protein for fiberboard application. Int. J. Adhes Adhes. 22: 267–272.

Zieleniewska, M., M. Leszczynski, M. Kuranska, A. Prociak, L. Szceopkowski, M. Krzyzowska and J. Ryszkowska. 2015. Preparation and characterisation of rigid polyurethane foams using a rapeseed oil-based polyol. Ind. Crop Prod. 74: 887–897.

Zirwer, D., K. Gast, H. Welfle, B. Schlesier and K. Schwenke. 1985. Secondary structure of globulins from plant seeds: a re-evaluation from circular dichroism measurements. Int. J. Biol. Macromols. 7(2): 105–108.

5

Wood Adhesives Containing Proteins and Carbohydrates

H.N. Cheng* and Zhongqi He

ABSTRACT

In recent years there has been resurgent interest in using biopolymers as sustainable and environmentally friendly ingredients in wood adhesive formulations. Among them, proteins and carbohydrates are the most commonly used. In this chapter, an overview is given of protein-based and carbohydrate-based wood adhesives. Included in the coverage are recent wood adhesives involving proteins, carbohydrates, glycoproteins, and protein/carbohydrate blends. A lot of interesting developments have been reported, particularly entailing modification reactions, use of additives, and blending. For example, soy protein is sometimes used as an ingredient for wood adhesives, and commercial wood adhesive products involving soy protein are available. It has been found that cottonseed protein added to soy protein can improve both its dry strength and its water resistance. In addition, cottonseed- and soy-protein based adhesives have been formulated with xylan, starch, or celluloses to determine the influence of polysaccharide fillers on protein-based adhesive properties. In some cases, adhesive strength is retained even when the cottonseed or soy protein is mixed with up to 75% polysaccharide.

Southern Regional Research Center, USDA Agricultural Research Service, 1100 Robert E. Lee Blvd., New Orleans, LA 70124, USA.
* Corresponding author: hn.cheng@ars.usda.gov

Introduction

Adhesives are needed for various wood applications. Prior to World War II, many agro-based biopolymers were used in adhesive formulations. Because of lower cost and better properties, synthetic adhesives mostly replaced natural polymer in the past 70 years (Lambuth, 1989). The global wood adhesives market was valued at $13.15 billion with a volume of 16,200 kilo tons in 2013 (Transparency Market Research, 2014). Most of the adhesives are based on urea-formaldehyde, melamine-urea-formaldehyde, phenol-formaldehyde resins, and polyurethanes. However, because of environmental concerns with formaldehyde, sustainability, and the desire to decrease the usage of petroleum-derived raw materials, there is a partial shift in the past 15 years towards the use of agro-based raw materials in wood adhesives (Frihart, 2011; Pizzi, 2006; Hemingway and Conner, 1989). Some product development strategies include: (1) partial replacement of petroleum derived adhesive systems with agro-based raw materials, (2) generation of derivatives or degradation products from agro-based materials, which are utilized as adhesive building blocks, and (3) direct use of agro-based materials, sometimes with additives, to enhance their effectiveness. Agro-based materials being considered include proteins, carbohydrates, lignin, and tannin.

The purpose of this article is to review recent developments in the use of the proteins and carbohydrates in wood adhesive formulations. Topics of interest include proteins, carbohydrates, glycoproteins, and protein/carbohydrate blends. A particular emphasis has been placed on work reported from the authors' laboratories.

Protein-Based Adhesives

Animal protein adhesives are known since antiquity; indeed the ancient civilizations in Egypt and China used animal glues for both wood and paintings (Lambuth, 1989). These were produced through hydrolysis of the collagen from skins, bones, tendons, and other tissues. These materials were common in woodworking glue but have now been replaced by synthetics.

Other animal proteins often used include blood proteins and casein (Detlefsen, 1989). Casein glues have moderate resistance to moisture but tend to be susceptible to microbial and fungal growth. Because casein's properties are not outstanding and the price is not cheap, their use in wood adhesive application today is limited; it is sometimes included in wood adhesive testing (Umemura et al., 2003). Blood glues are dark red or black in color and have moderate resistance to moisture. In a recent article (Lin and Gunasekaran, 2010), cow blood with alkali modification was tested as a wood adhesive. Various parameters were studied with respect to bonding strength and water resistance, including pH, the degree of hydrolysis, viscosity, water solubility, and curing time. Addition of sodium silicate at > 2% (v/v) improved the water resistance irrespective of the pH of the adhesive formulation. The adhesive bonding strength was independent of the pH and was comparable to that of phenol-formaldehyde in the dry condition, but somewhat lower in the wet state.

Vegetable proteins are being increasingly used in wood adhesive formulations. Soy protein is by far the most common (Qi et al., 2016; Kumar et al., 2002; Liu and Li, 2007; Sun and Bian, 1999). Relative to blood and casein glues, soy proteins have lower dry strength and moisture resistance, but recent formulations have improved the end-use performance (Frihart, 2011). Commercial wood adhesive products involving soy protein are now available (Frihart, 2011; Allen et al., 2010; Orr, 2007). Because soy protein is discussed extensively elsewhere in this book (Qi et al., 2016), it is not covered in detail here.

Several other vegetable proteins have been reported recently in the literature. Among them, cottonseed protein has received a fair amount of attention (Cheng et al., 2013; He et al., 2014a; 2014b; 2016a; Cheng et al., 2016b). Cottonseed protein isolate was found to be superior to soy protein isolate in adhesive strength and water resistance on maple veneers; the effects of several modifiers, including alkali, guanidine hydrochloride, sodium dodecyl sulfate, and urea were also investigated (Cheng et al., 2013). Successful attempts were made to use water or buffer washing of cottonseed meal, and the adhesive properties of these washed cottonseed meal were equal or even slightly better than the cottonseed protein isolate (He et al., 2014a; 2014b; 2014c; 2016a). A number of new modifiers were found that gave improved adhesive performance particularly to cottonseed protein (Cheng et al., 2016b).

Canola protein, together with sodium bisulfate, was recently studied as a wood adhesive (Li et al., 2012). The protein was extracted from defatted canola meal by using the alkali solubilization-acid precipitation method and modified with varying $NaHSO_3$ concentrations. As the concentration of $NaHSO_3$ increased, the recovery of canola protein increased, but the protein purity decreased. The canola protein exhibited good dry adhesive strength and water resistance. $NaHSO_3$ had a slight weakening effect on the adhesion performance of the canola protein but improved the handling and flowability of the protein.

Furthermore, spent hen protein (Wang and Wu, 2012), wheat gluten (D'Amico et al., 2013; Nordqvist et al., 2012), and distillers dry grains (Bandara et al., 2013) were also reported as wood adhesives. A comparative study of several proteins (mixed porcine and bovine blood, soy, peanut, blood/soy blend, blood/peanut blend) formulated with phenol-formaldehyde (PF) resins was carried out (Yang et al., 2006). In adhesive testing, the blood protein/PF performed almost as well as neat PF resins and met the requirement for exterior and interior grades of medium density fiberboard (MDF). PF formulations made with soy or peanut proteins only met the requirement for interior grade, but the addition of blood protein to soy or peanut protein improved their performance for exterior application as well.

As shown by the comparative protein study (Yang et al., 2006), polymer blending can be utilized to vary polymer properties. Indeed, some blends of proteins as adhesives have been studied a long time ago, e.g., casein and protein meal glues and soy protein/blood glues (Cone and Galber, 1934; Bradshaw and Dunham, 1931). Cone and Galber (1934) claimed that the oilseed residue flour glues might be modified to great advantage by the introduction of blood, either as the dry blood albumin or even as fresh blood. The mixture had a better spreading consistency than either seed protein or blood protein jellies. A similar strategy, co-precipitation of proteins from mixtures of raw materials, had been applied to improve protein gelling characteristics and functionality (Alu'datt et al., 2013). For example, Berardi and Cherry (1979) reported the preparation and composition of coprecipitated protein isolates from cottonseed, soybean, and peanut flours. The co-isolates improved the nutritional values of these proteins. Alu'datt et al. (2012) reported the preparation, characterization and properties of whey-soy proteins co-precipitate gels. Their oscillatory rheometric data indicated that protein concentration directly influenced gel rigidity, and protein gels obtained from the co-precipitates at 16% protein concentration showed superior gel strength and water holding capacity relative to the gel of either whey or soy protein isolate alone.

As noted earlier, cottonseed protein was found to have better dry strength and hot water resistance relative to soy protein (Cheng et al., 2013). If the use of soy protein is preferred, a promising line of approach is to blend cottonseed protein into soy protein (Cheng et al., 2016a). Blending with the appropriate weights of each protein was done in water, and the same procedure was used for adhesive application on maple veneer, adhesive curing, and mechanical testing. The two proteins mixed well in water, and nothing unusual was found during the process. The bonding results (Table 1) showed that the dry adhesive strength and hot water resistance of soy protein adhesive formulations were both improved with the addition of cottonseed protein. The water resistance was particularly notable such that even the addition of 25% cottonseed protein produced a noticeable improvement. More than that, the two proteins can be blended together in different ratios to give a range of adhesive properties and hot water

Table 1. Properties of adhesives from blends of soy protein (SP) and cottonseed protein (CSP)[a]. Adapted from Cheng et al., 2016a.

CSP %	SP %	Tensile strength[b] (MPa) of dry bonded wood strips	CSP %	SP %	Tensile strength[c] (MPa) after 2 cycles of water (63°C)– soaking and drying
100	0	2.58 ± 0.22 [A]	100	0	3.34 ± 0.23 [A]
80	20	2.30 ± 0.19 [B]	75	25	2.95 ± 0.49 [B]
60	40	1.99 ± 0.22 [C]	50	50	3.41 ± 0.28 [A]
40	60	1.88 ± 0.18 [CD]	25	75	2.86 ± 0.35 [B]
20	80	1.71 ± 0.15 [DE]	0	100	0.30 ± 0.17 [C]
0	100	1.52 ± 0.12 [E]			

[a] n = 10 for each treatment. Data in each column with the same letter indicates that the treatments are not significantly different at p = 0.05.
[b] Testing done on bonded wood composites based on 1.27 cm x 8.89 cm wood strips and 1.27 cm x 2.54 cm overlap between two strips.
[c] Testing done on bonded wood composites based on 2.54 cm x 8.89 cm wood strips and 2.54 cm x 2.54 cm overlap between two strips.

resistance, thus providing the flexibility in formulating different adhesives to meet different requirements.

Carbohydrate-based Adhesives

Carbohydrates have been used as adhesives since antiquity, but they are utilized today more in paper products than in wood (Baumann and Conner, 1994; Conner, 1989). Among carbohydrates, starch is the most often used. An old method is to dissolve starch in a hot alkaline solution, which is then cooled down to be applied to wood (Kennedy, 1989). Recently, several approaches have been adopted to improve the performance of starch-based wood adhesives. These include starch/poly(vinyl alcohol) blend (Imam et al., 1999; 2001), starch/tannin blend (Moubarik et al., 2009; 2010; 2011), starch reacted with isocyanates (Tan et al., 2011), starch grafted with vinyl acetate by itself (Wang et al., 2012), with urea (Wang et al., 2013), and with silica nanoparticles (Wang et al., 2011). These formulations showed varying degrees of improved dry adhesive strength and water resistance when compared to starch alone.

In addition, a number of other polysaccharides have been reported for wood adhesive applications. For example, konjac glucomannan and chitosan were investigated for their adhesive properties on plywood (Umemura et al., 2003). Konjac had good dry bond strength but water resistance was extremely low. Chitosan also had good dry bond strength and exhibited superior water resistance compared to casein and soy glues. When konjac and chitosan were blended, the dry adhesive strength was enhanced, and the bonding properties were better than those of casein and soy glues alone. At intermediate or high chitosan levels, the chitosan–konjac blends showed superior wet resistance to casein and soy glues.

Recently, several polysaccharide dispersions were prepared and evaluated as wood adhesives, including locust bean gum, guar gum, xanthan gum and tamarind gum (Norstrom et al., 2014). The best results were shown for locust bean gum. Water resistance was only indirectly tested through contact-angle measurements. The same research group also looked at xylan as possible wood adhesives (Norstrom et al., 2015), but they found it to have mediocre bonding performance, especially regarding water resistance. The addition of dispersing agents [poly(vinyl alcohol) or poly(vinyl amine)] and crosslinkers [glyoxal or hexa(methoxymethyl) melamine] was found to improve both dry adhesive strength and water resistance.

Several publications reported bacterial polysaccharides as wood adhesives. In one report, a high molecular weight polysaccharide produced from bacteria was tested on epoxy glass and manufactured woods (Combie et al., 2004). In a related publication (Haag et al., 2004), the same research group found the adhesive strength of their bacterial polysaccharide to be similar to that of pullulan, but it could be improved about 20% by the addition of Tween 80 and a rhamnolipid surfactant. In a follow-up work (Haag et al., 2006), they found another bacterial polysaccharide that gave good dry adhesive properties. Through acetylation, the effect of humidity was partly mitigated on both polysaccharides, but still their suggested use was for indoor furniture and cabinetry (Haag et al., 2006). Very recently, an exopolysaccharide isolated from *Bacillus magaterium* was found to have satisfactory dry adhesive strength on wood, but no water resistance data were reported (Kumar and Shah, 2015).

Two methods were reported that modified carbohydrates to facilitate their use in glues: one used phenol or ethylene glycol and sulfuric acid to form a liquefied product from cellulose, and another employed enzymes or micro-organisms to degrade the carbohydrates (Hamarneh, 2010). Several methods were devised to improve the water resistance of carbohydrate glues (Pizzi, 2006; Hamarneh, 2010). One approach was to modify urea-formaldehyde or phenol-formaldehyde resins with the addition of the carbohydrate (e.g., cellulose, starch, and hemicellulose). Another approach was to convert carbohydrates into furfuryl alcohol, 2,5-bis(hydroxylmethylfuran), or diglucosyl urea, and these degradation materials could then be transformed to an adhesive through the reaction with formaldehyde or diisocyanate.

In a recent work (Cheng et al., 2016a), it was shown that cottonseed proteins added to a carbohydrate adhesive could improve both the dry tensile strength of carbohydrate adhesive and the durability of the adhesion in a hot water test. This work will be described in the section below on protein/carbohydrate blends.

Glycoprotein Adhesives

In nature, cell surface glycoproteins exhibit adhesive properties (Von der Mark and Sorokin, 2002; Xu and Mosher, 2011). For example, the integrins can mediate adhesion to extracellular matrix and cell-cell interactions. The cadherins are glycoproteins present in tissues that are involved in cell adhesion together with Ca^{++} ions. The selectins are a family of cell adhesion molecules that bind to mucins. The immunoglobulins are involved in the recognition, binding, or adhesion processes of cells through antibodies. Mucins (commonly found in mucous membranes or saliva) contain glycoproteins that can cause the formation of a gel-like material that helps bioadhesion (Smart, 2005). The glue on spider web contains glycoproteins (Stellwagen et al., 2015). Likewise the glue in silk contains glycoproteins (Dutta et al., 2012). The structure and function of adhesive gels produced from invertebrates were reviewed by Smith (2002).

Despite their frequent occurrence in living systems, glycoproteins have not so far been independently exploited for wood adhesion. However, a main component of soy protein, β-conglycinin, is a glycoprotein that contains ca. 5% high mannose type of carbohydrate (Kimura et al., 1997; Mo and Sun, 2013), and it likely contributes to soy protein's adhesive strength. It may be noted that when a blend of protein and polysaccharide is pressed at 80–130°C for adhesive bonding, Maillard reaction products are produced, and these products have similar chemical structures as a glycoprotein in the sense that there is a covalent bond between the carbohydrate and the protein.

Protein/Carbohydrate Blends

A few studies have appeared on the use of protein/carbohydrate blends as wood adhesives. Chen et al. (2013) determined the effects of sucrose and glucose on the water resistance and bonding strength of soy-based adhesives with poplar veneers. They prepared adhesives containing the same amount of soy protein but different sucrose and glucose contents, and characterized the mechanisms of the influences of the sugars on the water resistance of soy protein-based adhesives. While an increased total carbohydrate content from defatted soy flour was not favorable, their data showed that the water absorptions and bonding strengths of soy protein improved with increasing contents of sucrose and glucose. The highest bonding strength was obtained when the carbohydrate in the adhesive contained 71% glucose. Fourier-transform infrared and X-ray photoelectron spectroscopies revealed that the cross-linking and hydrophobicities of these adhesives were enhanced by Maillard reactions between soy protein and sucrose and glucose. Thus, the authors proposed that the Maillard reactions of sucrose and glucose with soy protein decreased the hydrophilicity and improved the bonding strength of soy protein. This work provides an incentive to utilize defatted soy meal as wood adhesives as the price of soy flour is more attractive than soy protein isolate for commercial wood adhesives (Table 2).

Lorenz et al. (2015) examined commercial protein isolate, concentrate, and flour with different carbohydrate contents for their adhesive properties on maple veneers. They reported that dry strengths of all the adhesives were good (>5 MPa), but only the soy protein isolate had good wet strength as the isolate could provide 10 times the wet adhesive shear strength than the soy flour (Fig. 1). It is notable that the soy protein concentrate did not give better wet strength than soy flour in contrast to the assumption that

Table 2. Approximate price and compositions of commercial soy products in 2014. Compiled per Lorenz et al. (2015).

Product	Price ($/lb)	Composition (%)			
		Protein	Oil	Carbohydrate	Moisture
Whole bean	~ 0.23–0.25	36	18	36	10
Defatted meal	~ 0.25	48	0	44	< 10
Flour	~ 0.30–0.35	50	0	40	< 10
Protein concentrate	≥ 1.00	> 65	0	Up to 35	-[a]
Protein isolate	≥ 1.90	> 90	0	Up to 10	-

[a] No data available

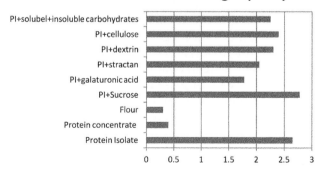

Wet strength (MPa)

Figure 1. Wet adhesive strength of soy products and the effect of added carbohydrates. The carbohydrates were added into 15% of soy protein isolate (PI) with 23% of carbohydrate that was equal to the carbohydrate content in the soy flour adhesive. Compiled per Lorenz et al. (2015).

removal of the soluble carbohydrates and low-molecular-weight proteins in preparing the concentrate should have improved the wet adhesive strength.

Thus, Lorenz et al. (2015) measured the effect of several soluble and insoluble external carbohydrates on the soy protein's wet adhesives strength (Fig. 1). They mixed the carbohydrates with soy protein at the concentration of total carbohydrates that occurs in soy flour, and the bonding performance of these mixtures was evaluated. Sucrose, which can move easily into the wood, increased slightly the protein isolate's wet adhesive strength. The observed effect of sucrose was consistent with that in Chen et al. (2013). Galacturonic acid and stractan (arabinogalactan) had the biggest negative effect while cellulose had a small effect on the wet adhesive strength of protein isolate. However, the wet adhesive strength of all the protein isolates with external carbohydrates was higher than that of the flour adhesive with same percentage of indigenous carbohydrates. In other words, the carbohydrate plus commercial isolate was still much better than soy flour or a commercial soy concentrate. Whereas the mechanism of the differential functions of indigenous and external carbohydrates in soy protein's adhesive strength is not clear (Lorentz et al., 2015), a possibility is that not all carbohydrates may equally affect the protein's adhesive performance and apparently, the carbohydrate composition of soy flour and protein concentrate differs from the carbohydrates added to soy protein isolate. The explanation is supported by the data of different effects of carbohydrates in Fig. 1 and Chen et al. (2013).

In cottonseed meal-related studies, He et al. (2014a) observed that higher protein content of cottonseed meal products did not necessarily lead to higher adhesive strength, implying other components could also contribute to the adhesive performance of cottonseed products. One of the components should be the insoluble carbohydrates as the water washed cottonseed meal showed better water resistance than unwashed cottonseed meal (He et al., 2014a; 2014b). Further characterization of these products indicated that both water soluble proteins and carbohydrates were removed during water washing, the water washed cottonseed meal products were less hydrophilic (He et al., 2014c; 2015). Thus, water soluble carbohydrates may contribute to poor water resistance components, and insoluble carbohydrates may increase the water resistance of cottonseed protein adhesives. Recently, He et al. (2016b) examined the functions of the water soluble and insoluble carbohydrates in soy meal in the same way. Removal of water soluble carbohydrates in soy meal increased its dry adhesive strength, but its water resistance was more similar to of that of unwashed soy meal than to that of soy protein isolate (Table 3). This result of no or little improvement of water resistance by water washing is in contrast to the result of cottonseed meal products (He et al., 2014a), but

Table 3. Dry and soaked adhesive strength and protein content of soy meal products. Adapted from He et al. (2016b).

	Dry strength (MPa)	Soaked strength (MPa)	Protein (%)
----------Bonded at 100°C----------			
Soy meal	2.40 ± 0.50	1.25 ± 0.19	49.6
Water washed meal	3.22 ± 0.72	1.52 ± 0.39	66.2
Lab-prepared protein isolate	3.51 ± 0.95	3.73 ± 0.62	91.9
Commercial protein isolate	4.16 ± 0.56	4.24 ± 0.37	86.3
---------Bonded 130°C-----------			
Soy meal	2.85 ± 0.51	2.48 ± 0.29	49.6
Water washed meal	3.25 ± 0.80	2.20 ± 0.29	66.2
Lab-prepared protein isolate	3.73 ± 0.62	3.67 ± 1.28	91.9
Commercial protein isolate	4.50 ± 0.42	4.95 ± 0.34	86.3

similar to the observation on ethanol-purified soy protein concentrate by Lorenz et al. (2015). Therefore, removal of either water or ethanol soluble carbohydrates would not improve the water resistance of soy meal-based adhesives. It is assumed that the different carbohydrate profiles in soy and cottonseed products was the factor in the different adhesive performances between the two types of oilseed-based products (He et al., 2016b).

Also similar to Lorenz et al. (2015), He et al. (2016b) observed better adhesive performance of commercial soy protein isolate than the lab-prepared protein isolate (Table 3). In reviewing the metal composition data, higher Ca and Mg contents in the lab-prepared protein isolate may have contributed to better adhesive performance than commercial soy protein isolate (He et al. 2016b). On the other hand, Lorenz et al. (2015) proposed that difference in the protein denaturation in protein isolation processes could be a large factor that contributes to the difference in adhesive performance between commercial and lab prepared soy protein isolate. It is certain that further elucidation of the mechanisms or causes of the differing adhesive performances would shed light on the adhesive mechanisms of the protein isolates, thus optimizing the use of these materials and their fractions for wood bonding.

Cheng et al. (2016a) reported on the blending of soy and cottonseed protein with other *polysaccharides*, e.g., xylan, cellulose, and starch. The results for protein-xylan blends are shown in Table 4. Xylan had only modest dry adhesive strength (1.12 MPa) and no hot water resistance. The addition of about 15% cottonseed protein considerably improved its dry strength,

Table 4. Adhesive properties of adhesives from blends of xylan and soy protein (SP) or cottonseed protein (CSP)[a]. Adapted from Cheng et al. (2016a).

Protein %	Xylan %	Tensile strength[b] (MPa) of dry bonded wood strips	Tensile strength[c] (MPa) after 2 cycles of water (63°C)-soaking and drying
0	100	1.12 ± 0.22 [E]	all failed
15 CSP	85	2.26 ± 0.28 [C]	0.20 ± 0.10 [C]
40 CSP	60	2.37 ± 0.18 [BC]	1.84 ± 0.30 [B]
60 CSP	40	2.87 ± 0.29 [A]	3.03 ± 0.38 [A]
80 CSP	20	2.81 ± 0.32 [AB]	3.34 ± 0.28 [A]
100 CSP	0	2.48 ± 0.25 [ABC]	3.34 ± 0.23 [A]
15 SP	85	1.34 ± 0.23 [DE]	all failed
40 SP	60	1.45 ± 0.33 [DE]	all failed
60 SP	40	1.49 ± 0.28 [DE]	all failed
80 SP	20	1.60 ± 0.18 [D]	0.28 ± 0.15 [C]
100 SP	0	1.49 ± 0.17 [DE]	0.30 ± 0.17 [C]

[a] $n = 10$ for each treatment. Data in each column with the same letter indicates that the treatments are not significantly different at $p = 0.05$.

[b] Testing done on bonded wood composites based on 1.27 cm x 8.89 cm wood strips and 1.27 cm x 2.54 cm overlap between two strips.

[c] Testing done on bonded wood composites based on 2.54 cm x 8.89 cm wood strips and 2.54 cm x 2.54 cm overlap between two strips.

almost equal to that of cottonseed protein by itself. For hot water resistance, the addition of about 50% cottonseed protein was needed to bring xylan's water resistance to the similar level as cottonseed protein itself. In contrast, 50% addition of soy protein provided some improvements in dry strength but no effect on hot water resistance.

An analogous set of data (Table 5) was reported for cellulose/protein blends (Cheng et al., 2016a). Cellulose powder (20 μm size) from Sigma Aldrich (Milwaukee, Wisconsin) and the protein were added to water and made into a suspension, which is then applied to the wood surface and cured. By itself, cellulose is a poor adhesive, with a dry tensile strength of 0.46 MPa and no hot water resistance. The addition of as low as 25% cottonseed protein provides remarkably improved dry tensile strength and hot water resistance. In contrast, the addition of 25% soy protein modestly increases the dry tensile strength and no hot water resistance. It was noted (Cheng et al., 2016a) that the performance properties of a protein in adhesion depend on the reaction of protein with the wood surface and on the formation of crosslinked networks among the denatured protein molecules during the heat bonding process. The polysaccharide added as the second component in the blend can dilute the protein molecules but may also react with the protein during heat bonding. As long as the

Table 5. Adhesive properties of adhesives from blends of cellulose and soy protein (SP) or cottonseed protein (CSP)[a]. Adapted from Cheng et al. (2016a).

Protein %	Cellulose %	Tensile strength[b] (MPa) of dry bonded wood strips	Tensile strength[c] (MPa) after 2 cycles of water (63°C)-soaking and drying
0	100	0.46 ± 0.08 [D]	all failed
25 CSP	75	2.36 ± 0.33 [A]	2.83 ± 0.50 [B]
50 CSP	50	2.54 ± 0.26 [A]	3.43 ± 0.23 [A]
75 CSP	25	2.54 ± 0.19 [A]	2.67 ± 0.50 [B]
100 CSP	0	2.55 ± 0.26 [A]	3.34 ± 0.23 [A]
25 SP	75	1.05 ± 0.17 [C]	all failed
50 SP	50	1.30 ± 0.26 [BC]	all failed
75 SP	25	1.54 ± 0.25 [B]	0.25 ± 0.10 [C]
100 SP	0	1.52 ± 0.12 [B]	0.30 ± 0.17 [C]

[a] $n = 10$ for each treatment. Data in each column with the same letter indicates that the treatments are not significantly different at $p = 0.05$.
[b] Testing done on bonded wood composites based on 1.27 cm x 8.89 cm wood strips and 1.27 cm x 2.54 cm overlap between two strips.
[c] Testing done on bonded wood composites based on 2.54 cm x 8.89 cm wood strips and 2.54 cm x 2.54 cm overlap between two strips.

denatured protein can form a continuous network during heat bonding and the polysaccharide can help with this bonding, the adhesive strength and water resistance properties can be maintained with up to a critical minimum level of protein. Only when the protein level become too low, such that the continuous networks are unable to form, is the adhesion markedly diminished.

Thus, for xylan, cellulose, and starch adhesives (Cheng et al., 2016a), the addition of 40–50% cottonseed protein can notably improve the adhesive properties of carbohydrate adhesive. From the point of view of protein-based adhesives, the addition of 50% less expensive polysaccharides can significantly lower the cost of the protein adhesive formulation.

Conclusions

A review is made of recent wood adhesives work involving proteins and carbohydrates. Because of environmental regulations regarding emission of organic compounds including formaldehyde, wood adhesives based on soy proteins have gained attention in the past 15 years. Some of the issues that need to be addressed for soy protein include lower cost, increased adhesive strength, and improved water resistance. Cottonseed protein has been found to be a viable alternative to soy protein, with stronger adhesive strength and hot water resistance when compared to soy protein. Carbohydrates have been used as wood adhesives in the past; recent developments have yielded several approaches that improved both adhesive strength and water resistance. In addition, two types of blends in wood adhesive formulations are included in this review: (1) blends of two proteins, and (2) blends of protein with carbohydrates. Both blending approaches have been shown to be promising, and further studies of the blends may be anticipated in the future.

Acknowledgements

The authors thank Drs. K. Thomas Klasson and Michael K. Dowd for helpful discussions and Catrina Ford and Dorselyn Chapital for expert experimental work. Mention of trade names or commercial products in this publication is solely for the purpose of providing specific information and does not imply recommendation or endorsement by the U.S. Department of Agriculture. USDA is an equal opportunity provider and employer.

Keywords: Blending, Carbohydrate, Cottonseed protein, Glycoprotein, Soy protein, Wood adhesive

References

Allen, A.J., J.J. Marcinko, T.A. Wagler and A.J. Sosnowick. 2010. Investigations of the molecular interactions of soy-based adhesives. Forest Prod. J. 60(6): 534–540.

Alu'datt, M.H., I. Alli and M. Nagadi. 2012. Preparation, characterization and properties of whey-soy proteins co-precipitates. Food Chem. 134: 294–300.

Alu'datt, M.H., G.J. Al-Rabadi, I. Alli, K. Ereifej, T. Rababah, M.N. Alhamad and P.J. Torley. 2013. Protein co-precipitates: A review of their preparation and functional properties. Food Bioprod. Process. 91: 327–335.

Bandara, N., L. Chen and J. Wu. 2013. Adhesive properties of modified triticale distillers grain proteins. International Journal of Adhesion and Adhesives. 44: 122–129.

Baumann, M.G.D. and A.H. Conner. 1994. Carbohydrate polymers as adhesives. *In*: A. Pizzi and K.L. Mittai (eds.). Handbook of Adhesive Technology. Marcel Dekker, New York, Chapter 15.

Berardi, L. and J. Cherry. 1979. Preparation and composition of coprecipitated protein isolates from cottonseed, soybean, and peanut flours. Cereal Chem. 56: 95–100.

Bradshaw, L. and H.V. Dunham. 1931. Process of making slow-setting casein glue, and dry base for use in such process. U.S. Patent 1,829,259.

Chen, N., Q. Lin, J. Rao and Q. Zeng. 2013. Water resistances and bonding strengths of soy-based adhesives containing different carbohydrates. Ind. Crop. Prod. 50: 44–49.

Cheng, H.N., M.K. Dowd and Z. He. 2013. Investigation of modified cottonseed protein adhesives for wood composites. Ind. Crops Prod. 46: 399–403.

Cheng, H.N., C. Ford, M.K. Dowd and Z. He. 2016a. Soy and cottonseed protein blends as wood adhesives. Ind. Crop. Prod. 85: 324–330.

Cheng, H.N., C. Ford, M.K. Dowd and Z. He. 2016b. Use of additives to enhance the properties of cottonseed protein as wood adhesives. Int. J. Adhes. Adhes. 68: 156–160.

Combie, J., A. Steel and R. Sweitzer. 2004. Adhesive designed by nature (and tested at Redstone Arsenal). Clean Techn. Environ. Policy. 6: 258–262.

Cone, C.N. and H. Galber. 1934. Method of making an adhesive and the product thereof. U.S. Patent 1,976,435.

Conner, A.H. 1989. Carbohydrates in adhesives. pp. 271–288. *In*: R.W. Hemingway, A.H. Conner and S.J. Branham (eds.). Introduction and Historical Perspectives. Washington, DC: American Chemical Society.

D'Amico, S., U. Muller and E. Berghofer. 2013. Effect of hydrolysis and denaturation of wheat gluten on adhesive bond strength of wood joints. J. Appl. Polym. Sci. 129: 2429–2434.

Detlefsen, W.D. 1989. Blood and casein adhesives for bonding wood. *In*: R.W. Hemingway, A.H. Conner and S.J. Branham (eds.). Washington, DC: American Chemical Society.

Dutta, S., R. Bharali, R. Devi and D. Devi. 2012. Purification and characterization of glue like sericin protein from a wild silkworm *Antheraea assamensis* helfer. Global J. Biosci. Biotechnol. 1: 229–233.

Frihart, C.R. 2011. Wood adhesives. Forest Prod. J. 61: 4–12.

Haag, A.P., G.G. Geesey and M.W. Mittleman. 2006. Bacterially derived wood adhesive: Bacterially derived biopolymers as wood adhesives. Int. J. Adhes. Adhesives 26: 177–183.

Haag, A.P., R.M. Maier, J. Combie and G.G. Geesey. 2004. Bacterially derived biopolymers as wood adhesives. Int. J. Adhes. Adhesives 24: 495–502.

Hamarneh, A.I.M. 2010. Novel wood adhesives from bio-based materials and polyketones. PhD Thesis, University of Groningen.

He, Z., H.N. Cheng, D.C. Chapital and M.K. Dowd. 2014a. Sequential fractionation of cottonseed meal to improve its wood adhesive properties. J. Amer. Oil Chem. Soc. 91: 151–158.

He, Z., D.C. Chapital, H.N. Cheng and M.K. Dowd. 2014b. Comparison of adhesive properties of water- and phosphate-buffer-washed cottonseed meals with cottonseed protein isolates on bonding maple and poplar veneers. Int. J. Adhes. Adhes. 50: 102–106.

He, Z., M. Uchimiya and H. Cao. 2014c. Intrinsic fluorescence excitation-emission matrix spectral features of cottonseed protein fractions and the effects of denaturants. J. Am. Oil Chem. Soc. 91: 1489–1497.

He, Z., H. Zhang and D.C. Olk. 2015. Chemical composition of defatted cottonseed and soy meal products. PLoS One 10(6): e0129933. DOI:10.1371/journal.pone.0129933.

He, Z., D.C. Chapital, H.N. Cheng and O.M. Olanya. 2016a. Adhesive properties of water-washed cottonseed meal on four types of wood. J. Adhes. Sci. Technol. 30: 2109–2119.

He, Z., D.C. Chapital and H.N. Cheng. 2016b. Comparison of the adhesive performances of soy meal, water washed meal fractions, and protein isolates. Modern Appl. Sci. 10(5): 112–120.

Hemingway, R.W. and A.H. Conner. 1989. Opportunities for future development of adhesives from renewable resources. pp. 487–494. *In*: R.W. Hemingway, A.H. Conner and S.J. Branham (eds.). Adhesives from Renewable Resources, Washington, DC: American Chemical Society.

Imam, S.H., S.H. Gordon, L. Mao and L. Chen. 2001. Environmentally friendly wood adhesive from a renewable plant polymer: Characteristics and optimization. Polym. Degrad. Stab. 73(3): 529–533.

Imam, S.H., L.J. Mao, L. Chen and R.V. Greene. 1999. Wood adhesive from crosslinked poly(vinyl alcohol) and partially gelatinized starch: Preparation and properties. Starch-Starke 51(6): 225–229.

Kennedy, H.M. 1989. Starch- and dextrin-base adhesives. pp. 326–336. *In*: R.W. Hemingway, A.H. Conner and S.J. Branham (eds.). Washington, DC: American Chemical Society.

Kimura, Y., A. Ohno and S. Takagi. 1997. Structural analysis of N-glycans of storage glycoproteins in soybean (Glycine max. L) seed. Biosci. Biotechnol. Biochem. 61: 1866–1871.

Kumar, S. and A.K. Shah. 2015. Characterization of an adhesive molecule from Bacillus megaterium ADE-0-1. Carbohydr. Polym. 117: 543–548.

Kumar, R., V. Choudhary, S. Mishra, I.K. Varma and B. Mattiason. 2002. Adhesives and plastics based on soy protein products. Ind. Crops Prod. 16: 155–172.

Lambuth, A.L. 1989. Adhesives from renewable resources: Historical perspective and wood industry needs. pp. 1–10. *In*: R.W. Hemingway, A.H. Conner and S.J. Branham (eds.). Adhesives from Renewable Resources, Washington, DC: American Chemical Society.

Li, N., G. Qi, X.S. Sun, M.J. Stamm and D. Wang. 2012. Physicochemical properties and adhesion performance of canola protein modified with sodium bisulfite. J. Am. Oil Chem. Soc. 89: 897–908.

Lin, H. and S. Gunasekaran. 2010. Cow blood adhesive: Characterization of physicochemical and adhesion properties. Int. J. Adhes. Adhesives 30: 139–144.

Liu, Y. and K. Li. 2007. Development and characterization of adhesives from soy protein for bonding wood. Int. J. Adhes. Adhes. 27: 59–67.

Lorenz, L., M. Birkeland, C. Daurio and C. Frihart. 2015. Soy flour adhesive strength compared to that of purified soy proteins. Forest Prod. J. 65: 26–30.

Mo, X. and X.S. Sun. 2013. Soy proteins as plywood adhesives: Formulation and characterization. J. Adhes. Sci. Technol. 27: 2014–2026.

Moubarik, A., N. Causse, T. Poumadere, A. Allal, A. Pizzi, F. Charrier et al. 2011. Shear refinement of formaldehyde-free corn starch and mimosa tannin (Acacia mearnsii) wood adhesives. J. Adhes. Sci. Technol. 25(14): 1701–1713.

Moubarik, A., B. Charrier, A. Allal, F. Charrier and A. Pizzi. 2010. Development and optimization of a new formaldehyde-free cornstarch and tannin wood adhesive. Eur. J. Wood Prod. 68(2): 167–177.

Moubarik, A., A. Pizzi, A. Allal, F. Charrier and B. Charrier. 2009. Cornstarch and tannin in phenol-formaldehyde resins for plywood production. Ind. Crops Prod. 30(2): 188–193.

Nordqvist, P., D. Thedjil, S. Khosravi, M. Lawther, E. Malmstrom and F. Khabbaz. 2012. Wheat gluten fractions as wood adhesives-glutenins versus gliadins. J. Appl. Polymer Sci. 123: 1530–1538.

Norstrom, E., L. Fogelstrom, P. Nordqvist, F. Khabbaz and E. Malmström. 2014. Gum dispersions as environmentally friendly wood adhesives. Ind. Crops Prod. 52: 736–744.

Norstrom, E., L. Fogelstrom, P. Nordqvist, F. Khabbaz and E. Malmström. 2015. Xylan–A green binder for wood adhesives. Eur. Polym. J. 67: 483–493.

Orr, L. 2007. Wood Adhesives – A Market Opportunity Study, Omni Tech International, Ltd., Midland, Michigan. http://www.soynewuses.org/downloads/reports/final_WoodAdhesivesMarketOpportunity.pdf.

Pizzi, A. 2006. Recent developments in eco-efficient bio-based adhesives for wood bonding: Opportunities and issues. J. Adhesion Sci. Technol. 20: 829–846.

Qi, G., N. Li, X.S. Sun and D. Wang. 2016. Adhesion properties of soy protein subunits and protein adhesive modification, Chapter X, this book.

Smart, J.D. 2005. The basics and underlying mechanisms of mucoadhesion. Advanced Drug Delivery Rev. 57(11): 1556–1568.

Smith, A.M. 2002. The structure and function of adhesive gels from invertebrates. Integr. Comp. Biol. 42: 1164–1171.

Stellwagen, S.D., B.D. Opell and M.E. Clouse. 2015. The impact of UVB radiation on the glycoprotein glue of orb-weaving spider capture thread. J. Exp. Biol. 218: 2675–2684. Doi: 10.1242/jeb.123067.

Sun, X.S. and K. Bian. 1999. Shear strength and water resistance of modified soy protein adhesives. J. Am. Oil Chem. Soc. 76: 977–980.

Tan, H., Y. Zhang and X. Weng. 2011. Preparation of the plywood using starch-based adhesives modified with blocked isocyanates. Procedia Engin. 15: 1171–1175.

Transparency Market Research. 2014. "Wood Adhesives and Binders Market–Global Industry Analysis, Size, Share, Growth, Trends and Forecast, 2014– 2020." Albany, NY.

Umemura, K., A. Inoue and S. Kawai. 2003. Development of new natural polymer-based wood adhesives I: dry bond strength and water resistance of konjac glucomannan, chitosan, and their composites. J. Wood Sci. 49: 221–226.

Von der Mark, K. and L. Sorokin. 2002. Adhesive Glycoproteins. *In*: P.M. Royce and B. Steinmann (eds.). Connective Tissue and Its Heritable Disorders: Molecular, Genetic, and Meidcal Aspects, 2nd Ed., Hoboken: John Wiley.

Wang, C. and J. Wu. 2012. Preparation and characterization of adhesive from spent hen proteins. Int. J. Adhes. Adhes. 36: 8–14.

Wang, Z., Z. Gu, Y. Hong, L. Cheng and Z. Li. 2011. Bonding strength and water resistance of starch-based wood adhesive improved by silica nanoparticles. Carbohydr. Polym. 86(1): 72–76.

Wang, Z., Z. Li, Z. Gu, Y. Hong and L. Cheng. 2012. Preparation, characterization and properties of starch-based wood adhesive. Carbohydr. Polym. 88(2): 699–706.

Wang, Z., Z. Gu, Z. Li, Y. Hong and L. Cheng. 2013. Effects of urea on freeze–thaw stability of starch-based wood adhesive. Carbohydr. Polym. 95: 397–403.

Xu, J. and D. Mosher. 2011. Fibronectin and other adhesive glycoproteins. pp. 41–75. *In*: R.P. Mecham (ed.). The Extracellular Matrix: An Overview. Berlin: Springer-Verlag.

Yang, I., M. Kuo, D.J. Myers and A. Pu. 2006. Comparison of protein-based adhesive resins for wood composites. J. Wood Sci. 52: 503–508.

6

Preparation and Utilization of Water Washed Cottonseed Meal as Wood Adhesives

Zhongqi He* and H.N. Cheng

ABSTRACT

Cotton fiber and cottonseed are the two major products of cotton crop. The ratio of fiber and cotton is 100/150. Cotton fiber represents 85–90% of cotton's total economic value. Thus, enhanced utilization of cottonseed products as industrial raw materials would greatly benefit cotton growers and processors. Defatted cottonseed meal is the residual fraction after oil crushing. Cottonseed protein isolate is a product of defatted cottonseed meal by alkali extraction and acid precipitation. Water washed cottonseed meal is a product from defatted cottonseed meal by simple washing procedure to remove water-soluble ingredients. In this chapter, we review the studies of preparation, characterization, and testing of water washed cottonseed meal as wood adhesives. Data indicate that the washed cottonseed meal serves as wood adhesives, better than the defatted meal, and comparable or even better in some cases, than protein isolate. Thus, water washing is an economic and environment-friendly method to produce cottonseed-meal based wood adhesives. Future research should be focused on the application of the washed cottonseed meal in production of oriented strand board, fiber board and particleboard as the composite wood industry is the primary consumer of wood adhesives.

Southern Regional Research Center, USDA Agricultural Research Service, 1100 Robert E. Lee Blvd., New Orleans, LA 70124, USA.
* Corresponding author: Zhongqi.He@ars.usda.gov

Introduction

Cotton (*Gossypium hirsutum* L.), as a major fiber source for the textile industry, is produced in more than 30 countries (Campbell et al., 2014). Cotton is America's number one value-added crop. Much of the cotton land area in the US is located in the southern and southeastern region (e.g., Georgia, Alabama, Arkansas, North Carolina, Mississippi, and Texas) (Bellaloui et al., 2015b; He et al., 2013c; Tazisong et al., 2013; Tewolde et al., 2015). Cotton crop is mainly harvested for the cotton bolls while the other biomass is left or burnt in the field (He et al., 2016a; Windeatt et al., 2014). Cotton fiber and cottonseed are the two major products of cotton bolls. The ratio of fiber and cotton is 100/150 (Bellaloui and Turley, 2013; Pettigrew and Dowd, 2014). Currently, however, cotton fiber represents 85–90% of cotton's total economic value (Campbell et al., 2014; NCPA, 2016). Thus, enhanced utilization of cottonseed products as industrial raw materials would greatly benefit cotton growers and processors.

Cottonseed value is determined by the value of the products produced. Cottonseed is processed to produce five parts, i.e., linter, hulls, crude oil, meal and waste (Fig. 1) (NCPA, 2016). Linters which are short fibers still clinging to the seed are one of the finest sources of cellulose and are used to produce a variety of things like plastics, rocket propellants, rayon, pharmaceutical emulsions, cosmetics, photography and X-ray film, upholstery and fine writing paper. Hulls, tough and protective coatings for the kernel, find their way mostly into the feed industry as a source of roughage for livestock. Cottonseed oil is seen by the food industry as a premium oil and, as such, is eagerly sought by prepared food makers. Thus, it is not commonly found bottled on grocery store shelves. Defatted cottonseed meal (CSM), the residual fraction after oil crushing, counted for 45% of cottonseed, is used in animal feeds and garden fertilizer (Broderick et al., 2013; Dowd, 2015; He et al., 2014e; Wanapat et al., 2013). Research effort has also been undertaken for the utilization of whole cottonseed and its products for bioenergy and related biochar production (He et al., 2016a; Putun, 2010). For this purpose, US National Cottonseed Products Association (NCPA, 2011) has petitioned to U.S. Environmental Protection Agency Fuels Programs Registration for renewable fuel pathway for biodiesel using cottonseed oil. The potential value-added products of CSM include, but are not limited to, bioplastics and films (Yue et al., 2012), superabsorbent hydrogel (Zhang et al., 2010) antioxidant meal hydrolysates (Gao et al., 2010), as well as bio-oil and biochar (Singh et al., 2014). These bio-based products differentially utilize the functional components (e.g., proteins, peptides, and carbohydrates) in CSM such that expensive purified fractions are not always necessary. In particular, wood adhesives are promising products from CSM as shown in this chapter.

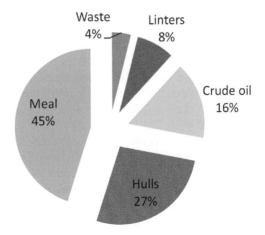

Figure 1. Composition of cottonseed. Compiled per NCPA (2016).

Historic Information on Cottonseed Meal as Wood Adhesives

As early as in the 1950s', cottonseed meal was studied as plywood glue (Hogan and Arthur Jr., 1951b). The hydraulic-pressed (45% protein, 6% oil), screw-pressed (46% protein, 4–5% oil) and hexane-extracted (56% protein, 1.5% oil) CSM preparations were tested to glue birch plywood veneers. Preparation of CSM glue consists of dispersing the meal (passed a 200-mesh screen) in an aqueous alkali solution and adding other chemicals to improve the tack, spreading characteristics and water resistance of the glue. Specifically, four parts of meal is suspended in three parts of water, and stirred for 15 min to produce a uniform paste. Then, the chemicals in solution are added to the meal-water paste. Following the reaction to produce the glue, additional water is mixed with the product to yield a dispersion having the desired solid content. Finally, the glue is thoroughly blended for 1 h to give a smooth and uniform dispersion. The glue dispersion is spread on the birch veneers by a plywood glue spreader. The plywood panels are assembled by cold (25°C) and/or hot (83–114°C) pressing for about 10 min, and conditioned for 6 d at 24°C and 32% relative humidity for dry adhesive strength test. Wet adhesives strength is measured after the glued specimens are further immersed in water for 48 h at 25°C. The results from these early studies indicate that cold pressing followed by hot pressing remarkably increases the wet shear strength of the bonded wood specimens whereas the dry strength is on the same level within the testing conditions. To use the meal dispersion in lower pH (7.5–7.8) conditions to minimize discoloration of the veneer, urea and formaldehyde are also added into the meal. Although the early data show satisfactory adhesive strength, the formulation with formaldehyde would be a health concern nowadays (Act, 2010). Among the three meals, hexane-extracted meal performed better than hydraulic- and screw-pressed meals. The difference is probably due to the fact that the protein in the last two meals is less soluble and more denatured than that in the hexane-extracted meal (Hogan and Arthur Jr., 1951b).

The hexane-extracted meal was further studied and compared with peanut meal and commercial casein for its adhesive properties under different environmental conditions (Hogan and Arthur Jr., 1951a; Hogan and Arthur Jr., 1952). The glued and conditioned birch plywood specimens were immersed in a testing reagent solution for 1–14 d, and their shear strength at break were determined immediately following the completion of soaking. Five organic reagents and six inorganic reagents were used in the chemical resistance test (Table 1). No matter what reagent was applied, soaking decreased the shear strength of tested wood veneers. The resistance of the bonding capability to inorganic reagents was all lower than that to water. However, the effects of acids, bases, and salts on the shear strength of the bonded woods was about equal. On the other hand, the resistance of the bonding capability to the five organic reagents was all greater than that to water. Furthermore Hogan and Arthur Jr. (1952) observed three types of effects among the organic reagents. There was an initial decrease in the

Table 1. Chemical resistance of cottonseed meal, peanut meal, and casein-based adhesives. Data are the shear strength at break (MPa) with birch wood veneers measured after they were soaked in the reagent solution for 1 (Phase I) and 3 or 4 d (Phase II). Compiled per Hogan and Arthur Jr. (1952).

Reagent	Cottonseed meal		Peanut meal		Casein	
	Phase I	Phase II	Phase I	Phase II	Phase I	Phase II
No soak	2.35	- [a]	2.21	-	2.54	-
H_2O	0.99	0.52	0.89	0.34	1.25	0.85
3% H_2SO_4	0.75	0.26	0.70	0.25	0.94	0.35
30% H_2SO_4	0.26	0.16	0.57	0.14	0.76	0.21
1% NaOH	0.47	0.25	0.42	0.21	0.55	0.28
10% NaOH	0.02	-	0.17	-	0.41	-
10% HNO_3	0.43	0.20	0.40	0.16	0.56	0.25
10% HCl	0.48	0.19	0.39	0.15	0.58	0.25
10% NaCl	0.39	0.19	0.35	0.16	0.63	0.25
10% Na_2SO_4	0.45	0.23	0.45	0.25	0.79	0.37
CCl_4	1.05	1.16	1.21	1.39	1.25	1.34
Benzene	1.27	1.38	1.14	1.24	1.54	1.48
Acetone	0.97	1.10	0.87	0.97	1.34	1.42
50% Ethanol	0.99	0.79	0.79	0.70	1.32	1.07
95% Ethanol	1.10	1.18	1.01	1.06	1.41	1.38
Methanol	0.87	0.92	0.81	0.87	0.94	1.00

[a] No data available

shear strength, followed by an increase with the wood specimens soaked with carbon tetrachloride and acetone solvent. The shear strength of the wood specimens soaked in aqueous ethanol and methanol changed more like that with water. That is, the shear strength of the bonds decreased with increasing soaking time until a minimum value was reached. The initial decrease in shear strength was observed with the wood specimens soaked in benzene; then there were no more significant changes in the shear strength in the soaking periods ranging from 1 to 14 days. The difference between inorganic and organic reagents suggests that ionic or valence forces are the primary attractive forces involved in the bond between the adhesive and wood surface.

In addition, these early viscosity studies (Arthur Jr. and Karon, 1948; Cheng and Arthur Jr., 1949) observed that the viscosity of the cottonseed protein dispersion tended to decrease as the dispersions were aged while the dispersion of other vegetable proteins increased in viscosity with aging. Thus, Hogan and Arthur Jr. (1951b) argued that the unique property of cottonseed protein make it possible to develop a cottonseed meal glue with a long "working life". In other words, a dispersion of CSM having low viscosity, at least, for several hours is likely. Thus, combined with additional data from interior and exterior accelerated cyclic service tests Hogan and Arthur Jr. (1951a; 1952) concluded that CSM glue is superior to peanut meal and comparable to commercial casein glue, and cottonseed meal glue can probably be used for bonding plywood designed for interior uses, for regions of comparatively low relative humidity and for bonding plywood used in building forms which may be discarded after several times of use. However, there were no more studies on cottonseed meal-based wood adhesives in the late 20th century as petroleum-based wood adhesives prevailed (Lambuth, 2003).

Preparation and Characterization of Water Washed Cottonseed Meal for Wood Adhesives

Background and Justification

With the increasing interest in plant seed meal-based products as renewable and environment-friendly wood adhesives, Cheng et al. (2013) examined the adhesive properties of cottonseed protein isolate (CSPI) and compared its performance with soy protein isolate (SPI). They tested the adhesives of the two protein products on maple veneer bonding with or without alkaline, guanidine hydrochloride, sodium dodecyl sulfate, and urea modification. Cheng et al. (2013) concluded that CSPI-based adhesives (with or without sodium dodecyl sulfate-SDS modification) are more effective than or at least comparable to the more widely-studied SPI wood adhesives (Chen et al., 2014; Qi et al., 2013; Tong et al., 2015).

CSPI is typically prepared from flour by alkaline extraction followed by acid precipitation (Berardi et al., 1969; He et al., 2013b; Zhang et al., 2009). The technique uses corrosive reagents and it makes protein isolate-based adhesives relatively expensive. In order to find an economic and environmentally friendly way to formulate cottonseed proteins into wood adhesives, He et al. (2014c) sequentially separated CSM into several fractions by extraction with either water and 1 M NaCl or with phosphate buffer (35 mM Na_2HPO_4/NaH_2PO_4, pH 7.5) and 1 M NaCl, and the various fractions tested for their adhesive properties when bonded to maple veneer. They (He et al., 2014c; 2014d) found that these water- and buffer-washed solid fraction (i.e., washed meal)-based adhesives was significantly improved, compared with the unwashed meal-based adhesive. Furthermore, the water resistance of these fractions (the wet shear strength around 1.5 MPa) was comparable to that of cottonseed protein isolate (> 90% protein) when the joints were glued at 100°C. For the purpose of cost-effectiveness, He and Chapital (2015) later focused on utilization of the water washed meal fraction (termed as water washed cottonseed meal-WCSM) for wood adhesives.

Major Components of Water Washed Cottonseed Meal

For understanding of the chemical composition and quality control, He et al. (2015) determined the contents of several key components in laboratory (5–13 g starting material)-produced WCSM and protein isolates. For comparison, these parameters in defatted soy meal and its protein isolates were also determined. Estimated from total N contents, the contents of crude protein in cottonseed and soy meal were 50.7, and 49.6% of the dry matter, respectively (Table 2). Water washing enriched the protein content

Table 2. Content (% of dry matter) of crude protein in cottonseed and soy meal and their products-water washed meal, protein isolate (PI), water-extracted protein (Pw) and alkali-extracted protein (Pa). Data are presented in average with standard error (n = 2 or 3) per total nitrogen (TN) and amino acid (AA) contents. Adapted from He et al. (2015).

	Cottonseed		Soy	
	TN-based	**AA-based**	**TN-based**	**AA-based**
Meal	50.7 ± 5.2	69.3 ± 8.0	49.6 ± 0.1	65.8 ± 1.7
Washed meal	74.5 ± 0.7	89.7 ± 2.3	66.2 ± 1.5	80.1 ± 2.9
PI	97.4 ± 2.5	108.5 ± 0.8	92.0 ± 0.2	106.6 ± 1.6
Pw	80.6 ± 0.3	97.8 ± 1.1	-[a]	-
Pa	104.0 ± 0.3	105.4 ± 2.7	-	-

[a] Not determined.

in the washed meals by 12 to 25%. The protein contents of both CSPI and cottonseed alkali–extracted protein (CSPa) neared 100%, while SPI contained about 92% protein. The protein content of cottonseed water-extracted protein isolate (CSPw) was 81%. The protein contents calculated from the sum of amino acids (AAs) were all higher than the corresponding values calculated from total N. Yet both modes of protein content calculation provided similar trends among the samples. This observation also implies that the conversion factor of 6.25 from N% to protein % was appropriate as a lower factor of 5.30 for cottonseed suggested in some works (Rhee, 2001) would lead even lower N-based protein content than AA-based value. He et al. (2015) further analyzed the amino acid (AA) profiles in these products. Whereas there were some differences in the AA contents between the meal and their products, it is difficult to derive a general trend. Thus, He et al. (2015) grouped the contents into 10 essential AAs and 11 non-essential proteinous and nonproteinous AAs (Table 3). Although the essential AA content did not always differ significantly ($P > 0.05$) among the fractions, it was generally lower in the washed meal, alkali-extracted and total protein isolates, compared to the whole defatted meal. When these AAs regrouped with their polar or nonpolar side chains He et al. (2015), found that CSPw contained more polar side chains than WCSM and CSPa. This observation explained a previous finding that CSPa was less hydrophilic than CSPw as revealed by fluorescence excitation-emission matrix spectroscopy He et al. (2014a). Hettiarachchy et al. (1995) report that hydrophobicity plays

Table 3. Contents (percent of protein) of essential amino acids (EAAs), non-essential and non proteinous amino acids (NAAs), amino acids with polar side chains (AAsP), and amino acids with nonpolar side chains (AAsN) in cottonseed (CS) and soy meals and their products-water washed meal, protein isolate (PI), water-extracted protein (PIw) and alkali-extracted protein (PIa). Data are presented in average (n = 2 or 3). The values with different letters in a column of the same type of meals are significantly different at $P \leq 0.05$. Adapted from He et al. (2015).

	EAAs	NAAs	AAsP	AAsN
	Cottonseed			
Meal	65.8 a	34.2 a	62.7 a	35.2 a
Washed meal	63.5 ab	36.5 ab	59.6 b	38.6 b
PI	58.8 c	41.2 c	58.8 bd	39.7 bd
Pw	61.0 ab	39.0 ab	69.5 c	29.1 c
Pa	58.2 c	41.8 c	56.7 d	42.2 d
	Soy			
Meal	60.9 a	39.1 a	60.9 a	37.4 a
Washed meal	59.0 b	41.0 b	59.0 a	38.9 b
PI	58.9 b	41.9 b	58.8 a	39.0 ab

an important role in wood adhesives. Thus, the better water resistance of the water insoluble fraction (i.e., WCSM)-based adhesives than the water soluble fraction of cottonseed meal (He et al., 2014c) could be at least partially attributed to the more non polar (i.e., hydrophobic) AAs in WCSM. Unlike cottonseed, there were not much differences observed in the AA profiling of the three soy products.

The Fourier transform infrared (FTIR) spectra of the three cottonseed products are shown in Fig. 2 (He et al., 2014d). There are three major bands related to protein structures, i.e., amide I (C=O stretching), II (CN stretching, NH bending), and III (CN stretching, NH bending) at 1653, 1532 and 1236 cm^{-1}, respectively (He et al., 2013b; Zhao et al., 2008). These protein-related bands are basically similar to each other, implying that the secondary structures of the proteins, thus the fundamental bonding modes, in these CSM-based adhesives are same. The broad band at 1070 cm^{-1} can be assigned mainly to carbohydrate as it is a major component in cottonseed products although phosphorous in nucleotide and phytate with less contents in cottonseed might also have some contribution to the band intensity at this region (He et al., 2006; He et al., 2013b; Yu and Irudayaraj, 2005). The relative band area is 85.5, 60.5, and 15.8, respectively, for CSM, WCSM, and CSPI, indicated the removal of carbohydrate during preparation of WCSM

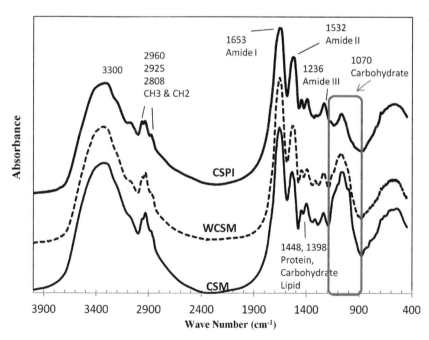

Figure 2. FTIR spectra of cottonseed meal (CSM), water-washed CM (WCSM), and protein isolate (CSPI). Adapted from He et al. (2014d).

and CSPI from CSM. Chen et al. (2013) also observed that the band area at 1055 cm^{-1} in soy flour-based adhesives decreased as the carbohydrate content decreased.

Pilot-scale Produced Water Washed Cottonseed Meal and Coproducts

For evaluating the operational parameters for production of large quantities of cottonseed meal-based products He et al. (2016d), further proposed and tested in the pilot scale [10 lb (4.54 kg) starting material] multiple-step procedure that can be used to produce CSPw and WCSM, and alternatively, CSPw, CSPa and alkali insoluble fraction (CSIR) (Fig. 3). In this pilot experiment, they also used the "real world" mill-produced CSM as the starting material that is not always the same in chemical composition as the laboratory-produced CSM (Table 4). Partly for this reason, in the pilot trial protein is enriched in WCSM to 46.3% from 34.1% in CSM (Table 5), lower than that in lab-produced WCSM (Table 2). The alkali–extracted residual fraction CSIR remains to have a protein content of 34.9%. The protein content of CSPw is 64.4% which was lower than the protein content of laboratory-made CSPw (Table 2). The product CSPa is basically protein-dominated,

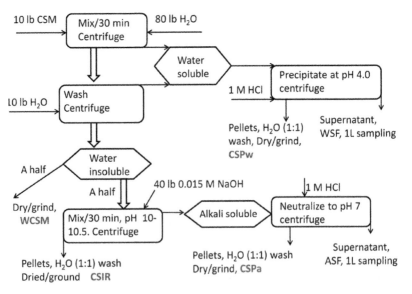

Figure 3. Pilot scale production of washed cottonseed meal product (WCSM) and co-product water soluble protein (CSPw), alternative products alkali soluble protein (CSPa) and alkali insoluble cottonseed meal residual fraction (CSIR) per 10 lb (4.54 kg) of the staring material cottonseed meal (CSM). Water (WSF) and alkali (ASF) soluble residual supernatant fractions were not collected. Reprinted from He et al. (2016d).

Table 4. Comparison of the major components in mill-produced (Mill-1 and Mill-2) and lab-prepared (Lab-1 and Lab-2) cottonseed meals. Reprinted from He et al. (2016d).

	Moisture	Protein	Oil	CF	ADF	NDF	ADL
	% of sample weight						
Mill-1	8.4	34.1	2.5	11.7	20.8	23.0	7.2
Mill-2	9.8	40.0	1.3	11.5	19.0	27.0	6.4
Lab-1	11.4	58.7	0.3	2.4	3.7	16.1	0.3
Lab-2	8.2	50.7	~1.0	3.3	5.5	11.9	1.0

more like the total protein isolate CSPI, which is also observed with the laboratory-made CSPa and CSPI (He et al., 2015; He et al., 2013b). Oil is an impurity (2.5%) in the starting material CSM (Table 4). There is still some oil in these products with the highest content (3.4%) in CSPw (Table 5). Data in Table 5 also show that fiber is a major component in WCSM and CSIR, but a minor (i.e., impurity) in protein fractions CSPw, CSPa, and CSPI. The contents of ADF, NDF, and ADL are all higher in the pilot-scale produced WCSM and CSIR than the two fractions from laboratory experiments (He et al., 2015) apparently due to the higher fiber content in the mill-produced starting material CSM. Those data also reveal that there are 17.6% and 8.4% of WCSM, and 24.6% and 10.2% of CSIR, respectively, as cellulose and hemicelluloses. The content of crude fiber (CF) is 11.7%, 16.0%, and 23.5% in CSM, WCSM and CSIR, respectively (Table 5), following the same orders of acid detergent fiber (ADF), neutral detergent fiber (NDF), and acid detergent lignin (ADL) in the three samples. Thus, water washing enriches not only protein component in WCSM and protein fractions CSPw and CSPa, but also fiber and lignin components in WCSM and CSIR. It is notable that the content of so called "crude fiber" is lower than ADF and NDF in both WCSM and CSIR. Similarly, higher content of CF than ADF and NDF is reported in cotton hull and burrs (Cheng and Biswas, 2011) and wheat milling fractions (Saunders and Hautala, 1979). This difference is due to the fact that CF measured the organic residues after both alkali and acid digestions so that portions of both the structural carbohydrates and lignin might have also been destroyed during the digestions (Saunders and Hautala, 1979).

In the pilot trial, contents of ash, six macroelements and seven micro elements of the starting material CSM and the products are also determined (Table 6). The ash content is lower in all products than in CSM, suggesting that washing reduces the mineral elements in CSM products. The contents of P and K, the highest two in the six macroelements also decrease in the same order of ash content, suggesting that P and K compounds are the major contributors of ash. The contents of the divalent metals Ca and Mg are lower

Table 5. Major components of pilot produced cottonseed meal products. Reprinted from He et al. (2016d).

	Moisture	Protein	Oil	CF	ADF	NDF	ADL	Cellulose	Hemicellulose
								----% of product weight----	
WCSM	8.6 ± 1.1	46.3 ± 2.2	1.0 ± 0.4	16.0 ± 2.2	27.0 ± 2.7	35.4 ± 3.8	9.4 ± 0.8	17.6 ± 2.0	8.4 ± 2.6
CSIR	10.7 ± 0.9	34.9 ± 2.5	0.5 ± 0.1	23.5 ± 2.1	38.5 ± 4.1	48.6 ± 6.7	13.9 ± 2.1	24.6 ± 2.0	10.2 ± 5.4
CSPw	8.9 ± 1.4	64.4 ± 2.6	3.4 ± 3.0	0.9 ± 0.4	2.5 ± 0.4	1.93 ± 0.5	0.9 ± 0.1	1.6 ± 0.3	ND[‡]
CSPa	8.2 ± 1.5	101.2 ± 2.0	0.1 ± 0.1	0.1 ± 0.0	0.3 ± 0.1	ND	ND	0.3 ± 0.1	ND
CSPI[†]	9.6 ± 0.2	94.8 ± 1.0	0.2 ± 0.1	1.0 ± 0.1	0.7 ± 0.1	0.1 ± 0.0	0.1 ± 0.0	0.6 ± 0.1	ND

[†] For comparison only. Produced separately by alkali extraction-acid precipitation with the same batch of the starting material CSM.
[‡] Not detected.

Table 6. Ash and selected mineral element contents in starting material CSM and the products. Reprinted from He et al. (2016d).

	Ash	P	Ca	K	Mg	Na	S	Fe	Zn	Cu	Mn	Ni	Al	B
	------% of product weight------							------ppm------						
CSM	7.15	1.48	0.26	1.80	0.72	0.24	0.52	181	77	11	31	1.6	149	21
WCSM	5.20a[+]	1.22α	0.27a	1.04a	0.67a	0.08c	0.46c	166b	100a	15β	50a	0.5αβ	130b	20a
CSIR	4.96a	1.02α	0.29a	0.82ab	0.66a	0.41a	0.39d	133bc	92a	10β	49a	0.3β	91b	20a
CSPw	4.64a	1.11αβ	0.10b	0.85ab	0.15b	0.07c	0.76a	362a	19c	40αβ	21b	61αβ	313a	20a
CSPa	1.25b	0.33β	0.07bc	0.16c	0.08bc	0.20b	0.60b	48c	54b	63αβ	12b	1.5α	89b	6b
CSPI[‡]	2.16b	0.49β	0.03c	0.33bc	0.05c	0.17bc	0.65b	69bc	27c	60α	8b	1.3β	79b	8b

[+] Values with the same letter in the same column are not significantly different at $\alpha = 0.05$. Lower case and Greek letters indicate groupings according to Ryan's Q and Games-Howell's tests, respectively.

[‡] For comparison only. Produced separately by alkali extraction-acid precipitation with the same batch of the starting material CSM.

in the three protein products CSPw, CSPa, and CSPI than CSM. These data indicate that these minerals in the meal and their products should be mainly present in the forms of K phytate (potassium inositol hexakisphosphate) bonding with P (i.e., $K_{12}C_6H_6O_{24}P_6$), with the secondary highest contributions of Mg and Ca phytate ($Mg_6C_6H_6O_{24}P_6$ and $Ca_6C_6H_6O_{24}P_6$) (Han, 1988; He et al., 2013a; 2006). The content of S is higher in the three protein products CSPw, CSPa, and CSPI than in CSM, WCSM, and CSIR. This observation indicates that the amino acids cysteine (cystine) and/or methionine in protein (He et al., 2014b; Voet and Voet, 1990) are the major S source in the products. Furthermore, the higher S content in CSPw than in CSPa and CSPI is consistent with the previous observation that the content of cysteine of CSPa is the highest among the three lab-produced protein fractions (He et al., 2014e). The contents of the micromineral elements are in the general range of cottonseed and meal products (Bellaloui and Turley, 2013; Bellaloui et al., 2015a; He et al., 2013c). Similar to those lab-produced products (He et al., 2014e) and due to the nature of micro abundance, the changes of these mineral contents are relatively small. A positive observation is that, without remarkable changes in these microminerals and other nutrients, CSIR could be used at least as animal feedstuffs as CSM does (Arieli, 1998; Bellaloui et al., 2015b) whereas its industrial utilization needs to be explored.

Adhesive Performance of Water Washed Cottonseed Meal

Preparation and Testing Protocol

He and Chapital (2015) list the general procedure to test the WCSM as wood adhesives. Briefly, WCSM (or the comparing CSM and CSPI) and water are mixed with a weight ratio of 3 to 25, and stirred by a magnetic bar for 2 h in a beaker. The adhesive slurry is then brushed onto one end of two wood veneer strips covering 25.4 mm (1.0") length (Fig. 4a). Two wood strips are bonded together by overlapping the tacky adhesive coated area (1.0" x 1.0") of two strips and hot-pressing for 20 min. After the bonded wood specimens are conditioned (Fig. 4b), the lap-shear strengths at break of the wood pairs are measured and termed as (dry) adhesive strength. The water resistance is performed by soaking the bonded wood specimens in water (Fig. 4c). Two measurements are used to evaluate the water resistance ability of these adhesives. The wet strength is the direct impact of water soaking on the adhesive strength. The wet adhesive strength is immediately measured after these wood specimens have been soaked at room temperature (23°C) for 2 d. The soaked strength measures how much adhesive strength can be recovered after the bonded wood specimens are re-dry. The soaked adhesive strength is measured after two cycles of soaking (4 h at 63°C) and drying (1–2 d). Alternatively, the soaked strength is measured on the wood

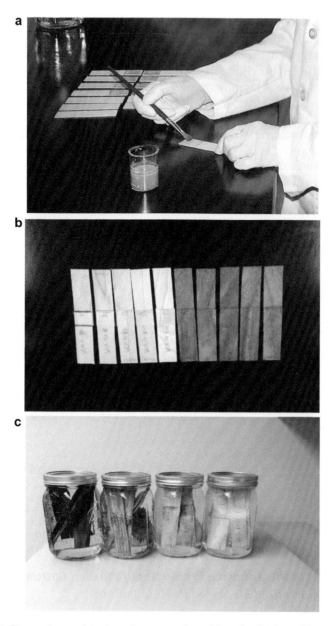

Figure 4. Preparation and testing of cottonseed-meal based adhesives. (a) Application of adhesive slurry onto wood strips. Courtesy of USDA-ARS AgResearch Magazine (http://agresearchmag.ars.usda.gov/2015/jun/cottonseed/); (b) Bonded wood pairs set aside for conditioning (the bonding area –25.4 x 25.4 mm) is shown between the red lines at the most left pair. Reprinted from He and Chapital (2015); (c) Water resistance experiments. Soaked strength is measured with the wood pairs after two cycles of soaking (4 h at 63°C) and drying (1–2 d), and wet strength is measured with the wood pairs immediately after soaking at 23°C for 2 d.

specimens that are soaked in water at 23°C for 2 d and conditioned at 23°C and 50% relative humidity for 7 days (He et al., 2016b). Figure 5 shows the different types of broken wood pairs during the tensile shear strength test. From the left to right, the breaking types are gradually changed from substrate (wood) failure, to cohesive failure, and to adhesive failure (delamination). The latter two failures are referred to as adhesive bond failures so that the shear strengths of broken wood pairs with such failures represent the bonding strength of the adhesive preparation under the testing conditions. On the other hand, one should be aware that the actual adhesive strength of the meal product observed with a wood failed pair could be higher than the value at break. More information on the measures and significance of adhesive strength, wood failure, and delamination can be found in Frihart and Hunt (2010).

Figure 5. Broken wood pairs during a tensile shear strength test. The left most pairs represent the wood failure mode. From the left to right, the breaking types are gradually changed from substrate (wood) failure, to cohesive failure, and to adhesive failure (delamination). Reprinted from He and Chapital (2015).

Comparison of the Adhesive Properties of Washed Cottonseed Meal with Untreated Meal and Protein Isolate

He et al. (2016b) compared the adhesive strength of CSM, WCSM, and CSPI on maple samples at the different pH conditions used for the adhesive slurry preparation (Table 7). The optimal adhesive strength of all three products is with their natural pH condition (i.e., pH 6.0 or so). At all pH conditions tested, the dry strength of the three products or adhesives is similar. However, soaking the bonded maple pairs lower the adhesive

Table 7. Shear strength (MPa) at break of maple wood strips bonded with adhesive slurries of cottonseed meal (CSM), water washed cottonseed meal (WCSM), and cottonseed protein isolate (CSPI) made at different pH. Data are presented in the format of average ± standard deviation (n = 5). Compiled from He et al. (2016b).

pH	Adhesive	Dry (MPa)	Wet (MPa)	Soaked (MPa)
4.5	CSM	2.77 ± 0.87	0.98 ± 0.25	1.96 ± 0.82
	WSCM	3.15 ± 1.46	2.12 ± 0.21	3.18 ± 1.20
	CSPI	1.78 ± 0.74	0.86 ± 0.53	1.81 ± 0.82
6.0	CSM	3.73 ± 0.49	1.20 ± 0.14	2.53 ± 0.48
	WSCM	3.82 ± 0.92	2.36 ± 0.13	4.04 ± 0.48
	CSPI	3.76 ± 0.80	2.15 ± 0.58	4.09 ± 0.41
7.5	CSM	2.87 ± 0.61	0.63 ± 0.21	1.60 ± 0.41
	WSCM	3.10 ± 0.79	1.30 ± 0.11	3.44 ± 0.52
	CSPI	3.82 ± 0.44	1.24 ± 0.23	2.95 ± 0.58
11.0	CSM	2.52 ± 0.22	0.52 ± 0.08	1.29 ± 0.10
	WSCM	3.05 ± 0.50	0.64 ± 0.19	2.52 ± 0.31
	CSPI	3.04 ± 0.54	1.24 ± 0.23	2.87 ± 0.20

strengths of all three products. Even though, the highest values of wet strength are still with the slurries prepared at pH 6.0. The pattern of the effect of pH is similar to that of the dry strength. With slurry pH at 4.5, 6.0 and 7.5, WCSM shows the higher wet strength than its corresponding WCM and CSPI. The relative decreases of shear strength due to water soaking, compared to the dry samples, are in the general order of WCSM < CSPI < WCM. The relative changes of WCSM (–31.9% and –38.2%) with pH 4.5 and 6.0 are not only smaller than the decrease of CSPI (corresponding values –51.9% and –42.7%), but also lower than the –52.9 to –63.5% changes of soy protein adhesives made at the same pH range (Wang et al., 2009). This observation implies that certain insoluble components in WCSM (He et al., 2015) are capable to improve the water resistance of the plant seed protein based adhesives. Drying partly recovers the lost adhesive strength of CSM as the decreases of shear strength at break of the dried samples are about 21% to 36% less than those of wetted samples. Drying recovers more adhesive strength for WCSM and CSPI. As a matter of fact, the adhesive strength of WCSM is fully recovered with three low pH conditions (4.5, 6.0 and 7.5). The full recovery of CSPI's adhesive strength is reached with two low pH conditions (4.5 and 6.0) after the bonded pairs are dried. It is noted that there are some data with more than 100% recovery (He et al., 2014d; Liu et al., 2010; Wang et al., 2009; Zhong et al., 2007). In addition to the high standard deviations of these data, it can be also due to the removal of some soluble components (including alkaline and sodium) during the soaking-drying cycles (Liu et al., 2010) and possible increase of other adhesion forces by substituting other attractive forces for ionic forces (Hogan and Arthur, 1952). It is also noticeable that the change of the dry, wet, and soaked strengths of the three cottonseed products from the optimal pH 6.0 is not sharper than the soy protein-based adhesive slurries affected by pH away from its optimal pH near 6.0 by Zhong et al. (2007). This difference indicates that cottonseed adhesive products, especially WSCM, are more effective in meeting pH requirements of bonding products.

The Effect of Wood Type

The adhesive properties of WCSM have been tested on bonding the veneers of five types of woods (i.e., maple, poplar, Douglas fir, walnut and white oak) (He et al., 2016b; 2014d; 2016c; 2014f). Wood type is also a factor to influence the adhesive properties of WCSM (Table 8). Under the testing conditions, the trend of the dry adhesive strength of WCSM on the five woods is Douglas fir > poplar > white oak > maple > walnut. Similarly, in the comparison of the adhesive strength of modified soy protein on five woods (walnut, cherry, soft maple, poplar and yellow pine), Kalapathy et al. (1995) report that soft maple gave the highest strength, inconsistent with the hardness order. By examination of the photomicrographs of the adhesive bonds in red oak, hybrid poplar, and Douglas-fir, Modzel et al. (2011) report that the penetration depth of the phenol–formaldehyde (PF) adhesive in the oak and poplar samples is approximately 400 μm, but only 100 μm in the Douglas fir sample. X-ray microtomography also shows that Douglas fir has lower permeability than either oak or poplar. Thus, the low permeability of Douglas fir would have higher interlocking with WCSM slurry on the interface than poplar, walnut, maple and white oak. Similar to the shear strength of the dry specimens, WCSM shows the greatest shear strength on wet Douglas fir and maple specimens. The wet strength of WCSM on the other three woods is at the same level around 1.90 MPa. Compared to the dry shear strength values, soaked-and-dried Douglas fir and maple specimens have recovered the shear strength that are lost in the wet test specimens to the same level of the dried specimens. In the process

Table 8. Shear strength (MPa) of five types of wood strips bonded at 100°C with water washed cottonseed meal. Compiled from He et al. (2016b; 2016c).

Wood	Dry	Wet	Soaked
Poplar	4.51 ± 0.64	1.97 ± 0.11	3.04 ± 0.74
Douglas fir	4.92 ± 1.13	2.34 ± 0.27	4.41 ± 0.49
Maple	3.82 ± 0.92	2.36 ± 0.13	4.04 ± 0.48
Walnut	3.69 ± 0.41	1.83 ± 0.14	2.49 ± 0.30
White oak	4.25 ± 0.81	1.90 ± 0.12	3.43 ± 0.34

of drying bonded specimens after hot water immersion for the other three wood types, the shear strength have partially recovered to 67%, 67%, and 81% of the corresponding dry strength for polar, walnut, and white oak samples. The inconsistent orders of dry, wet, and soaked strength among the five woods suggest that the two adhesive parameters (strength and water resistance) are affected by different wood properties.

Pot Life of Washed Cottonseed Meal and Protein Isolate

Pot life experiments indicate that storage of WCSM slurries up to 8 days does not change both dry and soaked adhesive strengths of WCSM slurries prepared at pH 6.0, 7.5, or 9.0 whereas the dry and soaked adhesives strengths of CSPI slurries decreased slightly with the same storage conditions (He et al., 2016b). Storage changes the color and odor of these slurries (Table 9). The newly prepared slurries of both WCSM and CSPI are almost odor-less with a general brownish color. The color of WCSM slurries gradually become dark as storage time increases. Odd smell begins appearing at day 4, and becomes stronger at day 8. The slurries of CSPI seem more mold-resistant as no odd smell appears at day 4. The odd odor is stronger in the storage slurries with a higher pH for both WCSM and CSPI. He et al. (2016b) argue that 2–4 days should be long enough

Table 9. Effect of storage time on physical appearance and apparent viscosity (mPa.s at shear rate of 10 s^{-1}) of the adhesive slurries of washed cottonseed meal (WCSM) and cottonseed protein isolate (CSPI) prepared at pH 6.0, 7.5 and 9.0. Adapted from He et al. (2016b).

pH	Storage time (day)			
	0	2	4	8
WCSM				
6.0	Brown (21)	Green/taupe (66)	Dark tan, musty (88)	Greenish brown, musty (196)
7.5	Light brown (48)	Light tan (127)	Tan, musty (166)	Yellowish tan, rotten egg smell (251)
9.0	Brown (134)	Light brown (379)	Tan, sour/sulfury (380)	Yellowish tan, strong rotten egg (436)
CSPI				
6.0	Brown (6)	Light brown (6)	Light tan (6)	Light tan (7)
7.5	Caramel (37)	Light tan (34)	Light tan (26)	Light tan, sour smell (19)
9.0	Dark brown (16)	Light tan (14)	Light tan (47)	Light tan, rotten egg smell (19)

for the general on-site use of WCSM and CSPI for wood bonding while synthetic reagents (Norstrom et al., 2014; Wang and Guo, 2014) or copper preservatives (Lambuth, 2003) may be added into the adhesive slurries if a longer storage time is needed.

The slurry of WCSM shows a higher viscosity than CSPI slurry under the same conditions. The viscosities of both WCSM and CSPI slurries increase at higher pH (Table 9). The viscosities of WCSM slurries prepared at all three pH conditions increase with storage time, ranged from near 200 mPa.s at pH 6.0 (d 8) to > 400 mPa.s at pH 9.0 (d 8). Storage time affects the viscosity of CSPI slurries less than that of WCSM slurries. The viscosity of CSPI slurries prepared at pH 6.0 is low and basically unchanged over storage time. The viscosity of CSPI slurries at pH 7.5 indeed decreases during the storage period. These data indicate that WCSM not only possesses the better adhesive performance, but also is more adjustable in its rheological properties than CSPI. The carbohydrate components in WCSM (He et al., 2015) should have contributed to the better rheological behaviors than CSPI as Cheng and Arthur Jr. (1949) observe that sucrose increase the viscosity of cottonseed protein dispersion and Wang and Guo (2014) report that sucrose made whey protein into a viscous flowable liquid instead of an unflowable slurry and paste. It is reported that a viscosity of 5000–25,000 mPa.s for most wood laminating purposes (both cold or hot press), over 50,000 mPa.s for mastic consistency wood laminating operations, and about 8000–20,000 mPa.s are optimal for no clamp cold press technique (Kumar et al., 2002). The viscosity data in Table 9 indicate that WCSM is more flexible than CSPI in setting relevant parameters for practical applications.

Conclusions

Defatted cottonseed meal and its protein isolate are both capable of serving as wood adhesives. However, the former shows poor water resistance while the latter is expensive to prepare. Washing is a cheap and environment-friendly way to improve the adhesive performance of cottonseed meal. Water washed cottonseed meal is prepared in both laboratory and pilot scales. The composition of the two products is basically same with more fiber found in pilot-produced washed meal. This difference is due to the high fiber content in the mill-produced defatted meal used for the starting material of the pilot trial.

The adhesive experimental data confirm that, compared to the defatted meal, the water resistance of water washed cottonseed meal is improved. More than that, in most cases, its water resistance is comparable, or even better than cottonseed protein isolate. The washed cottonseed meal is also better and more flexible in operational conditions and parameter setup than protein isolate. Future research should be focused on the application of the washed cottonseed meal in production of oriented strand board, fiber board and particleboard as the composite wood industry is the primary consumer of wood adhesives.

Keywords: Composition, Cottonseed, Maple, Meal, Protein isolate, Pilot, Soaking, Storage, Viscosity, Washing, Water resistance

References

Act. 2010. S. 1660 Formaldehyde standards for composite wood products act. http://www.gpo.gov/fdsys/pkg/BILLS-111s1660enr/pdf/BILLS-111s1660enr.pdf. (Accessed Oct. 15, 2015).

Arieli, A. 1998. Whole cottonseed in dairy cattle feeding: A review. Anim. Feed. Sci. Technol. 72: 97–110.

Arthur, Jr. J.C. and M.K. Karon. 1948. Preparation and properties of cottonseed protein dispersions. J. Am. Oil Chem. Soc. 25: 99–102.

Bellaloui, N. and R.B. Turley. 2013. Effects of fuzzless cottonseed phenotype on cottonseed nutrient composition in near isogenic cotton (Gossypium hirsutum L.) mutant lines under well-watered and water stress conditions. Front. Plant Sci. 4: 516. doi: 10.3389/fpls.2013.00516.

Bellaloui, N., S.R. Stetina and R.B. Turley. 2015a. Cottonseed protein, oil, and mineral status in near-isogenic Gossypium hirsutum cotton lines expressing fuzzy/linted and fuzzless/linted seed phenotypes under field conditions. Front. Plant Sci. 6: 137. http://dx.doi.org/10.3389%2Ffpls.2015.00137.

Bellaloui, N., R.B. Turley and S.R. Stetina. 2015b. Water stress and foliar boron application altered cell wall boron and seed nutrition in near-isogenic cotton lines expressing fuzzy and fuzzless seed phenotypes. PloS One 10: e0130759. Doi:10.1371/journal.pone.0130759.

Berardi, L.C., W.H. Martinez and C.J. Fernandez. 1969. Cottonseed protein isolates: Two step extraction procedure. Food Technol. 23: 75–82.

Broderick, G.A., T.M. Kerkman, H.M. Sullivan, M.K. Dowd and P.A. Funk. 2013. Effect of replacing soybean meal protein with protein from upland cottonseed, Pima cottonseed, or extruded Pima cottonseed on production of lactating dairy cows. J. Dairy Sci. 96: 2374–2386.

Campbell, B.T., D. Boykin, Z. Abdo and W.R. Meredith. 2014. Cotton. pp. 13–32. In: S. Smith, B. Diers, J. Specht and B. Carver (eds.). Yield Gains in Major U.S. Field Crops. American Society of Agronomy, Crop Science Society of America, Soil Science Society of America, Madison, WI.

Chen, M., Y. Chen, X. Zhou, B. Lu, M. He, S. Sun and X. Ling. 2014. Improving water resistance of soy-protein wood adhesive by using hydrophilic additives. BioResources 10: 41–54.

Chen, N., Q. Lin, J. Rao and Q. Zeng. 2013. Water resistances and bonding strengths of soy-based adhesives containing different carbohydrates. Ind. Crop. Prod. 50: 44–49.

Cheng, F.W. and J.C. Arthur Jr. 1949. Viscosity of cottonseed protein dispersions. J. Am. Oil Chem. Soc. 26: 147–150.

Cheng, H.N. and A. Biswas. 2011. Chemical modification of cotton-based natural materials: Products from carboxymethylation. Carbohydr. Polymer. 84: 1004–1010.

Cheng, H.N., M.K. Dowd and Z. He. 2013. Investigation of modified cottonseed protein adhesives for wood composites. Ind. Crop. Prod. 46: 399–403.

Dowd, M.K. 2015. Seed. pp. 745–781. In: D.D. Fang and R.G. Percy (eds.). Cotton. 2nd Ed. Agronomy Monograph 57. ASA, CSSA, and SSSA, Madison, WI.

Frihart, C.R. and C.G. Hunt. 2010. Adhesives with wood materials: bond formation and performance. pp. 10.01–10.24. Wood Handbook: Wood as an Engineering Material: General Technical Report FPL; GTR-190. US Dept. of Agriculture, Forest Service, Forest Products Laboratory, Madison, WI.

Gao, D., Y. Cao and H. Li. 2010. Antioxidant activity of peptide fractions derived from cottonseed protein hydrolysate. J. Sci. Food Agric. 90: 1855–1860.

Han, Y.W. 1988. Removal of phytic acid from soybean and cottonseed meals. J. Agric. Food Chem. 36: 1181–1183.

He, Z. and D.C. Chapital. 2015. Preparation and testing of plant seed meal-based wood adhesives. J. Vis. Exp. 97: e52557 doi: 10.3791/52557.

He, Z., C.W. Honeycutt, T. Zhang and P.M. Bertsch. 2006. Preparation and FT-IR characterization of metal phytate compounds. J. Environ. Qual. 35: 1319–1328.

He, Z., J. Zhong and H.N. Cheng. 2013a. Conformational change of metal phytates: solid state 1D ^{13}C and 2D ^{1}H-^{13}C NMR spectroscopic investigations. J. Food Agri. Environ. 11(1): 965–970.

He, Z., H. Cao, H.N. Cheng, H. Zou and J.F. Hunt. 2013b. Effects of vigorous blending on yield and quality of protein isolates extracted from cottonseed and soy flours. Modern Appl. Sci. 7(10): 79–88.

He, Z., M. Shankle, H. Zhang, T.R. Way, H. Tewolde and M. Uchimiya. 2013c. Mineral composition of cottonseed is affected by fertilization management practices. Agron. J. 105: 341–350.

He, Z., M. Uchimiya and H. Cao. 2014a. Intrinsic fluorescence excitation-emission matrix spectral features of cottonseed protein fractions and the effects of denaturants. J. Am. Oil Chem. Soc. 91: 1489–1497.

He, Z., D.C. Olk and H.M. Waldrip. 2014b. Soil amino compound and carbohydrate contents influenced by organic amendments. pp. 69–82. *In*: Z. He and H. Zhang (eds.). Applied Manure and Nutrient Chemistry for Sustainable Agriculture and Environment. Springer, Amsterdam, the Netherland.

He, Z., H.N. Cheng, D.C. Chapital and M.K. Dowd. 2014c. Sequential fractionation of cottonseed meal to improve its wood adhesive properties. J. Am. Oil Chem. Soc. 91: 151–158.

He, Z., D.C. Chapital, H.N. Cheng and M.K. Dowd. 2014d. Comparison of adhesive properties of water- and phosphate buffer-washed cottonseed meals with cottonseed protein isolate on maple and poplar veneers. Int. J. Adhes. Adhes. 50: 102–106.

He, Z., H. Zhang, D.C. Olk, M. Shankle, T.R. Way and H. Tewolde. 2014e. Protein and fiber profiles of cottonseed from upland cotton with different fertilizations. Modern Appl. Sci. 8(4): 97–105.

He, Z., D.C. Chapital, H.N. Cheng, K.T. Klasson, M.O. Olanya and J. Uknalis. 2014f. Application of tung oil to improve adhesion strength and water resistance of cottonseed meal and protein adhesives on maple veneer. Ind. Crop. Prod. 61: 398–402.

He, Z., H. Zhang and D.C. Olk. 2015. Chemical composition of defatted cottonseed and soy meal products. PLoS One 10(6): e0129933. DOI:10.1371/journal.pone.0129933.

He, Z., M. Uchimiya and M. Guo. 2016a. Production and characterization of biochar from agricultural byproducts—overview and utilization of cotton biomass residues. *In*: M. Guo et al. (eds.). Agricultural and Environmental Applications of Biochar: Advances and Barriers. Soil Sci. Soc. Am., Madison, WI.

He, Z., D.C. Chapital and H.N. Cheng. 2016b. Effects of pH and storage time on the adhesive and rheological properties of cottonseed meal-based products. J. Appl. Polymer Sci. in press.

He, Z., D.C. Chapital, H.N. Cheng and M.O. Olanya. 2016c. Adhesive properties of water washed cottonseed meal on four types of wood. J. Adhes. Sci. Technol. in press.

He, Z., K.T. Klasson, D. Wang, N. Li, H. Zhang, D. Zhang and T.C. Wedegaertner. 2016d. Pilot-scale production of washed cottonseed meal and co-products. Modern Appl. Sci. 10(2): 25–33.

Hettiarachchy, N.S., U. Kalapathy and D.J. Myers. 1995. Alkali-modified soy protein with improved adhesive and hydrophobic properties. J. Am. Oil Chem. Soc. 72: 1461–1464.

Hogan, J.T. and J.C. Arthur Jr. 1951a. Cottonseed and peanut meal glues: permanence of plywood glue joint as determined by interior and exterior accelerated cyclic service tests. J. Am. Oil Chem. Soc. 28: 272–274.

Hogan, J.T. and J.C. Arthur Jr. 1951b. Preparation and utilization of cottonseed meal glue for plywood. J. Am. Oil Chem. Soc. 28: 20–23.

Hogan, J.T. and J.C. Arthur Jr. 1952. Cottonseed and peanut meal glues: resistance of plywood bonds to chemical reagents. J. Am. Oil Chem. Soc. 29: 16–18.

Kalapathy, U., N.S. Hettiarachchy, D. Myers and M.A. Hanna. 1995. Modification of soy proteins and their adhesive properties on woods. J. Am. Oil Chem. Soc. 72: 507–510.

Kumar, R., V. Choudhary, S. Mishra, I.K. Varma and B. Mattiason. 2002. Adhesives and plastics based on soy protein products. Ind. Crop. Prod. 16: 155–172.

Lambuth, A.L. 2003. Protein adhesives for wood. pp. 457–478. In: A. Pizzi and K.L. Mittal (eds.). Handbook of Adhesive Technology, 2nd Ed. Marcel Dekker, Inc., New York, N.Y.

Liu, D., H. Chen, P.R. Chang, Q. Wu, K. Li and L. Guan. 2010. Biomimetic soy protein nanocomposites with calcium carbonate crystalline arrays for use as wood adhesive. Bioresour. Technol. 101: 6235–6241.

Modzel, G., F. Kamke and F. De Carlo. 2011. Comparative analysis of a wood: adhesive bondline. Wood Sci. Technol. 45: 147–158.

NCPA. 2011. National Cottonseed Products Association Petition for Renewable Fuel Pathway for Biodiesel Using Cottonseed Oil. http://www.cottonseed.com/Whatsnew/Cottonseed%20-%20RFS2%20-%20Petition%20-%2012-12-11.pdf. Access date: 10/01/2015 [Online] (verified Oct. 1.).

NCPA. 2016. National Cottonseed Products Association-The Products. http://www.cottonseed.com/aboutncpa/TheProducts.asp [Online].

Norstrom, E., L. Fogelstrom, P. Nordqvist, F. Khabbaz and E. Malmstrom. 2014. Gum dispersions as environmentally friendly wood adhesives. Ind. Crop. Prod. 52: 736–744.

Pettigrew, W.T. and M.K. Dowd. 2014. Nitrogen fertility and irrigation effects on cottonseed composition. J. Cotton Sci. 18: 410–419.

Putun, E. 2010. Catalytic pyrolysis of biomass: effect of pyrolysis temperature, sweeping gas flow rate and MgO catalyst. Energy 35: 2761–2766.

Qi, G., N. Li, D. Wang and X.S. Sun. 2013. Adhesion and physicochemical properties of soy protein modified by sodium bisulfite. J. Am. Oil Chem. Soc. 90: 1917–1926.

Rhee, K.C. 2001. Determination of total nitrogen. Current protocols in food analytical chemistry: B:B1:B1.2.

Saunders, R.M. and E. Hautala. 1979. Relationships among crude fiber, neutral detergent fiber, *in vitro* dietary fiber, and *in vivo* (rats) dietary fiber in wheat foods. Am. J. Clin. Nutr. 32: 1189–1191.

Singh, V., A. Soni, S. Kumar and R. Singh. 2014. Characterization of liquid product obtained by pyrolysis of cottonseed de-oiled cake. Journal of Biobased Materials and Bioenergy 8: 338–343.

Tazisong, I.A., Z. He and Z.N. Senwo. 2013. Inorganic and enzymatically hydrolyzable organic phosphorus of Alabama Decatur silt loam soils cropped with upland cotton. Soil Sci. 178: 231–239.

Tewolde, H., M.W. Shankle, T.R. Way, A. Adeli, J.P. brooks and Z. He. 2015. Managing fall-applied poultry litter with cover crop and subsurface band placement in no-till cotton. Agron. J. 107: 449–458.

Tong, X., X. Luo and Y. Li. 2015. Development of blend films from soy meal protein and crude glycerol-based waterborne polyurethane. Ind. Crop. Prod. 67: 11–17.

Voet, D. and J. Voet. 1990. Chapter 4. Amino acids. pp. 59–74. Biochemistry. John Wiley & Son, Inc., New York, N.Y.

Wanapat, M., N. Anantasook, P. Rowlinson, R. Pilajun and P. Gunun. 2013. Effect of carbohydrate sources and levels of cotton seed meal in concentrate on feed intake, nutrient digestibility, rumen fermentation and microbial protein synthesis in young dairy bulls. Asian-Aus. J. Anim. Sci. 26: 529–536.

Wang, D., X.S. Sun, G. Yang and Y. Wang. 2009. Improved water resistance of soy protein adhesive at isoelectric point. Trans. ASABE 52: 173–177.

Wang, G. and M. Guo. 2014. Property and storage stability of whey protein-sucrose based safe paper glue. J. Appl. Poly. Sci. 131.

Windeatt, J.H., A.B. Ross, P.T. Williams, P.M. Forster, M.A. Nahil and S. Singh. 2014. Characteristics of biochars from crop residues: Potential for carbon sequestration and soil amendment. J. Environ. Manag. 146: 189–197.

Yu, C. and J. Irudayaraj. 2005. Spectroscopic characterization of microorganisms by Fourier transform infrared microspectroscopy. Biopolymers. 77: 368–377.

Yue, H.-B., Y.-D. Cui, P.S. Shuttleworth and J.H. Clark. 2012. Preparation and characterization of bioplastics made from cottonseed protein. Green Chem. 14: 2009–2016.

Zhang, B., Y. Cui, G. Yin, X. Li and X. Zhou. 2009. Alkaline extraction method of cottonseed protein isolate. Modern Appl. Sci. 3(3): 77–82.

Zhang, B., Y. Cui, G. Yin, X. Li and Y. You. 2010. Synthesis and swelling properties of hydrolyzed cottonseed protein composite superabsorbent hydrogel. Int. J. Polym. Mat. Polym. Biomat. 59: 1018–1032.

Zhao, X., F. Chen, W. Xue and L. Lee. 2008. FTIR spectra studies on the secondary structures of 7S and 11S globulins from soybean proteins using AOT reverse micellar extraction. Food hydrocoll. 22: 568–575.

Zhong, Z., X.S. Sun and D. Wang. 2007. Isoelectric pH of polyamide-epichlorohydrin modified soy protein improved water resistance and adhesion properties. J. Appl. Polym. Sci. 130: 2261–2270.

7

Comparative Evaluation of Rice Bran- and Corn Starch-modified Urea Formaldehyde Adhesives on Improvements of Environmental Performance of Agro-based Composites

Altaf H. Basta,[1,*] *Houssni El-Saied*[1] and *Jerrold E. Winandy*[2]

ABSTRACT

Investigations have continued for production of high performance agro-based composites using environmentally acceptable approaches. The possibility of preparing high performance agro-based composites using non toxic rice bran-modified urea formaldehyde (RB-based UF) adhesive systems is evaluated in comparison with that synthesized from starch-based modified UF adhesive system (St-based UF). This chapter focuses on the research of incorporating rice bran into urea-formaldehyde using denaturalized RB in either a slurry (wet) or dry form. It also investigates optimizing the nitrogen content of modified starch systems. The evaluations include assessment of the role of RB- and modified starch-based adhesives

[1] Cellulose and Paper Dept., National Research Centre, Dokki-12622, Cairo, Egypt.
[2] Dept. of Bioproducts and Biosystems Engineering, University of Minnesota, St. Paul, MN 55108 USA.
* Corresponding author: altaf_basta2004@yahoo.com

in reducing emissions of free-HCHO content and its effect on viscosity and gel time of modified UF adhesives. It also includes investigations of the physical, mechanical and water resistance properties of sugarcane bagasse particleboard composites. The results obtained show that both adhesive systems exhibit improved environmental performance (reduction in HCHO emission is ~53%) over a commercially HCHO-based adhesive (UF). Further, they provided particleboards that meet the requirements of grade H-3 per particle board grade requirements from NPA (ANSI A208, 1999).

Introduction

Formaldehyde-based adhesives, especially urea-formaldehyde adhesives, are used worldwide in manufacturing lignocellulosic composites intended for interior applications. They are used because they have good adhesion properties and are of low cost. During the curing of formaldehyde-based adhesives, especially urea-formaldehyde adhesives in the hot press stage, the free formaldehyde may react with chemical constituents of the lignocelluloses, incorporated into the adhesive polymers, off-gas into the air, or destroy via the Cannizzaro reaction (a chemical reaction that converts formaldehyde into methyl alcohol and formic acid). In addition, during storage or in-service use of lignocellulosic composites, especially when exposed to elevated humidity or temperatures, measurable levels of free formaldehyde can be generated due to hydrolysis, isomerisation and decomposition of formaldehyde-resin (Deppe, 1982).

Numerous investigators (Gosselin et al., 1984; Blair et al., 1986; Collin et al., 1988; Nunn et al., 1990) studied the health effects of formaldehyde. Epidemiology studies have inferred possible association between formaldehyde exposure and increasing risk of cancer. Exposure to extremely high concentrations of HCHO in air can produce spasms and edema of the larynx. Individual susceptibility to the irritating effects of airborne formaldehyde can increase with repeated exposures (Beane Freeman et al., 2009). Formaldehyde emission properties of different wood based composite panels and methods of determination were studied in detail (Kim and Kim, 2005; Roffael, 2006; Costa et al., 2012; Salem et al., 2012).

Attempts have been made to protect the environment or to minimize gaseous toxicants by utilizing phenolic or amino resins as wood-product adhesives. These formaldehyde-free adhesives are thermosetting or cross-linked type of polymers, e.g., isocyanate binders and polyvinyl alcohol, or use natural products, e.g., natural tannin and proteinaceous adhesives (Dix and Marutzk, 1988; Tomita, 1997; 1999; Tohmura et al., 2005; Kim, 2009; Pizzi, 2010; Roffael, 2007; 2011; Li et al., 2011; Guangyan et al., 2012; Navarretea et al., 2013). Unfortunately, the poor water resistance of boards bonded by biobased adhesives (especially proteinaceous adhesives) make them unacceptable in the marketplace. Rice bran (RB)-isocyanate adhesive was reported to be a good adhesive agent for agricultural byproducts bonding, and improved the water resistance property of final products (Pan et al., 2006; Wang et al., 2008; 2010).

Changing the surface properties of natural lignocellulosic fibres by different grafting techniques in producing green composites was investigated extensively by Thakur et al. (2012a,b,c; 2014). Additional *in situ* grafting of agro-fibres with free styrene containing polyester was studied by El-Saied et al. (2012). Alkali pre-treatments of the cellulosic fibers lead to increase the degree of grafting the vinyl monomers onto fibers (Mansour et al., 1991). Addition of small percentages of nano-clay has also been reported to improve the performance of UF adhesives for plywood and particleboards (Lei et al., 2007).

We have reported an improvement in the bondability of sugarcane bagasse fibres with commercial UF when adding inorganic and organic materials (biobased polymers), such as nitrogen-containing starch derivatives as HCHO-scavenger, or rice bran (Basta et al., 2004; 2006a; 2006b; 2011; 2014a). Our studies indicated the potential for performance enhancement of rice straw-based composites via pre-treatment process together with using UF-scavenger adhesive system (Basta et al., 2014b), or using free-formaldehyde adhesive based on polyalcoholic polymers (Basta et al., 2013).

In this chapter, we synthesized and compared the results of our two published articles (Basta et al., 2011; 2014a). We assessed the relative effectiveness of each process and bio-based polymer (denaturized rice-bran or modified corn starch) as additives to urea-formaldehyde adhesives. The performance of both synthesized modified UF adhesive and agro-based panels was evaluated in comparison with commercial UF for production of agro-composites that fulfill the requirements of particle board grade H-3 NPA (ANSI A208, 1999).

Experimental

Raw Materials and Commercial Adhesives

- Sugarcane bagasse is an agricultural waste product with 41% α-cellulose, 28% pentosans, 21% lignin and 2% ash. It was used to manufacture lignocellulosic, particleboard-type composites. The raw sugarcane bagasse used in this study was provided by the Kom Ombo Sugar Company of Upper Egypt.
- Commercial urea formaldehyde (UF) with a resin content of 60% was the "control" adhesive in this study. This resin was supplied from Speria Misr Company.

- RB flour contained a 10% moisture and is a by-product after the rice oil is extracted by hexane. It was supplied from Damietta Company of Damietta City, Egypt.

Preparation and Characterization of Adhesive Systems

Rice Bran-based Modified UF Adhesive System

The experimental adhesive systems were prepared by replacing the commercial urea-formaldehyde adhesive with different percentages (10%, 20%, and 30%) of denaturized rice bran. The effects of pH's of denaturized RB, and processed form of the RB (i.e., wet or dry), as well as the methods of incorporating RB with the UF adhesive synthesis were evaluated for optimizing the rice bran-relevant parameters in manufacturing environment-friendly wood adhesives.

For denaturalization treatment process, RB flour with 10% moisture was screened. Only the RB passing through a 100-mesh screen was further modified by alkali treatment. In this process a calculated amount of RB was put in a beaker with distilled water and the pH values of the resulting slurry were adjusted to 9.0, 10.0, or 11.0 with NaOH solution. The solid content of slurry was controlled to 18 percent and then gelatinized in a shaker water bath at 60°C for 2 h. The resulting slurry of RB is denoted as "wet RB". Some wet RB was then oven-dried in metal trays at 75°C for at least 24 h to moisture content of approximately 10% to produce "dry RB". The "dry RB" was further milled for 3 min in a stein Laboratory Mill and sieved through the US 100 mesh screens. Both the wet RB and the dry RB adhesive systems were then prepared and equilibrated for 1 h to achieve adequate adhesive hydration before they were used for making the particleboard.

Synthesis of Starch-based Modified Urea-formaldehyde Adhesives

In an attempt to improve the performance of urea-formaldehyde, an inexpensive and commercially accepted adhesive for wood composites, acrylamide (AM)-containing starch derivatives were used to modify UF during its synthesis. Maize starch from Miser Co. and glucose were modified by acrylamide using a previously developed method (Basta et al., 2006b). Prehydrolysis of the starch at different concentrations of hydrochloric acid, at different times and temperatures, and at differing AM-to-starch ratios provided five starch derivatives with different nitrogen content (195.7–554.0 m atom/100 g starch). The nitrogen content of these prepared biopolymers was determined by using Kjeldahl digestion method. The conditions of preparing these starch-based biopolymers are listed in Table 1.

Table 1. Conditions of preparation of starch-based biopolymers.

Conditions of Hydrolysis			AM/Starch	Matom N/100 g St.)
HCl, M	Time, hrs	Temp.		
0.5	2	60	-	-
1	2	60	-	-
1.5	2	60	-	-
1.5	4	60	-	-
1.5	2	80	-	-
-	-	-	0.5	196
-	-	-	1	441
0.5	2	60	0.5	237
1	2	60	0.5	294
1.5	2	60	0.5	377
1.5	4	60	0.5	460
1.5	2	80	0.5	487
1.5	2	60	1	554

Amide-containing starch (AM-St.)-modified UF adhesive was synthesized by incorporating 5% AM-Starch derivative (based on urea), acquiring different nitrogen contents with paraformaldehyde (PF) and urea (U), and using 2-stage process (alkaline-acidic). An amine-free starch-modified UF adhesive control was also prepared for comparison using a three-neck round bottom flask with a reflux condenser. In this process, the PF/U control with a PF/U-ratio of 1.7:1 was prepared in 120 ml H_2O adjusted to pH 8.0 with 50% NaOH solution. The mixture under continuous mechanical stirring was heated to 96°C in 45 min, and was held at this temperature for 30 min. In the second stage, the pH of the mixture was adjusted to pH 5.5 with acetic acid and subsequently the mixture was boiled at 96°C for 30 minutes. Finally the reaction mixture was concentrated by removing the water under vacuum and maintaining a pH of ~8.0.

Characterization of Adhesive System

To evaluate the investigated modified UF adhesives the following characteristics were estimated: Free-HCHO (% using sodium sulphite method), gel time (sec), bond strength, and viscosity, in comparison with the properties of commercial UF (EN 1243, 1999; Cetin and Ozmen, 2002). Viscosity of the unmodified and modified UF adhesives were measured at 25°C using a Brookfield Programmable DV-II Viscometer, with spindle # 3. Each test was replicated three times.

Composites preparation and tests

Sugarcane bagasse-based particleboard composites were prepared by blending the ground raw bagasse (3 mm diameter) with 10% of its dry weight commercial UF or 12% starch-based modified UF adhesive systems. 1% ammonium sulfate (w/w) based on the solid weight of UF was used as a curing catalyst, according to the procedures outlined in the wood handbook (Youngquist, 1999). Different nitrogen- and free-nitrogen content containing starch derivatives were incorporated during synthesis of UF. With regard to RB/UF adhesive systems, different parameters were applied, e.g., percentage of RB replaced the UF (10%, 20% or 30%), pH of denaturalization of RB (pH = 9, 10 or 11), method of incorporating RB (wet, or dry form).

Press conditions for the resinated bagasse were 200 bar at 160°C for 5 min. The target thickness of the composite boards produced was ~7.4–7.9 mm. The pressed particleboards were trimmed to avoid edge effects and then cut into various sizes for properties evaluation.

All composites prepared using the various adhesives were subjected to mechanical and physical tests. The modulus of rupture (MOR), modulus of elasticity (MOE) and internal bond (IB) were estimated according to ASTM D1037-06a (2006) and ASTM-D3043 (2006) using universal Instron testing machine (Model 1122; Instron Corporation, Conton, MA). Water absorption and thickness swelling percentages were also estimated according to the ASTM D1037-06a (2006), and to the GDR standard, TGL 11-373. Data are the averages of four replicate measurements.

Results and Discussion

Characteristics of RB-based modified UF adhesive systems

Figure 1a shows free-HCHO content of RB incorporating UF adhesive systems, in comparison to commercial standard UF adhesives. It is clear that, incorporating RB as replacing parts (10–30%) to UF showed major advantages in reducing the free-HCHO emissions of the adhesive. The pH-value of denaturalization of RB and its form (wet or dry) also has an effect on influencing the increase or decrease in system efficiency as a function of RB type and replacement percentage. The efficiency of reducing the free-HCHO emissions in adhesive systems ranges from 20.2% to ~52.9%. The best reduction was observed in the case of incorporating 30% wet RB denaturalized at pH 10–11.

For the viscosity performance of RB-modified UF-adhesives, Fig. 2a shows that replacement of UF with different percentages of denaturalized RB in dry form provides adhesives with significant higher viscosities compared to commercial UF. The viscosity of dry form DB modified resins was 1.1–8.4 times higher than the controls. Despite variations in solid

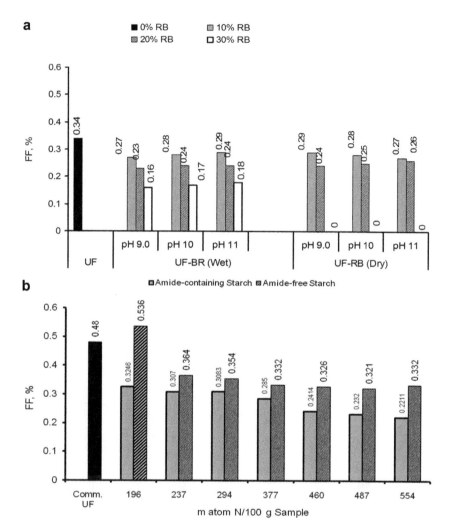

Figure 1. Effect of UF modifying methods on free-HCHO of produced adhesive systems.
a. Rice-bran modifications
b. Starch modifications

content, an increase in viscosity is noticed. We were not able to determine the viscosity at 30% incorporating percentage due to its gel-like form (Fig. 2a). Incorporating denaturalized RB in wet form, produced a decrease in UF viscosity from 279 to 126 mPa.sec. This trend can be related to both the dilution values and pH of denaturalization of RB. Moreover, for incorporating the slurry of denaturalized RB, this decrease in UF viscosity also may result in the decrease in the rate of condensation reactions, than

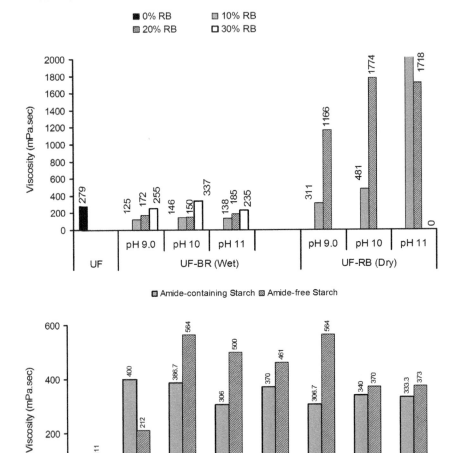

Figure 2. Effect of UF modifying methods on the viscosity of the adhesive systems.
a. Rice-bran modifications
b. Starch modifications

its incorporating in dry form [UF-RB (dry)]. This would appear to decrease the rate of emission of formaldehyde for the process. This pointview is strengthened by the gel time values at all pH's of denaturalization of RB, whereas the lower gel time values are observed with increasing the viscosity of adhesive (Figs. 2a and 3a).

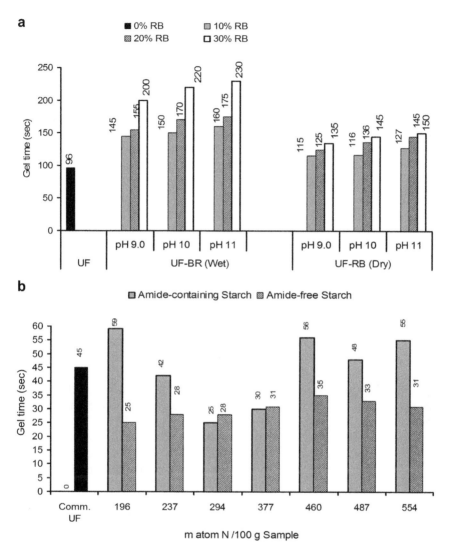

Figure 3. Effect of UF modifying methods on gel time of the adhesive systems.
a. Rice-bran modifications
b. Starch modifications

Characteristics of AM-Modified UF Adhesives

The efficiency of starch modifiers on reduction of the HCHO emissions is illustrated in Fig. 1b. Incorporating starch-based biopolymer during synthesis of UF greatly reduced the environmental impact, through reducing the free-HCHO emitted from the particleboard/UF adhesive system. The free-HCHO emissions decreased when using both amide-free and amide-containing starch. The efficiency of reducing the free-HCHO emissions ranges from 32.3% to ~54%. The best reduction in free-HCHO emissions was achieved when composites were made using modified UF-resin, with higher level of nitrogen amide-containing starch (554 m atom N/100 g).

The use of starch-based modifier provides UF adhesives with higher viscosities (Fig. 2b). Amide-free starches provide adhesives with generally higher viscosities compared to amide-containing starches despite variations in the individual solid content (~59.5 to 63.4%) of matched adhesive systems. Increasing the nitrogen content of amide-containing starch from 196–554 m atom N/100 g starch produced a decrease in UF viscosity from 400–330 mPa.sec. This trend generally parallels that of values of free-HCHO content in modified UF adhesives (Fig. 1).

Comparison of viscosity for the UF adhesives made from amide-free starches with those made from amide-containing starches shows a clear relation between gel time and viscosity (Figs. 2b and 3b). The increase in viscosity accompanied by the decrease in gel time confirms the findings of Osemeahon and Barminas (2006) that the viscosity seems to represent the gel point of resins. However, for amide-containing starch-UF adhesives, the viscosity and gel time values are not clearly correlated. This is probably due to the change in the content of amide groups which influence its chemical ability to capture the free-HCHO molecules produced from condensation step of synthesis of UF. In other words, at relatively higher nitrogen content of amide-containing starch (e.g., 554 m atom N/100 g St.), the amide groups may retard the formation of free-HCHO molecules in condensation step and consequently increased the gel time.

In summary, the amide-containing starch (554 m atom N/100 g St.) or incorporating 30% denaturized RB (at pH 8–11) would provide adhesives which have minimal free-HCHO emissions while still maintaining acceptable viscosity and gel time. Thus, such adhesive systems are less toxic for environmental concerns.

Mechanical and Water Resistance Properties of Bagasse-based Composites

RB-UF Adhesive Systems

For static bending properties (MOR & MOE), Figs. 4a and 5a show that all three percentages (10, 20 and 30%) of incorporating RB slurry (wet form) to UF tended to increase the modulus of rupture property of sugarcane bagasse composites when compared with any of the dry forms of RB, or the commercial "UF control" adhesive (Fig. 4a). All tested adhesive systems provided improvement in MOE (Fig. 5a). Replacing 10–20% UF

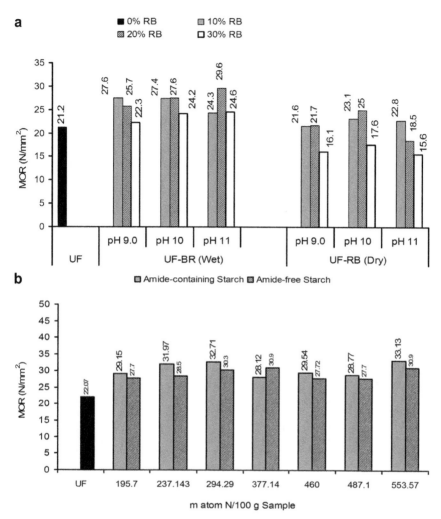

Figure 4. Effect of UF modifying methods on the modulus of rupture (MOR) of particleboards.
a. Rice-bran modifications
b. Starch modifications

by denaturalized RB slurry (at pH 9–10) resulted in composites with better static bending properties. MOR was generally improved by about 25% and MOE by about 32%, both compared to the UF control. The improvements at relatively lower replacement percentages may be ascribed to the viscosity values of modified UF adhesives. Low viscous RB-UF gives rise to penetration of adhesive through bagasse fibers and facilitates bonding agro-fibers during hot pressing.

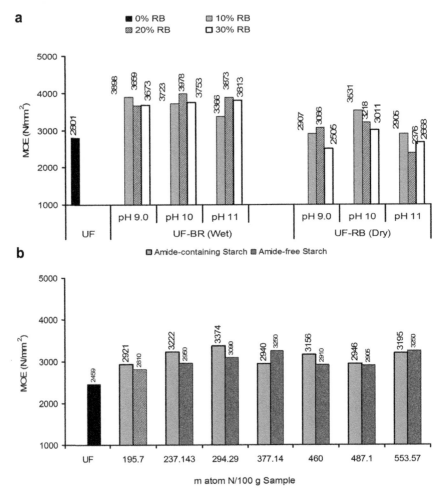

Figure 5. Effect of UF modifying methods on the modulus of elasticity (MOE) of particleboards.

a. Rice-bran modifications

b. Starch modifications

It is evident that there was no benefit in increasing MOR when increasing the percentage of RB in the wet slurry or by any incorporation of denaturalized RB in its dry form. This observation is probably related to deficiency in water content. Water in the RB slurry leads to increased fiber flexibility, induced through converting water to steam, which in turn allows for more fiber response to shape during the hot pressing stage.

As with the trend noted for MOR and MOE properties, the method of incorporating the RB with the UF adhesive noticeably affected the internal bond (IB) of particleboards (Fig. 6a). At all pH levels, addition of denaturalized RB and with all process for incorporating the RB to UF provided improvement in IB when compared to commercial UF with one exception (i.e., 30% dry DB which decreased IB at all pH levels).

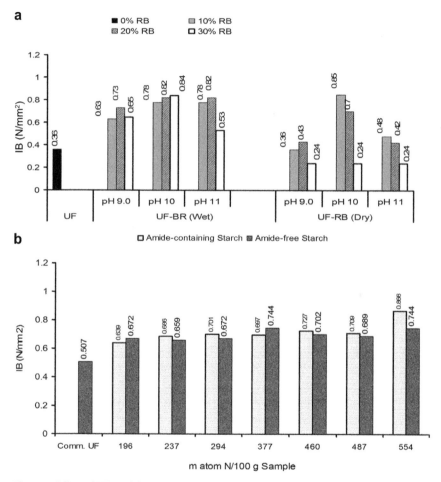

Figure 6. Effect of UF modifying methods on the internal bond (IB) of particleboards.
a. Rice-bran modifications
b. Starch modifications

For water swelling property (thickness swelling), Fig. 7 shows that using the RB (wet form) adhesive systems offers great improvements in the water resistance properties of the bagasse composites (thickness swell (TS) = 4–13%) compared to commercial UF (TS = 17.5%). The only dry-form DB system that improved TS was replacement of 10% of UF in dry form RB

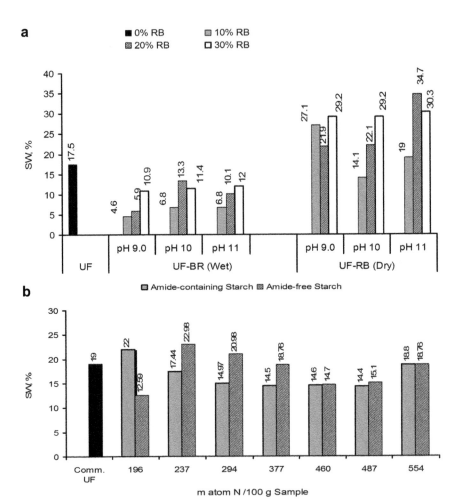

Figure 7. Effect of UF modifying methods on the thickness swelling (SW) of particleboards.
a. Rice-bran modifications
b. Starch modifications

when pre-denaturalized at pH 10 (SW = 14%). All other dry-form RB systems increased SW when compared to UF controls. A possible explanation of this observation was the lower viscosity of modified UF adhesive. The low viscosity enhances adhesive penetration through the fibers for enhanced bond formation (under heat and pressing) between protein groups (amide, amino) and methylol (CH_2OH groups) groups of UF with hydroxyl groups of bagasse fibers. This cross-linking action renders both RB and bagasse fibers resist to swell.

AM-Modified UF Adhesive Systems

The nitrogen content of AM-St. derivatives had a profound effect on mechanical and water resistance properties of final boards (Figs. 4b–7b). From Figs. 4b and 5b it is clear that the two types of UF resins synthesized using either amide-free or amide-containing starches tended to increase the static bending properties (MOR & MOE) of sugarcane fiber bagasse composites, compared with commercial UF adhesives. The modified UF synthesized with amide-containing starches resulted in composites with better static bending properties than UF resins using amine-free starch modifiers. The values of MOR and MOE of composites made by amide-containing starch-UF adhesives varied from 28 to 33 N/mm^2 and from 2850 to 3374 N/mm^2 (improved by 17.9 and 18.4%), respectively. With amide-free starch-UF adhesives the values for MOR and MOE varied from 27.7-to 30.9 N/mm^2 and from 2810 to 3250 N/mm^2 (improvement by 14% & 11.6%), respectively. The values of MOR and MOE of composites resulted from unmodified commercial UF were 22.07 and 2459 N/mm^2, respectively. The MOR and MOE of composites panels made by amide-containing starch modified UFs tended to be about 10 ± 3% higher than the other UF resin systems. The observation is probably related to the effects of viscosity of modified UF adhesives. Low viscous UF gives rise to lower molecular weight molecules (??) which favors molecular chain mobility that enhancing flexibility of polymer network in composites produced (Osemeahon and Barminas, 2006).

Similar to the trend of MOR and MOE properties, the internal bond (IB) of modified-UF is affected by nitrogen content of amide-containing starch modifier (Fig. 6). Incorporating high-nitrogen starch (554 m atom/100 g Sample) greatly improved IB (0.866 N/mm^2; i.e., improved by about 70%), compared with a nitrogen content of other modifiers (0.64–0.73 N/mm^2). In general, amide-free starches have less effect in improving the internal bond of modified UF (0.67–0.704 N/mm^2; improvement%: 31.4%–38%), compared with that produced from commercial UF (0.51 N/mm^2).

Water swelling property (SW) of the produced modified UF-based composites varied with amide-containing systems, and it was improved when m atom/100 g contents are 460 and 554. For amide-free systems improving in SW was observed when m atom/100 g content > 377 (Fig. 7b). Sugarcane fiber bagasse composites bonded by amide-free or amide-containing biopolymers generally tended to exhibit less water resistance (increased SW), compared with commercial UF (SW = 17.4%). Variations in the water swelling resistance between UF resins made with amide-containing starch and amide-free starches were noted. This may be ascribed to the competing reaction of amide groups (hydrophilic side) of biopolymers and urea with formaldehyde during adhesives synthesized.

It is interesting to note that water swelling data of most boards meet the requirement of standard specifications (≤ 25% by ANSI).

In summary, when comparing the properties of particle boards made from RB- and starch-based modified UF adhesives, both modified UF adhesive systems provided agro-fiber composites that meet or exceed the physical/mechanical property requirements of high grade particleboards (H-3), especially with respect to static bending values (MOR 23.5 N/mm^2, and MOE 2,750 N/mm^2). Incorporating 20% RB slurry, denaturized at pH 9–10, to modify UF, provided higher improvement in MOR, MOE, IB and SW properties of the produced board than that produced using high performance amide-containing starch derivatives (554 m atom N/100 g sample) as UF modifier. Moreover, the water swelling results of boards made from either modified UF adhesive meet the requirement of standard specifications (≤ 25% by ANSI).

Acknowledgements

The authors are grateful to Kom Ombo Co. for Particleboards, Upper Egypt for providing the raw materials and equipment for the experimental study dealing with RB-based modified UF-adhesive system.

The research work dealing with AM-modified UF adhesive system was carried out under the grant from Egypt-U.S. Science and Technology Program and sponsored by the Egyptian Ministry of Scientific Research and the U.S. Department of State with Contact/Agreement No. 280.

Keywords: Modified-HCHO-based adhesive, Urea-formaldehyde, Biopolymer-based adhesive, Protein-based adhesive, Agro-based Composite, Environment, Free formaldehyde

References

ANSI A208, 1999. American National Standardization Institute. Particleboards. ANSI 208.1.

ASTM D1037-06a. 2006. Standard Test Methods for Evaluating Properties of Wood-Base Fiber and Particle Panel Materials. ASTM Book of Standards. West Conshohocken, PA.

ASTM D3043-06. 2006. Standard Test Methods for structural panels in flexure. ASTM Book of Standards. West Conshohocken, PA.

Basta, A.H., H. El Saied, R.H. Gobran and M.Z. Sultan. 2004. Enhancing environmental performance of formaldehyde-based adhesive in lignocellulosic composites. Polym.-Plast. Technol. Eng. 43: 819–843.

Basta, A.H., H. El Saied, R.H. Gobran and M.Z. Sultan. 2006a. Enhancing environmental performance of formaldehyde-based adhesive in lignocellulosic composites. Pt. II. Pigment Resin Technol. 34: 12–21.

Basta, A.H., H. El Saied, R.H. Gobran and M.Z. Sultan. 2006b. Environmental performance of formaldehyde-based adhesives in lignocellulosic composites. Pt. III. Des. Monomers and Polm. 9(4): 325–347.

Basta, A.H., H. El Saied, J.E. Winandy and R. Sabo. 2011. Preformed amide-containing biopolymer for improving the environmental performance of synthesized urea–formaldehyde in Agro-fiber Composites. J. Polym. Environ. 19: 405–412.

Basta, A.H., H. El-Saied and V.F. Lofty. 2013. Performance of rice straw-based composites using environmentally friendly polyalcoholic polymers-based adhesive system. Pigment Resin Technol. 42(1): 24–33.

Basta, A.H., H. El-Saied and E.M. Deffallah. 2014a. Optimising the process for production of high performance bagasse-based composites from rice bran-UF adhesive system. Pigment Resin Technol. 43(4): 212–218.

Basta, A.H., H. El-Saied and V.F. Lofty. 2014b. Performance assessment of deashed and dewaxed rice straw on improving the quality of RS-based composites. RSC Adv. 4(42): 21794–21801.

Beane Freeman, L., A. Blair, J.H. Lubin et al. 2009. Mortality from lymphohematopoietic malignancies among workers in formaldehyde industries, The National Cancer Institute cohort. J. National Cancer Institute. 101(10): 751–761.

Blair, A., P. Stewart, M. O'Berg, W. Gaffey, J. Walrath, J. Ward, R. Bales, S. Kaplan and D. Cubit. 1986. Mortality among industrial worker exposed to formaldehyde. J. Natl. Cancer Inst. 76: 1071–1084.

Cetin, N.S. and N. Ozmen. 2002. Preparation and characterization of phenol–formaldehyde adhesives modified with enzymatic hydrolysis lignin. Int. J. Adhes. Adhes. 22: 481–486.

Collin, J.J., J.C. Caporossi and H.M.D. Utidjiam. 1988. Formaldehyde exposure and nasopharyngeal Cancer: Re-examination of the National Cancer Institute study and an update of one plant. Nat. Cancer Inst. 80: 376–377.

Costa, N.A., J. Pereir, J. Ferra, P. Cruz, J. Martins et al. 2013. Scavengers for achieving zero formaldehyde emission of wood-based panels. Wood Sci. Technol. 47: 1261–1272.

Deppe, H.-J. 1982. Emission von Organischen Substanzen aus Spanplatten. pp. 91. *In*: K. Aurand, B. Seifert and J. Wegner (eds.). Luftqualitat in Innenraumen. Stuttgart, Fischer.

Dix, B. and R. Marutzk. 1988. Tannin formaldehyde resin. Holz Roh-Werkt. 46(1): 19–24.

El-Saied, H., A.H. Basta, M.E. Hassanen, H. Korte and A. Helal. 2012. Behaviour of rice-byproducts and optimizing the conditions for production of high performance natural fiber polymer composites. J. Polym. Env. 20(3): 838–047.

EN 1243. (1999).The European Standard, Determination of free formaldehyde in amino and amidoformaldehyde condensates. Brussels: European Committee for Standardization.

Gosselin, R.E., R.P. Smith and H.C. Hodge. 1984. Clinical toxicology of commercial products, 5th Ed., PP. II-196-III-198. William and Wilkins, Baltimore.

Guangyan, Q., L. Ningbo, W. Donghai and S.S. Xiuzhi. 2012. Physicochemical properties of soy protein adhesives obtained by *in situ* sodium bisulfite modification during acid precipitation. J. Am. Oil Chem. Soc. 89(2): 301–312.

Kim, S. and H.-J. Kim. 2005. Comparison of standard methods and gas chromatography method in determination of formaldehyde emission from MDF bonded with formaldehyde-based resins. Bioresource Technol. 96: 1457–1464.

Kim, S. 2009. Environment-friendly adhesives for surface bonding of wood-based flooring using natural tannin to reduce formaldehyde and TVOC emission. Bioresource Technol. 100: 744–748.

Lei, H., G. Du, A. Pizzi and A. Celzard. 2007. Influence of nanoclay on urea-formaldehyde resins for wood Adhesives and It's Model. Appl. Polym. Sci. 106: 3958–3966.

Li, X., Z. Cai, J.E. Winandy and A.H. Basta. 2011. Effect of oxalic acid and steam pretreatment on the primary properties of UF-bonded rice straw particleboards. Ind. Crops Prod. 33: 665–669.

Mansour, O.Y., Z.A. Nagieb and A.H. Basta. 1991. Graft polymerization of some vinyl monomers onto alkali-treated cellulose. J. Appl. Polym. Sci. 43(6): 1147–1158.

Navarretea, P., A. Pizziab, H. Paschc, K. Roded and L. Delmotte. 2013. Characterization of two maritime pine tannins as wood adhesives. J. Adhes. Sci. and Technol. 27(22): 2462–2479.

Nunn, A.J., A.A. Craigen, J.H. Darbyhire et al. 1990. Six-Year follow-up of lung function in men occupationally exposed to formaldehyde. Br. J. Ind. Med. 47: 747–752.

Osemeahon, S.A. and J.T. Barminas. 2006. Properties of a low viscosity urea-formaldehyde resin prepared through a new synthetic Route. Bull. Pure. Appl. Sci. 250: 67–76.

Pan, Z., A. Cathcart and D. Wang. 2006. Properties of particleboard bond with rice bran and polymeric methylene diphenyl diisocyanate adhesives. Ind. Crop. Prod. 23: 40–45.

Pizzi, A. 2010. Emulsion polymer isocyanates as wood adhesive: A Review. J. Adhes. Sci. Technol. 24(8-10): 1357–1381.

Roffael, E. 2006. Volatile organic compounds and formaldehyde in nature, wood and wood based panels. Holz als Roh-und Werkstoff. 64: 144–149.

Roffael, E. 2007. Verfahren zur Herstellung von Holzfaserplatten mit verbesserter Formaldehydemission, hoher Feuchtebeständigkeit und Hydrolyseresistenz. DE 102007054123 B4, erteilt am 15.3.2012.

Roffael, E., C. Behn, D. Krug, A. Weber, C. Hartwig-Gerth and G. Gräfe. 2011. UF- und PMDI-Doppelbeleimung bei Faserplatten. Holz-Zentralblatt 137: 1216–1217.

Salem, M.Z.M., M. Böhm, J. Srba and J. Beránková. 2012. Evaluation of formaldehyde emission from different types of wood-based panels and flooring materials using different standard test methods. Build. Environ. 49: 86–96.

Thakur, V.K., A.S. Singha and M.K. Thakur. 2012a. Graft copolymerization of methylacrylate onto cellulosic biofibers: Synthesis, characterization and applications. J. Polym. Environ. 20(1): 164–174.

Thakur, V.K., A.S. Singha and M.K. Thakur. 2012b. Surface modification of natural polymers to impart low water absorbency. Int. J. Polym. Anal. 17: 133–143.

Thakur, V.K., A.S. Singha and M.K. Thakur. 2012c. In-air Graft copolymerization of Ethyl acrylate onto natural cellulosic polymers. Int. J. Polym. Anal. 17: 48–60.

Thakur, V.K., M.K. Thakur and R.K. Gupta. 2014. Graft copolymers of natural cellulose for green composites. Carbohydr. Polym. 104: 87–93.

Tohmura, S.-I., G.-Y. Li and T.-F. Qin. 2005. Preparation and characterization of wood polyalcohol-based isocyanate adhesives. Appl. Polym. Sci. 98(2): 791–795.

Tomita, B. 1997. New resin system from lignin, Sci. Technol. Polm. Adv. Mater. [Proc. Int. Conf. front. Polym. Adv. Mater.], 4th 1997 (Pub. 1998), pp. 747–750.

Tomita, B. 1999. Trend and future of wood adhesive. Mokuzai Kenkyu Shiryo. 25: 1–30.

Wang, W., X. Zhang and X. Li. 2008. A novel natural adhesive from rice bran. Pigment Resin Technol. 37: 229–233.

Wang, W., X. Zhang, F. Li and C. Qi. 2010. Rice bran adhesive modified with potassium permanganate and poly (vinyl alcohol). Pigment Resin Technol. 39: 355–358.

Youngquist, J.A. 1999. Wood-based composites and panel products. Wood handbook: wood as an engineering material. Madison, WI: USDA Forest Service, Forest Products Laboratory. General Technical Report FPL; GTR-113: Pages 10.1–10.31.

8

Tannins for Wood Adhesives, Foams and Composites

*Nicolas Brosse** and *Antonio Pizzi*

ABSTRACT

Tannin is a renewable resource that is "coming of age" in several fields different from their usual classical application, namely hide tanning to produce heavy duty leather. Since 1970s, tannin-based adhesives for wood industry have been produced and commercialized. More recently, new promising tannin-based materials such as biocomposites and rigid foams have also been developed. In this chapter, we present an updated overview of tannin research regarding the industrial extraction methods, the reactivity and the applications. The coverage also includes the results and perspectives in the field of condensed tannins extraction and utilization. Recent research efforts on new material applications are also presented: biocomposites composed of natural fibers (flax, hemp) and of tannin-based resin and also tannin-based rigid foams with excellent performance and characteristics. A new potential source of condensed tannins (the grape pomace) is proposed and the first results on the utilization of grape pomace as a source of tannins for wood adhesives are presented.

Introduction

Tannin or tannins have been used industrially for almost 50 years. They can be used as adhesives for interior and exterior wood bonding so that

Laboratoire d'Etude et de Recherche sur le MAteriau Bois, Faculté des Sciences et Techniques, Université de Lorraine, Bld des Aiguillettes, BP 70239, F-54506 Vandoeuvre-lès-Nancy, France.
* Corresponding author: nicolas.brosse@univ-lorraine.fr

they are excellent alternatives to the petroleum—derived phenolic resins for the production of particleboard, plywood, glulam and finger jointing.

The word "tannin" defines two classes of phenolic chemical compounds:

- hydrolysable tannins which are mixtures of simple sugars and ester of glucose with gallic and digallic acids;
- condensed tannins consisting of flavonoid units with varying degrees of condensation.

The structure of the flavonoid constituting the main monomer of condensed tannins may be represented as follows (Fig. 1).

This flavonoid unit is repeated 2 to 11 times in mimosa tannin, with an average degree of polymerization of 4 to 5, and up to 30 times for pine tannins, with an average degree of polymerization of 6 to 7 for their soluble extract fraction (Fechtal and Riedl, 1993; Thompson and Pizzi, 1995).

Condensed tannins (also called as proanthocyanidins, polyflavonoid tannins, catechol-type tannins, pyrocatecollic type tannins, non-hydrolyzable tannins or flavolans) represent more than 90% of the total world production of commercial tannins and are yielded by extraction from wood substance, bark, leaves, and fruits. Other components of the extraction solutions are sugars, pectines, and other polymeric carbohydrates, amino acids, and other substances. At the industrial scale, tannins are extracted from the bark of various trees through simple water-based procedures. Comprehensive reviews have already been published describing the chemistry of condensed tannins and their reactivity toward formaldehyde and the technology of industrial tannins adhesive formulation (Pizzi, 1983; 1994; 2003; 2006).

The ability of tannin-based materials to make significant impact as a substitute for polymeric materials depends on the availability of low-price and high-quality tannin fractions in large quantities. Grape is one of the world's largest fruit crops and grape pomace is one of the wastes generated by grape juice and wine-making processes. This pomace consists of skins,

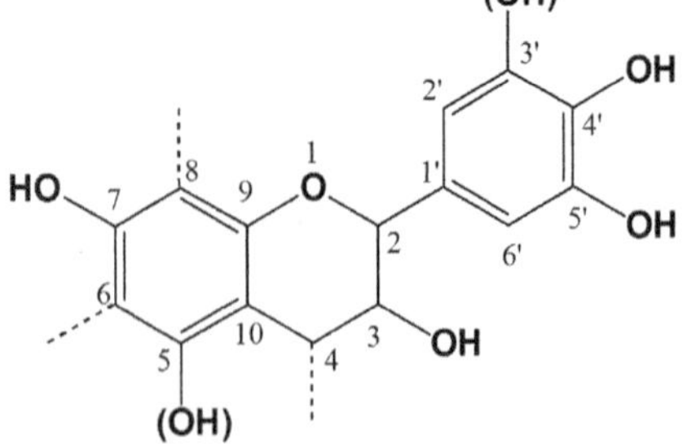

Figure 1. A general structure of the monomer of condensed tannins.

seeds, and in certain cases, of some stems, with the skins and seeds making up the major part. Because of a low extraction during winemaking, the solid residues still retain high levels of condensed tannins. Only small amounts of these by-products are up-graded for recovery of the phenolics to be used as natural health remedies, food supplements, or as novel nutrifunctional food ingredients. Therefore, grape pomace potentially could constitute an abundant and relatively inexpensive source for tannin adhesive production.

Composites have been an area of growing interest and a subject of active research. Biocomposites composed of natural fibres and an oil-derived thermoplastic matrix are well known for car door's interiors and other applications. However, for the same type of applications, composites also using natural matrix while still presenting high performance have not been developed and commercialised. In this context, there is a need for 100% green biodegradable biocomposites and the tannin-based resins are potential candidates for such applications.

History of Tannin Extraction

Leather tanning has been used for centuries, millennia in fact, by immersing hides in pits in which tree bark or wood rich in tannin, such as oak, had been left in. This method took up to one full year to produce good leather. However, the actual tannin extraction industry is relatively more recent. It started in Lyon, France and in northern Italy in the 1850s to satisfy the need for black dyes for silk clothes. As the fashion of women silk blouses vaned about ten years later, the multitude of small chestnut tannin extraction factories that had sprung up underwent a dramatic change of fortune, many going into bankruptcy and closing down, others combining to present enough critical mass to find an alternative use for tannin extract (Calleri, 1989).

The surviving producers, not many, managed to convince the leather manufacturers that, by dissolving tannin extract in the manufacturing pits, leather could be manufactured in just one month instead of the average of one year employed with the traditional bark bath technology. As a consequence the tannin extraction industry started its second life for an application quite different from the original one. The advantages of tannin extracts in leather-making, and even further, time saving allowed by their use, were such that the industry underwent rapid expansion and prospered. Shortness of materials in Europe to satisfy the rapidly growing demand for tannin extracts for leather forced the opening of factories in far away countries and the use of new types of tannins. Thus, in the early 1900s tannin factories using quebracho tannin from South America and Mimosa tannin from Southern and Central Africa started extraction in industrial quantities, and exported to the main northern hemisphere markets.

The two World Wars gave considerable impulse to the expansion of tannin extraction, after all armies marched on shoes with leather soles. To give a typical quantitative example, in 1946, just after the end of World War II, a major producer such as South Africa manufactured about 110 thousand metric tons of dry tannin extract solids. However, this year was the zenith of the use of vegetable tannins for leather making. The use of rubber soles in everyday shoes, which were cheaper and more readily available, as well as difficulty in supplying tannins to certain countries, led to the decrease in the tannin extract markets. By the beginning of the 1970s total production was down to 72000 tons/year. This was followed by a further decrease in tannin production in a period in which hides supply dwindled due to the decrease/difficulties of cattle farming in the 1960s and 1970s. Finally, the considerable shift in customers taste in shoes which came about with the comfortable sport and leisure shoes "boom" gave a final negative impulse to the use of tannins in leather making. It is interesting that today the same country that produced 110 thousand tons of tannin extract in 1946 produces only 42 thousand tons of tannin extract, only half of which are still used for leather manufacture.

As a consequence of the steady dwindling in tannin sales for leather, in the 1960s and 1970s the industry started to desperately look for new applications for these materials. After all, they had survived the collapse of the silk dyes market more than a century earlier, and a new lease of life could perhaps be found in other fields. Many applications were tried, from varnish primers for metals which effectively were in use in Britain for some time during the 1960s and 1970s, to antipollution flocculating agents that were successful for about 15 years in the 1970s and 1980s before being superseded by better synthetic materials. Tannin-based ore flotation agents are still in use, at about 600 tons/yearly of tannins, especially in a couple of feldspar mines in southern Africa. Furthermore, fluidifying agents for drilling muds and superplasticizing additives for cement were developed. To the knowledge of the author, about 70–80 tons per year of tannin-based cement additives were and are still used. The main use found, however, was for tannin adhesives for wood panels and other wood products. Production started in 1973 and reached 4500 dry tons/year of tannin by 1978, and is now around 25000 dry tons/year.

The application of tannins as wood adhesives contributed towards saving a few tannin extraction factories, stabilized the situation in some major producing countries, but did not translate into an immediate and definitive rescue of the industry. Worldwide, leather tanning still consumes more tannin. This difficulty has been due to the low prices that oil-derived synthetic adhesives enjoyed relatively to tannins. The first oil crisis of 1974 at least convinced a few southern hemisphere industrialists (in Australia, South Africa and New Zealand) to use tannin adhesives in order to avoid the difficulties of supply of synthetic adhesives. The phenomenon persisted, but with a couple of still existing exceptions, it did not take off in Europe or North America. After all, synthetic adhesives after 1974 became cheap again, in abundant supply, and were strongly "pushed" by all the big chemical companies. Considerable renewed interest in these materials started again after the year 2000 due to two factors: (i) the recent marked increase in oil prices that increased disproportionally the market price of all synthetic adhesives, and thus favoring natural materials. Tannin for instance, is now much cheaper than phenol and not far from the price of the cheapest adhesive of them all, urea-formaldehyde resins. (ii) The recent severe tightening of formaldehyde emission regulations, mainly the introduction of the extremely severe Japanese standard. A regulation that is now starting to spill over in other countries too (JIS, 1994).

Although tannin adhesives are now fast becoming an interesting industrial proposition as an alternative to synthetic adhesives, the use of tannins is expanding rapidly to an even more interesting field, namely the pharmaceutical/medicine field. Thus, a fourth market transformation has started and it appears set to overtake tannins transformation into wood adhesives, at least by monetary value addition. Thus, the therapeutic virtues of the addition of tannins in wine (consequence of the so-called "French Paradox"), the use of tannins to cure some gastrointestinal diseases, the increasing use of tannins as food supplements in North America and research on the utility of tannins in a multitude of diseases, even grave ones such as cancer and virus-induced diseases, are in full swing. They are because the value added to the base cost of tannins is considerable. Tannins for use for human consumption must be heavily purified. The price is approx 40 to 50 times higher than the cost of tannin for leather.

Industrial Extraction Procedures

From tree bark: Tannins can be extracted from bark and wood of trees at different temperatures and under slightly different conditions, but industrially are always extracted by countercurrent extraction using hot water. In general, the chips of bark are charged in batteries of 6 or 8 connected stainless steel autoclaves, each containing between 10 and 20 tons of bark or wood chips. For some tannins such as mimosa tannin extract the extraction of the bark chips is often done with just hot water at between 70°C and 90°C. The lower temperature is the preferred one because as the temperature increases the yield of extract increases but not the yield of tannins, as other materials, in particular carbohydrates, are extracted more. Thus, if the extract is prepared for adhesives use it is good practice to use temperatures not higher than 70°–75°C. In the case of other tannins such as quebracho tannin from wood chips and pine tannin from bark chips some chemicals are added. In general 2% sodium sulphite or bisulphite and 0.5% of sodium bicarbonates are used for the extraction. These conditions facilitate the extraction of the high molecular weight oligomers by increasing their water solubility.

More recently, for pine bark tannins and quebracho tannins, small percentages of urea are added to the extraction medium together to bisulphite and bicarbonate to increase the extraction yield by avoiding internal rearrangements to insoluble compounds. Such an approach has resulted in up to 40% increase in extraction yield tested at industrial level (Sealy-Fisher and Pizzi, 1992).

From grape industry: Grape is one of the most grown fruit crops in the world with more than 60 million tons produced annually. Viniculture is an important agricultural activity in a lot of countries in southern Europe like in Spain, Italy and France and produces huge amounts of grape pomace. After extraction in the distilleries of wide range of products (ethanol, grape seed oil, anthocyanins and tartrate), the remaining pomace is currently not upgraded. In France, about 700 000 to 1000 000 tons of dry grape pomace are produced each year. Considering the growing demand for green materials and components, agricultural by-products like pomace have an obvious potential as a renewable starting material. This pomace consists of skins, seeds, and in certain case of some stems, with the skins and seeds making up the major part (Zocca et al., 2007). Because of a low extraction during winemaking, the solid residues still retain high levels of condensed tannins. However, the majority of the grape pomace is currently used for composting or discarded in open areas potentially causing environmental problem. Only small amounts of these by-products are up-graded by different solvent extraction methods for recovery of the phenolics to use as natural health remedies, food supplements, or as novel nutrifunctional food ingredients (Shrikhande, 2000). Therefore, grape pomace potentially constitutes a very abundant and relatively inexpensive source for tannin adhesive production. However, this source of tannins is currently not utilized at the industrial scale.

The chemical composition of grape pomace samples originating from red and white winemaking from different areas of production of French vineyard was established. There are differences in composition depending on the pomace samples origin, but some general conclusions can be drawn (Rondeau et al., 2013). Pomace samples contained 20 to 46% w/w of sugars, mainly glucans and xyloglucans with low pectinaceous polysaccharides content and 20 to 51% of condensed tannins.

The optimization of the extraction of phenolics from solid residues from the wine industry in water medium in the presence of sodium carbonate using response surface methodology based on three variables central composite design was performed (Brahim et al., 2014). It was demonstrated that from grape pomace, the maximum of phenolics can be recovered in water medium, but at high temperature (100°C–120°C) with large amounts of basic reagents like carbonate (compared to tree bark). The utilization of harsh conditions for the extraction of tannins from grape pomace was rationalized by the fact that the easily hydrolysable fraction of condensed tannins was removed during the winemaking and distillery processes. Most of the extraction trials were performed at the lab scale but the process was also transferred to an industrial scale (unpublished results).

Chemical Structure of Condensed Tannins

In condensed tannins from mimosa bark and grape pomace (Lan et al., 2011a,b,c) the main polyphenolic pattern is represented by flavonoid analogs based on resorcinol A rings and pyrogallol B rings (I, Fig. 2). These constitute about 70% of the tannins. The secondary but parallel pattern is based on resorcinol A rings and catechol B rings (II, Fig. 2, Pizzi, 1983; Roux, 1965). These tannins represent about 25% of the total of mimosa bark tannin fraction.

The remaining part of the condensed tannin extracts are the "non-tannins". They may be subdivided into carbohydrates, hydrocolloid gums, and small amino and imino acid fractions (Pizzi, 1983; Roux, 1965). For bark extracts, hydrocolloid gums vary in concentration from 3 to 6% and contribute significantly to the viscosity of the extract despite their low concentration (Pizzi, 1983; Roux, 1965). For pomace extracts, relatively high carbohydrate contents (around 50%) were detected from the ^{13}C NMR spectra and low Stiasny numbers (around 0.5–0.6) were obtained (Lan et

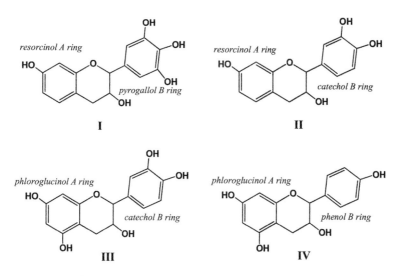

Figure 2. Main phenolic patterns in condensed tannins.

al., 2011a,b,c; 2012a,b). Similar flavonoid A and B ring patterns also exist in quebracho wood extract (*Schinopsis Balansae*, and *Lorentzii*) (King and White, 1957; Roux and Paulus, 1961; King et al., 1961), but no phloroglucinol A ring pattern, or probably a much lower quantity of it, exists in quebracho extract (King et al., 1961; Roux et al., 1975; Clark-Lewis and Roux, 1959). Similar patterns to wattle (mimosa) and quebracho are followed by hemlock and Douglas fir bark extracts. Completely different patterns and relationships do instead exist in the case of pine tannins (Roux et al., 1975; Clark-Lewis and Roux, 1959; Hemigway and McGraw, 1976) which present instead only two main patterns: one represented by flavonoid analogs based on phloroglucinol A rings and catechol B rings (III, Fig. 2, Hemigway and McGraw, 1976; Porter, 1974). The other pattern, present in much lower proportion, is represented by phloroglucinol A rings and phenol B rings (IV, Fig. 2, Hemigway and McGraw, 1976; Porter, 1974). The A rings of pine tannins then possess only the phloroglucinol type of structure, much more reactive toward formaldehyde than a resorcinol-type structure, with important consequences in the use of these tannins for adhesives.

Reactivity Condensed Tannins

The nucleophilic centers on the A ring of a flavonoid unit tend to be more reactive, than those found on the B ring (see Figs. 1 and 2). This is due to the vicinal hydroxyl substituents causing general activation in the B ring without any localized effects such as those found in the A ring.

Formaldehyde is generally the aldehyde used in the preparation, setting, and curing of tannin adhesives. It is normally added to the tannin extract solution at the required pH, preferably in its polymeric form of paraformaldehyde, which is capable of fairly rapid depolymerization under alkaline conditions, and as urea-formalin concentrates. Hexamethylenetetramine (hexamine) may also be added to resins due to its potential formaldehyde releasing action under heat. Hexamine is, however, unstable in acid medium (Megson, 1958) but becomes more stable with increased pH values. Hence under alkaline conditions the liberation of formaldehyde might not be as rapid and as efficient as wanted. Also, it has been fairly widely reported, with a few notable exceptions (Pizzi and Scharfetter, 1977), that bonds formed with hexamine as hardener are not as boil resistant (Pizzi and Stephanou, 1994) as those formed by paraformaldehyde. The reaction of formaldehyde with tannins may be controlled by the addition of alcohols to the system. Under these circumstances some of the formaldehyde is stabilized by the formation of hemiacetals [e.g., $CH_2(OH)(OCH_3)$] if methanol is used (Pizzi, 1983; Scharfetter et al., 1977). When the adhesive is cured at an elevated temperature, the alcohol is driven off at a fairly constant rate and formaldehyde is progressively released from the hermiacetal. This ensures that less formaldehyde is volatilized when the reactants reach curing temperature and that the pot life of the adhesive is extended. Other aldehydes have also been substituted for formaldehyde (Pizzi, 1983; Pizzi and Scharfetter, 1978; Plomley, 1966; Pizzi and Scharfetter, 1977). To-day, alternative aldehydes such as glyoxal, glutaraldehyde and others are experimented with, showing some success.

Formaldehyde reacts with tannins to produce polymerization through methylene bridge linkages at reactive positions on the flavonoid molecules, mainly the A rings. The reactive positions of the A rings are one of positions 6 or 8 (according to the type of tannin) of all the flavonoid units and both positions 6 and 8 of the upper terminal flavonoid units. The A rings of mimosa and quebracho tannins show reactivity toward formaldehyde comparable to that of resorcinol (Roux, 1965; King and White, 1957; Roux and Paulus, 1961). Assuming the reactivity of phenol to be 1 and that of resorcinol to be 10, the A rings have a reactivity of 8 to 9. However, because of their size and shape, the tannin molecules become immobile at a low level of condensation with formaldehyde, so that the available reactive sites are too far apart for further methylene bridge formation. The result may be an incomplete polymerization and therefore the production of a weak material. Bridging agents with longer molecules should be capable of bridging the distances that are too long for methylene bridges.

Pyrogallol or catechol B rings are by comparison unreactive and may be activated by anion formation only at relatively high pH (Roux et al., 1975). Hence the B rings do not participate in the reaction except at high pH values (pH 10), where the reactivity toward formaldehyde of the A rings is so high that the tannin-formaldehyde adhesives prepared have unacceptably short pot lives (Pizzi and Scharfetter, 1977). In general tannin adhesives practice, only the A rings are used to cross-link the network. With regard to the pH dependence of the reaction with formaldehyde, it is generally accepted that the reaction rate of wattle tannins with formaldehyde is slowest in the pH range 4.0 to 4.5 (Plomley, 1966); for pine tannins, the range is between 3.3 and 3.9.

In the reaction of polyflavonoid tannins with formaldehyde two competitive reactions are present:

1. The reaction of the aldehyde with tannin and with low-molecular-weight tannin-aldehyde condensates, which are responsible for the aldehyde consumption.
2. The liberation of aldehyde during the polycondensation reactions, probably due to the passage of unstable $-CH_2-O-CH_2-$ ether bridges initially formed to $-CH_2$-linked. The aldehyde formed is then available for further reaction.

In the case of some tannins, namely quebracho tannin, a third reaction of importance is present:

3. The simultaneous hydrolysis of some interflavonoid bonds, hence a depolymerization reaction, partly counteracting and hence slowing down hardening (Pizzi and Stephanou, 1994; Garnier et al., 2001; Pasch et al., 2001). Notwithstanding that the two major industrial polyflavonoid tannins which exist, namely mimosa and quebracho tannins, are very similar and both composed of mixed prorobinetinidins (resorcinol A rings and pyrogallol B rings, I in Fig. 2) and profisetinidins (resorcinol A rings and catechol B rings, II in Fig. 2), one could not explain this anomalous behaviour of quebracho tannin. It has now been possible to determine by both NMR (Pizzi and Stephanou, 1994) and particularly by laser desorption mass spectrometry (MALDI-TOF) for mimosa and quebracho tannins and some of their modified derivatives (Pasch et al., 2001) that: (i) mimosa tannin is predominantly composed of prorobinetinidins while quebracho is predominantly composed of profisetinidins, that (ii) mimosa tannin is heavily branched due to the presence of considerable proportions of "angular" units in its structure while quebracho tannin is almost completely linear (Pasch et al., 2001). This latter structural difference is the one which contributes to the considerable differences in viscosity of water solutions of the two tannins and which (iii) induces the interflavonoid link of quebracho to be more easily hydrolysable, due to the linear structure of this tannin, confirming NMR findings (Pizzi and Stephanou, 1994; Pasch et al., 2001) that this tannin is subject to polymerisation/depolymerisation equilibria. This also showed that the decrease of viscosity due to acid/base treatments to yield tannin adhesive intermediates does also depend in quebracho from a certain level of hydrolysis of the tannin itself and not only of the carbohydrates present in the extract. This tannin hydrolysis does not appear to occur in mimosa tannin in which the interflavonoid link is completely stable to hydrolysis.

It is interesting to note that while -CH_2-O-CH_2-ether bridged compounds have been isolated for the phenol-formaldehyde (Megson, 1958) reaction, their existance for fast-reacting phenols such as resorcinol and phloroglucinol (see Fig. 2) has been postulated, but they have not been isolated, as these two phenols have always been considered too reactive with formaldehyde. They are detected by a surge in the concentration of formaldehyde observed in kinetic curves due to methylene ether bridges decomposition (Porter, 1974).

When heated in the presence of strong mineral acids, condensed tannins are subject to two competitive reactions. One is degradative leading to lower-molecular-weight products, and the second is condensative as a result of hydrolysis of heterocyclic rings (p-hydroxybenzyl ether links)

(Roux et al., 1975). The p-hydroxybenzylcarbonium ions created condense randomly with nucleophilic centers on other tannin units to form "phlobaphenes" or "tanner's red" (Roux et al., 1975; Brown and Cummings, 1958; Brown and Cummings, 1961; Freudenberd and Alonso de Larna, 1958). Other modes of condensation (e.g., free radical coupling of B ring catechol units) cannot be excluded in the presence of atmospheric oxygen. In predominantly aqueous conditions, phlobaphene formation or formation of insoluble condensates predominates. The formation of these degradation products were also observed from grape pomace using high severity conditions (Lan et al., 2012a). These reactions, characteristic of tannins and not of synthetic phenolic resins, must be taken into account when formulating tannin adhesives.

Sulfitation of tannin is one of the oldest and most useful reactions in flavonoid chemistry. Slightly sulfited water is sometimes used to increase tannin extraction from the bark and grape pomace (Lan et al., 2012a,b) containing it. In certain types of adhesives, the total effect of sulfitation, while affording the important advantages of higher concentration of tannin phenolics in adhesive applications due to enhanced solubility and decreased viscosity, and of higher moisture retention by the tannin resins, allowing slower adhesive film dry-out, hence longer assembly times (Pizzi, 1979a), also represents a distinct disadvantage in that sulfonate groups promote sensitivity to moisture with adhesive deterioration and bad water resistance of the cured glue line even with adequate cross-linking (Pizzi, 1979b; Dalton, 1950; 1953; Parrish, 1958).

In recent years the importance of the marked colloidal nature of tannin extract solutions has come to the fore (Garnier et al., 2001; Pizzi and Stephanou, 1994; Merlin and Pizzi, 1996; Masson et al., 1996a,b; 1997; Garcia et al., 1997; Pizzi and Stephanou, 1993; Garnier et al., 2001). It is the presence of both polymeric carbohydrates in the extract as well as of the higher molecular fraction of the polyphenolic tannins which determines the colloidal state of tannin extract solutions in water (Pizzi and Stephanou, 1994a,b). The realization of the existence of the tannin in this particular state affects many of the reactions which lead to the formation and curing of tannin adhesives, to the point that reactions not thought possible in solution become, instead, not only possible but the favored ones (Pizzi and Stephanou, 1994ab), while reactions mooted to be of determinant importance, when found on models not in colloidal state, have in reality been shown to be inconsequential to tannin adhesives and their tannin applications (Pizzi, 1994; Pizzi and Stephanou, 1993).

Technology of Industrial Tannin Adhesives

The purity of vegetable tannin extracts varies considerably. Commercial wattle bark extracts normally contain 70 to 80% active phenolic ingredients.

From grape pomace, the phenolic content is lower (around 30 to 40%). The non-tannins fraction, consisting mainly of simple sugars and high-molecular-weight hydrocolloid gums, does not participate in the resin formation with formaldehyde. Sugars reduce the strength and water resistance in direct proportion to the amount added. Their effect is a mere dilution effect of the adhesive resin solids, with consequent proportional worsening of adhesive properties. The hydrocolloid gums, instead, have a much more marked effect on both original strength and water resistance of the adhesive (Pizzi, 1983; Scharfetter et al., 1977; Pizzi, 1978). If it is assumed that the non tannins in tannin extracts have a similar influence on adhesive properties, it can be expected that unfortified tannin-formaldehyde networks can achieve only 70 to 80% of the performance shown by synthetic adhesives.

In many glued wood products, the demands on the glue line are so high that unmodified tannin adhesives are unsuitable. The possibility of refining extracts has proved fruitless largely because the intimate association between the various constituents makes industrial fractionation difficult. Fortification is in many cases the most practical approach to reducing the effect of impurities. Fortification generally consists of co-polymerization of the tannin with phenolic or aminoplastic resins (Pizzi and Scharfetter, 1978; Scharfetter et al., 1977). It can be carried out during manufacture of the adhesive resin, during glue mix assembly, just before use, or during adhesive use. If added in sufficient quantity, various synthetic resins have been found effective in reducing the non tannin fraction to below 20% and in overcoming other structural problems (Pizzi and Scharfetter, 1978; Scharfetter et al., 1977). The main resins used are phenol-formaldehyde and urea-formaldehyde resols with a medium to high methylol group content. These resins can fulfill the functions of hardeners, fortifiers, or both. Generally, they are used as fortifiers between 10 and 20% of total adhesive solids, and paraformaldehyde is used as a hardener. Such an approach is the favorite one for marine-grade plywood adhesives. These fortifiers are particularly suitable for the resorcinolic types of condensed tannins, such as mimosa. They can be copolymerized with the tannins during resin manufacture, during use, or both (Pizzi, 1983; 1994; Pizzi and Scharfetter, 1978; Scharfetter et al., 1977). Copolymerization and curing are based on the condensation of the tannin with the methylol groups carried by the synthetic resin. Since tannin molecules are generally large, the rate of molecular growth in relation to the rate of linkage is high, so that tannin adhesives generally tend to have, fast gelling and curing times and shorter pot lives than those of synthetic phenolic adhesives. From the point of view of reactivity, phloroglucinol tannins such as pine tannins are much faster than mainly resorcinol tannins such as mimosa. The usual ways of slowing them down and, for instance, to lengthen adhesive pot life are:

1. To add alcohols to the adhesive mix to form hemiacetals with formaldehyde and therefore act as retardants of the tannin-formaldehyde reaction.
2. To adjust the adhesive's pH to have the required pot-life and rate of curing.
3. To use hexamine as hardener, which under the current conditions gives a very long pot life at ambient temperature but still fast curing time at higher temperatures.

The viscosity of bark extracts is strongly dependent on concentration. The viscosity increases very rapidly above a concentration of 50%. Compared to synthetic resins, tannin extracts are more viscous at the concentrations normally required in adhesives. The high viscosity of aqueous solutions of condensed tannins is due to the following causes, in order of importance:

1. *Presence of high-molecular-weight hydrocolloid gums in the tannin extract.* The viscosity is directly proportional to the amount of gums present in the extract (Pizzi, 1978; Garnier et al., 2001).
2. *Tannin-tannin, tannin-gum, and gum-gum hydrogen bonds.* Aqueous tannin extract solutions are not true solutions, but rather, colloidal suspensions in which water access to all parts of the molecules present is very slow. As a consequence, it is difficult to eliminate intermolecular hydrogen bonds by dilution only (Pizzi, 1978; Garnier et al., 2001).
3. *Presence of high-molecular-weight tannins in the extract* (Pasch et al., 2001; Pizzi, 1978; Garnier et al., 2001).

The high viscosity of tannin extracts solutions has also been correlated with the proportion of very high molecular weight tannins present in the extract. This effect is not well defined. In most adhesive applications such as in plywood adhesives, the viscosity is not critical and can be manipulated by dilution.

In the case of particleboard adhesives decrease of viscosity is, instead, an important prerequisite. When reacted with formaldehyde, unmodified condensed tannins give adhesives having characteristics that do not suit particleboard manufacture: namely, high viscosity, low strength, and poor water resistance. The most commonly used process to eliminate these disadvantages in the preparation of tannin-based particleboard adhesives consists of a series of subsequent acid and alkaline treatment of the tannin extract, causing hydrolysis of the gums to simple sugars and some tannin structural changes, thus improving viscosity, strength, and water resistance of the unfortified tannin-formaldehyde adhesive (Pizzi, 1983; 1994). Furthermore, such treatments may cause partial rearrangement of the flavonoid molecules that causes liberation of some resorcinol *in situ* in the tannin, rendering it more reactive, allowing better cross-linking with formaldehyde, and, ultimately, yielding an adhesive which without addition of any fortifier resins gives truly excellent performance for exterior-grade particleboard (Pizzi, 1983; 1994; Calleri et al., 1989).

This modification cannot be carried out too extensively, but only to a limited extent, to avoid precipitation of the tannin from solution by the formation of "phlobaphenes".

Particular gluing and pressing techniques have been developed for tannin particleboard adhesives (Pizzi, 1978b; 1979a) to achieve pressing times much faster than those traditionally obtained with synthetic phenol-formaldehyde adhesives, although recent advances in synthetic phenol-formaldehyde resins have markedly limited such an advantage (Pizzi, 1994; Zhao et al., 1999). Pressing times of 7 s/min of panel thickness have been achieved and press times of 9 s/min at 190 to 200°C press temperature are in daily operation: these are pressing times becoming comparable to what is obtainable with urea-formaldehyde or melamine-formaldehyde resins at the same pressing temperatures. The success of these simple types of particleboard adhesives relies heavily on industrial application technology rather than just on the preparation technology of the adhesive itself (Pizzi, 1978a; 1978b; Pizzi et al., 1994). A considerable advantage is the much higher moisture content of the resinated chips acceptable with these adhesives than with any of the synthetic phenolic and aminoresin adhesives. In the case of wood particleboard and of oriented strandboard (OSB) panels the technology so developed allows hot-pressing at moisture contents of around 24% against values of 12% for traditional synthetic adhesives, and presents other advantages as well (Pizzi, 1978a; Pizzi et al., 1994; Pichelin et al., 2001). The results obtained with industrial unfortified tannin-formaldehyde based panels before and after 2 h boiling are given in Table 1.

The best adhesive formulation for phloroglucinolic tannins such as pine tannin extracts was initially, a comparatively new adhesive formulation that is also capable of giving excellent results when using resorcinolic tannins such as a wattle tannin extract (Pizzi, 1980; 1981; 1982; Pizzi and Walton, 1992). The adhesive gluemix consisted only of a mix of an unmodified tannin extract 50% solution to which has been added paraformaldehyde and polymeric nonemulsifiable 4,4′-diphenylmethane diisocyanate (commercial pMDI) (Pizzi, 1980; 1981; 1982; Pizzi and Walton, 1992). The proportion of tannin extract solids to pMDI can be as high as 70:30 based

Table 1. Unfortified tanninformaldehyde adhesives obtained by acid-alkali, treatment, for exterior grade particleboard: example of industrial board results.

| Density | Swelling after a 2 h boil | | IB[a] dry | IB[a] after 2 h boil | Cyclic test V313 |
| | Measured wet | Measured dry | | | |
(g/cm³)	(%)	(%)	(kg/cm²)	(kg/cm²)	(%)
0.700	11.0	0.0	13.0	9.0	3.0

[a]IB = Internal Bond strength

on mass, but can be much lower in pMDI content. This adhesive is based on the following peculiar mechanism, by which the MDI, in water, is hardly deactivated to polyureas (Pizzi and Walton, 1992). Today, after 10 years industrial production of particleboard in Chile using tannin adhesives the use of pMDI has been phased out, the panels giving acceptable results for particleboard and MDF panels with just the pine-tannin-based adhesive (Valenzuela et al., 2012).

The properties of the particleboard manufactured with this system using pine tannin adhesives are listed in Table 2. The results obtainable with this system are then quite good and not too different from the result obtainable with some of the other tannin adhesives already described. In the case of phloroglucinolic tannin extracts being used, no pH adjustment of the solution is needed. One point that was given close consideration is the deactivating effect of water on the isocyanate group of pMDI. It has been found that the amount of deactivation by water of this group when in a concentrated solution (50% or over) of a phenol is much lower than previously thought (Pizzi, 1980; 1981; 1982; Pizzi and Walton, 1992). This is the reason that aqueous tannin extract solutions and pMDI can be reacted without substantial pMDI deactivation by the water present.

Grape tannins, extracted from pomace using different bases for the extraction (NaOH, Na_2CO_3 and $NaHCO_3$) were used to press one layer particleboard (Lan et al., 2011a,b). In this study, two different formulations (the nonfortified tannin adhesive and fortified with diisocyanate adhesive) were analyzed. The results are shown in Table 3. The IB strengths demonstrate that the nature of the extraction reagent (NaOH, Na_2CO_3 or $NaHCO_3$) greatly impacts the value of panel IB strength. It can be seen that formulations using NaOH for the tannins extraction yielded low panel performances and did not satisfy the IB strength results required (0.35 MPa) even with addition of pMDI in the formulation. In the same way, the utilization of $NaHCO_3$ for the tannins extraction led to an adhesive bearing poor properties; however, with addition of 20% of pMDI in the formulation, the adhesive yielded a good internal bond strength (IB = 0.36). The best results were obtained from Na_2CO_3 extracts and very good internal bond strength results (IB = 0.45 and 0.68 respectively) were obtained. A resin formulation composed of 95% of pomace extracts solution, 5% formaldehyde and without pMDI, yielded good internal bond strength which passed

Table 2. Properties of particleboard manufactured using pine tannin adhesives.

| Density (g/cm^3) | Swelling after a 2-h boil | | IB^a dry (kg/cm^2) | IB^a after 2h boil (kg/cm^2) |
	Measured wet (%)	Measured dry (%)		
0.690	15.0	4.3	8.4	4.3

^{a}IB = Internal Bond strength

Table 3. Extraction of tannins from grape pomace, adhesive formulations and particle board testing.

Tannins extraction conditions	Tannins/H_2CO/ pMDI	Density Kg/m^3	IB^a (MPa)
NaOH	20/1/0	696.3	0.23
NaOH	20/1/5	693.7	0.28
NaHCO$_3$	20/1/0	692.6	0.26
NaHCO$_3$	20/1/5	708.5	0.36
Na$_2$CO$_3$	20/1/0	706.1	0.45
Na$_2$CO$_3$	20/1/5	710.4	0.68

^{a}IB = Internal Bond strength

Figure 3. Particle boards before sanding, produced with tannins extracted from grape pomace.

relevant international standard specifications for interior-grade panels (IB > 0.35 MPa) (Lan et al., 2011a,b; 2012a,b).

The quest to decrease or completely eliminate formaldehyde emission from wood panels bonded with adhesives, although not really necessary in tannin adhesives due to their very low emission (as most phenolic adhesives), has nonetheless promoted research to further improve formaldehyde emission. This has centered on two lines of investigations: (i) tannins autocondensation, and (ii) the use of a hardener that does not emit at all, simply because no aldehyde has been added to the tannin. Methylolated nitroparaffins, and in particular the simpler and least expensive exponent of their class, namely trishydroxymethyl nitromethane (Valenzuela et al., 2012; Trosa and Pizzi, 2001), function well as hardeners of a variety of tannin-based adhesives while affording considerable side advantages to the adhesive and to the bonded wood joint. In panel products such as particleboard, medium density fiberboard and plywood, the joint performance which is obtained is of the exterior/marine grade type, while a very advantageous and very considerable lengthening in glue-mix pot-life is obtained. Furthermore, the use of this hardener is coupled with a marked reduction in formaldehyde emission. This is so low to be just limited to the formaldehyde generated by heating of the wood substrate (and slightly less, thus functioning as a mild depressant of emission from the wood itself). Furthermore, trishydroxymethyl nitromethane can be mixed in any proportion with traditional formaldehyde-based hardeners for tannin adhesives, its proportional substitution of such hardeners inducing a proportionally marked decrease in the formaldehyde emission of the wood panel without affecting the exterior/marine grade performance of the panel. Medium density fibreboard (MDF) industrial plant trials confirmed all the properties reported above and the trial conditions and results are reported (Trosa, 1999; Trosa and Pizzi, 2001). A cheaper but equally effective alternative to hydroxymethylated nitroparaffins is the use of hexamine as tannin hardener. This sometimes causes problems of early agglomeration in some tannins (Pichelin et al., 1999) and a better solution proposed to overcame such a problem was to use a mix of formaldehyde coupled with an ammonium salt.

Natural Fibres/Natural Matrices High Tech Laminates

Biocomposites are and have been an area of growing interest and a subject of active research for quite sometime now. This is due to both environmental concerns as well as to the foreseen future scarcity of oil and oil-derived products. Biocomposites using natural fibres and oil-derived polymer matrices have now existed and have been available commercially for quite sometime. Thus, composites from natural fibres plus polypropylene and other oil-derived thermoplastic matrices for car doors interiors and other applications are well known and are used, although not as extensively as could be wished. However, for the same type of applications, composites also using natural matrices, while still presenting high performance, are talked about, but in reality have not been developed nor commercialised. This is due to the difficulty in finding matrices of natural origin capable of imparting all the required performance to the resulting composites. For example, starch-bonded biocomposites suffer from poor water and moisture resistance, etc.

Composites of good performance formed from non-woven mats of flax and hemp fibres and natural resin matrices have been prepared (Fig. 4).

Both higher density thin composites as well as lower density thicker composites have been prepared (Pizzi et al., 2009). Two natural matrices types were used: (i) commercial mimosa flavonoid tannin extract with 5% hexamine added as hardener, and (ii) a mix of mimosa tannin+hexamine with glyoxalated organosolv lignin of low molecular weight, these two resins mixed 50/50 by solids content weight. The composites prepared were tested for modulus of elasticity (MOE) in bending and in tension and for maximum breaking strength in tension. Some of the mats were corona treated and the optimum length of corona treatment determined to improve the composites MOEs and breaking strength. These were related to the morphology of the treated fibre. Thermomechanical analysis, Brinell surface hardness and contact angle tests were also carried out with good results. The composites made with the mix of tannin and lignin resins as a matrix remained thermoplastic after a first pressing. The flat sheets prepared after the first pressing were then thermoformed into the shape wanted.

The composites prepared were of two types: (i) thin high density composites (1.2 mm thickness) and (ii) lower density 8 mm thick composites from non-woven mats of flax and hemp fibres and tanninn+hexamine alone or tannin+hexamine/glyoxalated lignin mixes. These composites were prepared for different potential applications. The first for covering other materials and the second to be tested as the back rebatable flap in cars. The latter are generally made in polypropylene. The average modulus of elasticity (MOE) in bending and in tension as well as the breaking load results which were obtained for the composites prepared were simply

Figure 4. Example of hemp non-woven mat and of the high resin composite produced by impregnating it with a tannin based resin (Lacoste et al., 2015).

exceptional (Pizzi et al., 2009), with resins load of around 30%–35%. More recent results have considerably improved the composite performance by achieving resin load contents equal to or in excess of 50% by weight of the composite. An example of a molded product, a skateboard, made exclusively from this composite fibrous material, of 98% natural content, is shown in Fig. 5.

Figure 5. Skateboard composed exclusively of a non-woven hemp fibre mat impregnated with a tannin+hexamine resin, pressed at 190°C.

Tannins Foams

Tannin-based rigid foams were developed for the first time in 1994 (Meikleham and Pizzi, 1994). They have been shown to be good thermal insulating materials (Tondi et al., 2008; Tondi and Pizzi, 2009; Tondi et al., 2009a,b,c). They have been shown to be apt to use for foaming in the cavity of hollow-core wooden doors and other wooden cavities as a thermal insulation material of good quality and based at the 95% level on natural materials. Their thermal insulation capacity is comparable to that of totally synthetic, oil-derived foams, such as polyurethanes, but with the advantage that they do not burn and thus do not emit toxic gases on burning (Celzard et al., 2011). Exposure to the weather of wooden hollow core structures filled with tannin-based foams confirm that the wood in contact with the foam is not affected by the acidity of the foam due to the already previously demonstrated incorporation of the acid hardener used by coreaction with the hardened polymer network. Moreover, alkaline-based foams have also been developed (Basso et al., 2014a,b,c).

Extensive physical characterisation tests have been done on tannin-based rigid foams (Tondi et al., 2009a,b,c; Li et al., 2012). The switch to the use of phosphoric acid as a catalyst has allowed the production of foams which not only do not burn, but which do not glow red at all. Time of resistance to flame increased approximately 5 times. Resistance to strong acid and strong alkaline environment has been tested with good results. Other properties such as solvent absorption, permeability and compression strength have shown that these foams have high affinity for water and they are anisotropic material. Hence tannin-based rigid foams have different behaviour in mechanical and permeability tests. Pycnometry test determines the skeletal density (1.55 g/cm^3) of these foams and physical derivative properties such as porosity and surface area have been calculated. The properties reported in this paper make tannin-based rigid foams, constituted by 95% natural material, a suitable foam for industrial insulation applications. Furthermore, when an aldehyde was needed, alternative aldehydes other than formaldehyde were successfully used (Lacoste et al., 2013).

Open cells structure tannin-furanic foams prepared using two different types of tannins have been shown to give good sound absorption/acoustic insulation characteristics at frequencies equal and higher than 1000 Hz with coefficients of acoustic absorption of 0.8–1.0. Their acoustic absorption coefficient was lower at 0.35–0.4 at lower frequencies (250–500 Hz). The more open cells floral foams of higher porosity, lower tortuosity, and lower resistivity have been shown to perform slightly better than the other types tested (Lacoste et al., 2015).

The foaming dynamics of several different tannin-based foams have been followed by simultaneously monitoring the variation of temperature, foam rising rate, internal foam pressure and dielectric polarisation, this latter being a direct measure of setting and curing of a thermosetting foam. This approach to foam monitoring described well the process and possible characteristics of the foam prepared. It constitutes an invaluable tool for foam formulation. For self-blowing foams it was shown that coordination between the foaming action and resin cross-linking must be respected. Thus, the findings indicate that to prepare a proper hardened foam, cross-linking must start at the same time the temperature and the internal pressure reach their maximum, and not before (Basso et al., 2013a,b).

One question often asked is if materials such as condensed flavonoid tannins can be used for polyurethane foams rather than just phenolic type foams. This can and has been achieved by several routes: (i) by oxypropylating or oxybutylating the hydroxy groups of the tannin and reacting this afterwards with a polyisocyanate (Basso et al., 2013c), or (ii) by reacting the same mixture in which an aldehyde such as glyoxal can react with the flavonoid and allowing the polyisocyanate to react simultaneosly with the hydroxybenzyl alcohol formed to form the polyurethanes (Basso et al., 2014a,b). This latter route is the easiest one to follow and has been already tried successfully industrially (Basso et al., 2014a) and is based on all kinetic findings already proven and used in wood adhesives practice since the early 1990s (Pizzi and Walton, 1992).

Conclusion and Prospects

This chapter has focused on the condensed tannins industrial extraction methods and utilizations. It also includes the results and perspectives of the recent studies in these fields. Although condensed tannins have been industrially used for very long time, with the turn of the 21st century a revival is obseved with an extension of their industrial use to green materials like adhesives, resins for composite and foams. The future of the green materials discussed in this chapter should accelerate for economic and environmental reasons.

Keywords: Condensed tannins, Adhesive, Resin, Particle board, Composite, Foam, Bark, Grape pomace, Wood industry

References

Basso, M.C., S. Giovando, A. Pizzi, A. Celzard and V. Fierro. 2013a. Tannin/furanic foams without blowing agents and formaldehyde. Ind. Crops Prods. 49: 17–22.

Basso, M.C., A. Pizzi and A. Celzard. 2013b. Influence of formulation on the dynamics of preparation of tannin-based foams. Ind. Crops Prod. 51: 396–400.

Basso, M.C., S. Giovando, A. Pizzi and A. Celzard. 2013c. Tannin foam surfactant effects. BioResources 8(4): 5807–5816.

Basso, M.C., S. Giovando, A. Pizzi, M.C. Lagel and A. Celzard. 2014a. Alkaline tannin rigid foams. J. Renewable Mat. 2(3): 182–185.

Basso, M.C., A. Pizzi, C. Lacoste, L. Delmotte, F.-A. Al-Marzouki, S. Abdalla and A. Celzard. 2014b. MALDI-TOF and ^{13}C NMR analysis of tannin-furanic-polyurethane foams adapted for industrial continuous lines application. Polymers 6: 2985–3004.

Basso, M.C., S. Giovando, A. Pizzi, H. Pach, N. Pretorius, L. Delmotte and A. Celzard. 2014c. Flexible-elastic copolymerized polyurethane-tannin foams. J. Appl. Polym. Sci. 131(13): DOI 10.1002/app.40499.

Brahim, M., F. Gambier and N. Brosse. 2014. Optimization of polyphenols extraction from grape residues in water medium. Ind. Crops Prod. 52: 18–22.

Brown, R. and W. Cummings. 1958. Polymerisation of flavans. Part II. The condensation of 4'-methoxyflavan with phenols. J. Chem. Soc. 4302–4305.

Brown, R. and W. Cummings. 1961. Polymerisation of flavans. Part IV. The condensation of flavan-4-ols with phenols. J. Chem. Soc. 3677–3682.

Calleri, L. 1989. Le Fabbriche Italiane de Estratto di Castagno, Silva, S. Michele Mondovi (CN), Italy CARB – California Air Resources Board, Composite Wood Products Public Workshop, June 20, 2006.

Celzard, A., V. Fierro, G. Amaral-Labat, A. Pizzi and J. Torero. 2011. Flammability assessment of tannin-based cellular materials. Polymer Degrad. Stab. 96: 477–482.

Clark-Lewis, J.W. and D.G. Roux. 1959. Natural occurrence of enantiomorphous leucoanthocyanidian: (+)-mollisacacidin (gleditsin) and quebracho(–)-leucofisetinidin. J. Chem. Soc. 1402–1406.

Dalton, L.K. 1950. Tannin-formaldehyde resins as adhesives for wood. Aust. J. Appl. Sci. 1: 54–70.

Dalton, L.K. 1953. Resins from sulphited tannins as adhesives for wood. Aust. J. Appl. Sci. 4: 54–70.

Fechtal, M. and B. Riedl. 1993. Use of eucalyptus and *Acacia mollissima* bark extract-formaldehyde adhesives in particleboard manufacture. Holzforschung. 47: 349–357.

Freudenberg, K. and J.M. Alonso de Larna. 1958. Annalen 612: 78.

Garcia, R., A. Pizzi and A. Merlin. 1997. Ionic polycondensation effects on the radical autocondensation of polyflavonoid tannins: An ESR study. J. Appl. Polymer Sci. 65: 2623–2633.

Garnier, S., A. Pizzi, O.C. Vorster and L. Halasz. 2001. Comparative rheological characteristics of industrial polyflavonoid tannin extracts. J. Appl. Polymer Sci. 81(7): 1634–1642.

Hemingway, R.W. and G.W. McGraw. 1976. Appl. Polymer Symp. 28.

JIS, Japanese Standards Association. 1994. Particleboard. JIS A 5908, Tokyo, Japan.

King, H.G.C. and T. White. 1957. Tannins and polyphenols of Schinopsis (quebracho) species: their genesis and interrelation. J. Soc. Leather Traders' Chem. 41: 368–384.

King, H.G.C., T. White and R.B. Huges. 1961. The occurrence of 2-benzyl-2-hydroxycoumaran-3-ones in quebracho tannin extract. J. Chem. Soc. 3234–3239.

Lacoste, C., M.C. Basso, A. Pizzi, M.-P. Laborie, D. Garcia and A. Celzard. 2013. Bioresourced pine tannin/furanic foams with glyoxal and glutaraldehyde. Ind. Crops Prod. 45(2): 401–405.

Lacoste, C., M.C. Basso, A. Pizzi, A. Celzard, E. Ella Bang, N. Gallon, B. Charrier, M.-P. Laborie and D. Garcia. 2015. Pine (*P. pinaster*) and quebracho (*S. lorentzii*) tannin-based foams as green acoustic absorbers. Ind. Crops Prod. 67: 70–73.

Lan, P., N. Brosse, L. Chrusciel, P. Navarette and A. Pizzi. 2011a. Extraction of condensed tannins from grape pomace for use as wood adhesives. Ind. Crops Prod. 33: 253–257.

Lan, P., R. El Hage, A. Pizzi, Z.-D. Guo and N. Brosse. 2011b. Extraction of polyphenolics from lignocellulosic materials and agricultural byproducts for the formulation of resin for wood adhesives. J. Biobased Mater. Bioener. 5: 460–465.

Lan, P., A. Pizzi, Z.-D. Guo and N. Brosse. 2011c. Condensed tannins extraction from grape pomace: Characterization and utilization as wood adhesives for wood particleboard. Ind. Crop. Prod. 34: 907–914.

Lan, P., A. Pizzi, Z.-D. Guo and N. Brosse. 2012a. Condensed tannins from grape pomace: Characterization by FTIR and MALDI TOF and production of environment friendly wood adhesive. Ind. Crops Prod. 40: 13–20.

Lan, P., F. Gambier, A. Pizzi, Z.-D. Guo and N. Brosse. 2012b. Wood adhesives from agricultural byproducts: lignins and tannins for the elaboration of particleboards. Cell Chem. Technol. 46(7-8): 457–462.

Li, X., M.C. Basso, V. Fierro, A. Pizzi and A. Celzard. 2012. Chemical modification of tannin/furanic rigid foams by isocyanates and polyurethanes. Maderas. 14(3): 257–265.

Masson, E., A. Merlin and A. Pizzi. 1996a. Comparative kinetics of induced radical autocondensation of polyflavonoid tannins. I. Modified and nonmodified tannins. J. Appl. Polymer Sci. 60: 263–269.

Masson, E., A. Pizzi and A. Merlin. 1996b. Comparative kinetics of the induced radical autocondensation of polyflavonoid tannins. III. Micellar reactions vs. cellulose surface catalysis. J. Appl. Polymer Sci. 60: 1655–1664.

Masson, E., A. Pizzi and A. Merlin. 1997. Comparative kinetics of the induced radical autocondensation of polyflavonoid tannins. II. Flavonoid units effects. J. Appl. Polymer Sci. 64: 243–265.

Megson, N.J.L. 1958. Phenolic Resins Chemistry, Butterworth, Sevenoaks, Kent, England.

Meikleham, N. and A. Pizzi. 1994. Acid and alkali-setting tannin-based rigid foams. J. Appl. Polymer Sci. 53: 1547–1556.

Merlin, A. and A. Pizzi. 1996. An ESR study of the silica-induced autocondensation of polyflavonoid tannins. J. Appl. Polymer Sci. 59: 945–952.

Parrish, J.R. 1958. African Forest Ass. 32: 26.

Pasch, H., A. Pizzi and K. Rode. 2001. MALDI-TOF Mass Spectrometry of Polyflavonoid Tannins. Polymer. 42(18): 7531.

Pichelin, F., C. Kamoun and A. Pizzi. 1999. Hexamine hardener behaviour: effects on wood glueing, tannin and other wood adhesives. Holz Roh Werkstoff. 57(5): 305–317.

Pichelin, F., A. Pizzi, A. Frühwald and P. Triboulot. 2001. A shear test for structural adhesives used in the consolidation of old timber. Holz Roh Werkstoff. 59(4): 256–265.

Pizzi, A. and H.O. Scharfetter. 1977. Warm-setting wattle adhesives for wood laminating, CSIR Special Report HOUT 138, Pretoria, South Africa.

Pizzi, A. 1978a. Wattle-based adhesives for particleboard. Forest Products J. 28(12): 42–47.

Pizzi, A. 1978b. Utilizing wattle-based adhesives in making particleboard. Adhesives Age. 21(9): 32–33.

Pizzi, A. and H.O. Scharfetter. 1978. The chemistry and development of tannin-based weather- and boil-proof cold-setting and fast-setting adhesives for wood. J. Appl. Polymer Sci. 22: 1945–1954.

Pizzi, A. 1979a. Glue blenders effect on particleboard using wattle tannin adhesives. Holzforschung Holzverwertung. 31(4): 85–86.

Pizzi, A. 1979b. Sulphited tannins for exterior wood adhesives. Colloid Polymer Sci. 257: 37–40.

Pizzi, A. 1980. Exterior wood adhesives by MDI crosslinking of polyflavonoid tannin B rings. J. Appl. Polymer Sci. 25: 2123–2127.

Pizzi, A. 1981. A universal formulation for tannin adhesives for particleboard. J. Macromol. Sci. Chem. A 16(7): 1243–1250.

Pizzi, A. 1982. Pine tannin adhesives for particleboard. Holz Roh Werkstoff. 40: 293.

Pizzi, A. 1983. In: A. Pizzi (ed.). Wood Adhesives Chemistry and Technology, Vol. I. Marcel Dekker, New York, Chap. 4.

Pizzi, A. and T. Walton. 1992. Non-emulsifiable, water-based diisocyanate adhesives. Holzforschung. 46: 541.

Pizzi, A., J. Valenzuela and C. Westermeyer. 1993. Non-emulsifiables, water-based, diisocyanate adhesives for exterior plywood, Part 2: industrial application. Holzforschung. 47: 68–71.

Pizzi, A. and A. Stephanou. 1993. A comparative C[13] NMR study of polyflavonoid tannin extracts for phenolic polycondensates. J. Appl. Polymer Sci. 50: 2105–2113.

Pizzi, A. 1994. Advanced Wood Adhesive Technology. Marcel Dekker, New York.

Pizzi, A. and A. Stephanou. 1994a. A ^{13}C NMR study of polyflavonoid tannin adhesive intermediates. I. Noncolloidal performance determining rearrangements. J. Appl. Polymer Sci. 51: 2109–2124.

Pizzi, A. and A. Stephanou. 1994b. A ^{13}C NMR study of polyflavonoid tannin adhesive intermediates. II. Colloidal state reactions. J. Appl. Polymer Sci. 51: 2125–2130.

Pizzi, A. and A. Stephanou. 1994. Phenol-formaldehyde wood adhesives under very alkaline conditions. Part II. Esters curing acceleration, its mechanism and applied results. Holzforschung. 48(2): 150.

Pizzi, A., J. Valenzuela and C. Westermeyer. 1994. Exterior plywood resins formulated from Pinus pinaster bark. Holz Roh Werkstoff. 52: 311–315.

Pizzi, A. 2003. Natural phenolic adhesives I: Tannin. Handbook of Adhesive Technology (2nd Edition, Revised and Expanded). 573–587.

Pizzi, A. 2006. Recent developments in eco-efficient bio-based adhesives for wood bonding: opportunities and issues. J. Adhesion Sci. Technol. 20(8): 829–846.

Pizzi, A., R. Kueny, F. Lecoanet, B. Massetau, D. Carpentier, A. Krebs, F. Loiseau, S. Molina and M. Ragoubi. 2009. Tannin-based rigid foams: Characterization and modification. Ind. Crop. Prod. 30: 235–240.

Plomley, K.F. 1966. Paper 39, Division of Australian Forest Products Technology.

Porter, L.J. 1974. Extractives of Pinus radiata bark. N.Z.J. Sci. 17: 213–218.

Rondeau, P., F. Gambier, F. Jolibert and N. Brosse. 2013. Compositions and chemical variability of grape pomaces from French vineyard. Ind. Crops Prod. 43: 251–254.

Rossouw, D.T., A. Pizzi and G. McGillivray. 1980. The kinetics of condensation of phenolics polyflavonoid tannins with aldehydes. J. Polymer Sci. Chem. Ed. 18: 3323–3343.

Roux, D.G. and E. Paulus. 1961. Condensed tannins. 8. The isolation and distribution of interrelated heartwood components of Schinopsis spp. Biochem. J. 78: 785–789.

Roux, D.G. 1965. Modern Applications of Mimosa Extract, Leather Industries Research Institute, Grahamstown, South Africa. 34–41.

Roux, D.G., D. Ferreira, H.K.L. Hundt and E. Malan. 1975. Structure, stereochemistry and reactivity of natural condensed tannins as basis for their extended industrial utilization. Appl. Polymer Symp. 28: 335.

Scharfetter, O., A. Pizzi and D.T. Rossouw. 1977. Laminating cold-set adhesives based on tannins, IUFRO Conference on Wood Gluing, Merida, Venezuela, pp. 335–353.

Sealy-Fisher, V.J. and A. Pizzi. 1992. Increased pine tannins extraction and wood adhesives development by phlobaphenes minimization. Holz Roh Werkstof. 50: 212–220.

Shrikhande, A.J. 2000. Wine by-products with health benefits. Food Res. Int. 33: 469–474.

Thompson, D. and A. Pizzi. 1995. Simple ^{13}C-NMR methods for quantitative determinations of polyflavonoid tannin characteristics. J. Appl. Polym. Sci. 55: 107–112.

Tondi, G., A. Pizzi and R. Olives. 2008. Natural tannin-based rigid foams as insulation for doors and wall panels. Maderas Ciencia y Tecnologia. 10(3): 219–227.

Tondi, G. and A. Pizzi. 2009. Tannin-based rigid foams: Characterization and modification. Ind. Crop. Prod. 29(1-2): 356–363.

Tondi, G., V. Fierro, A. Pizzi and A. Celzard. 2009a. Tannin-based carbon foam. Carbon 47(6): 1480–1492.

Tondi, G., V. Fierro, A. Pizzi and A. Celzard. 2009b. Erratum to "Tannin-based carbon foams". Carbon 47: 2761.

Tondi, G., W. Zhao, A. Pizzi, V. Fierro and A. Celzard. 2009c. Tannin-based rigid foams: a survey of chemical and physical properties. Bioresour. Technol. 100: 5162–5169.

Trosa, A. 1999. Doctoral thesis, University Henri Poincaré – Nancy 1, Nancy, France.

Trosa, A. and A. Pizzi. 2001. A no-aldehyde emission hardener for tannin-based wood adhesives for exterior panels. Holz Roh Werkstoff. 56(4): 229–233.

Valenzuela, J., E. Von Leyser, A. Pizzi, C. Westermeyer and B. Gorrini. 2012. Industrial production of pine tannin-bonded particleboard and MDF. Eur. J. Wood Wood Prod. 70(5): 735–740.

Zhao, C., A. Pizzi and S. Garnier. 1999. Fast advancement and hardening acceleration of low-condensation alkaline PF resins by esters and copolymerized urea. J. Appl. Polymer Sci. 74: 359–378.

Zocca, F., G. Lomolino, A. Curioni, P. Spettoli and A. Lante. 2007. Detection of pectinmethylesterase activity in presence of methanol during grape pomace storage. Food Chem. 102: 59–65.

9

Utilization of Citric Acid in Wood Bonding

Zhongqi He[1],* and *Kenji Umemura*[2]

ABSTRACT

Citric acid (CA) is a weak organic acid. It exists most notably in citrus fruits, after which it is named. As a commodity chemical, CA is produced on a large scale by fermentation. In this chapter, we first briefly review the applied research and methods for commercial production of CA. Then we synthesize and discuss the recent research on CA as a wood adhesive additive. For the purpose of wood bonding, CA can serve as a minor additive, functioning as a cross-linking catalyst, a cross-linking agent, or a dispersing agent to improve the adhesive strength and/or operational properties (such as viscosity). CA, with or without sucrose, can also be a major adhesive component in manufacturing fiber and particleboards with a variety of lignocellulose raw materials. The formation of ester linkages between CA and lignocellulose molecules (raw materials and/ or sucrose) developed adhesiveness and contributed to the good physical properties of the particleboards made with CA. These laboratory research accomplishments demonstrate that application of CA as a renewable natural adhesive compound for wood and lignocellulosic composites is possible. Further effort is on optimizing practical manufacturing conditions for specific types of wood-based molding and medium or low density lignocellulosic particleboards, using CA as an adhesive additive.

[1] Southern Regional Research Center, USDA Agricultural Research Service, 1100 Robert E. Lee Blvd., New Orleans, LA 70124, USA.

[2] Laboratory of Sustainable Materials, Research Institute for Sustainable Humanosphere, Kyoto University, Japan.

* Corresponding author: Zhongqi.He@ars.usda.gov

Introduction

Citric acid (CA, 2-hydroxy-1,2,3-propanetricarboxylic acid, β-hydroxytricarballyic acid, 3-carboxy-3-hydroxypentanedioic acid, 3-carboxy-3-hydroxypentane-1,5-dioic acid) is a weak organic acid with the formula $C_6H_8O_7$ (Fig. 1). CA is widely distributed in plants, and in animal tissues and fluids as it is a primary metabolic product and is formed in the tricarboxylic acid cycle. It exists most notably in citrus fruits after which it is named. Lemons and limes have particularly high concentrations of the acid; it can constitute as much as 8% of the dry weight of these fruits (about 47 g/L in the juices) (Penniston et al., 2008). The CA concentrations in citrus fruits range from 0.005 mol/L for oranges and grapefruits to 0.30 mol/L in lemons and limes. Within species, these values vary depending on the cultivar and the growth circumstances. As a commodity chemical, CA is produced on a large scale by fermentation (Angumeenal and Venkappayya, 2013). In 2007, worldwide annual production was approximately 1.6 x 10⁶ Mg (Berovic and Legisa, 2007).

At room temperature, CA is a white hygroscopic crystalline powder either in an anhydrous form or as a monohydrate. The anhydrous form crystallizes from hot water, while the monohydrate forms when citric acid is crystallized from cold water. Monohydrate crystals lose water in dry air or when heated at about 40 to 50°C. When heated above 175°C, it decomposes through the loss of carbon dioxide and water. CA is a slightly stronger acid than typical carboxylic acids because the anion can be stabilized by intramolecular hydrogen-bonding from other protic groups on CA. The pH values of 1 M, 0.5 M, and 0.1 M CA are around 1.6, 1.7, and 2.1, respectively. Thus, CA is widely used as an acidulant in beverages, confectionery, effervescent salts, in pharmaceutical syrups, elixirs, in effervescent powders and tablets, to adjust the pH of foods, and as synergistic antioxidant in processing cheese. It is also used in beverages, jellies, jams, preserves and candy to provide tartness. In the manufacture of alkyd resins, its esterized form is used as plasticizer and foam inhibitor. In analytical chemistry, it is used to determine citrate-soluble phosphate, and as a reagent for albumin, mucin, glucose and bile pigment (Windholtz, 1983). CA is an excellent chelating agent, binding metals. It is used to remove

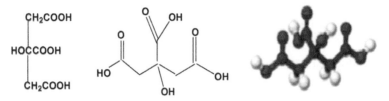

Figure 1. The formula and structure of citric acid.

limescale from boilers and evaporators. It can be used to soften water, which makes it useful in soaps and laundry detergents. By chelating the metals in hard water, it lets these cleaners produce foam and work better without need for water softening. In the industry, it is used to dissolve rust from steel (Verhoff and Bauweleers, 2014). In materials science, the citrate-gel method is a process similar to the sol-gel method, which is a method for producing solid materials from small molecules (Zayat and Levy, 2000). During the synthetic process, metal salts or alkoxides are introduced into a citric acid solution. The formation of citric complexes is believed to balance the difference in individual behaviour of ions in solution, which results in a better distribution of ions and prevents the separation of components at later process stages. The polycondensation of ethylene glycol and CA starts above 100°C, resulting in polymer citrate gel formation.

Production of Citric Acid

CA was first isolated in 1784 from lemon juice in the crystalline form (Verhoff and Bauweleers, 2014). It was first commercially-produced in England around 1826 from imported Italian lemons (lemons contain 7–9% CA). Lemon juice remained the commercial source of CA until 1919, when the first industrial process using *Aspergillus niger* started in Belgium. Currently, the extraction of CA is limited to some small factories in Mexico and Africa (Max et al., 2010). CA could be chemically synthesized from glycerol, symmetrical dicloroacetone, and other materials. However, chemical methods have so far proved uncompetitive. About 99% of world production of CA occurs via microbial processes, which can be carried out using surface or submerged cultures (Max et al., 2010).

Whereas many microorganisms can be employed to produce CA, *A. niger* is the main industrial producer. The theoretical yield is 112 g of anhydrous CA per 100 g of sucrose. However, in practice, due to losses during trophophase, the yield of CA often does not exceed 70% of the theoretical yield on carbon source (Max et al., 2010). Thus, despite a long and successful history of producing CA, effective conversion of low cost substrates to CA by fermentation is still an active research area (Angumeenal and Venkappayya, 2013).

Fermentation of a substrate to CA is directly related to the quality and quantity of the sugar source. In general, only the sugars that are quickly assimilated by the microorganism allow high final yield of CA. Among the synthetic substrates tried so far, sucrose was the best candidate followed by glucose, fructose and galactose (Vandenberghe et al., 1999). To make CA production cost effective, numerous renewable plant and fruit biomass rich in sugar content are studied as substrates.

Molasses is the effluent from sugar industry and is the non crystallisable residue remaining after sucrose isolation. It has been considered as a waste product or byproduct of the sugar industry due to its low price compared to other sugar sources and the presence of minerals, organic and inorganic compounds. Molasses needs pretreatment (such as by ammonium oxalate followed by treatment with diammonium phosphate) as the organic and inorganic components present in molasses may inhibit the fermentation processes (Angumeenal and Venkappayya, 2005). Other inexpensive agro-industrial wastes studied and/or used as substrates in fermentation include apple pomace, carob pod, carrot waste, coffee husk, corn cobs, grape pomace, kiwi fruit peel, kumara, orange waste, date syrup, pineapple waste, banana extract, potato chips waste and pumpkin (Angumeenal and Venkappayya, 2013, and references therein).

Finally, CA recovery from the fermentation media involves the precipitation of oxalic acid, possibly in the form of calcium oxalate at low pH, and subsequent separation from the medium containing the mycelium through rotating filters or centrifuges. CA is then precipitated at pH 7.2 and 70–90°C and recovered by filtration and drying. The crude product may be dissolved with sulfuric acid, treated with charcoal or ion exchange resins, and again crystallized as anhydrous CA (above 40°C) or as a monohydrate (below 36.5°C) (Max et al., 2010).

Citric Acid as a Wood Adhesive Additive

Serving as a Cross-linking Catalyst

Iman et al. (1999) reported the development of a starch-based wood adhesive where starch was chemically crosslinked with polyvinyl alcohol (PVOH) using hexamethoxymethylmelamine (Cymel 303 and 323) as a crosslinking agent and CA as a catalyst. Imam et al. (2001) further reported the characteristics and optimization of starch and polyvinyl alcohol (PVOH)-based cross-linked adhesive suitable for wood-to-wood bonding in interior applications. To prepare 1 kg of such an adhesive, 94.6 g of PVOH powder was completely dissolved into 536.3 g of deionized H_2O at about 80°C. The solution was then cooled to 50°C and 258.5 g of corn starch (12.3% moisture content) was added. The resulting slurry was stirred for 5 min and the temperature was slowly raised to 63°C. Subsequently, the solution was cooled to 25°C and 58.8 g of hexamethoxymethylmelamine cross-linker was added. Volumes were adjusted for loss of water due to evaporation and solid content (either 27 or 37%). CA (1.89 g/kg) and latex (5–7%) were added immediately prior to adhesive application. In this process, CA served as a catalyst for cross-linking reaction. Three-layered plywood specimens were prepared from 1.5–2.0 mm thick birch veneers with this cross-linked adhesive preparation. Their experimental data indicated that the crosslinked adhesive with 27% or less solid content appeared to have the best flow properties. The ideal curing temperature and curing time for wood bonding (also cross-linking) were 175°C and 15 min. The optimal concentrations for cross-linker, latex and CA were 15, 7 and 0.9%, respectively. Similarly, Sridach et al. (2013) investigated the effect of CA, PVOH, and starch ratio on the properties of cross-linked PVOH/pure tapioca starch adhesives, and showed again that the adhesive strength significantly increased when CA was used as a catalyst in the cross-linking reaction. Whereas the conditions of the cross-linking reaction were same as those in Imam et al. (2001) (i.e., 175°C and 15 min), CA content was lower (0.24%) in Sridach et al. (2013). Recently, followed the method of Iman et al. (2001), CA was also used as the cross-linking catalyst in sodium silicate wood adhesive modified by PVOH (Liu et al., 2015).

Serving as a cross-linking agent

In addition to serving as the catalyst of cross-linking reactions, CA can be a cross-linking agent itself. At temperatures above 100°C, the two carboxylic acid groups on CA can form an anhydride through loss of a water molecule. This anhydride then reacts with hydroxyl groups on the polysaccharide or protein to form an ester linkage (Feng et al., 2014; Chiou et al., 2013). Thus, CA has been used to cross-link starch and protein for agricultural biomass based plastics, fibers and films (Azeredo et al., 2015; Newson et al., 2014; Reddy et al., 2008; 2012; Shi et al., 2008). So CA is used as an adhesive additive in improvement of the adhesive performance (Table 1). Yang et al. (1997) reported ester cross-linking of cotton cellulose by polymeric carboxylic acids and CA. This work shows that CA esterifies the anhydride intermediate of two polymers of maleic acid (i.e., the homopolymer and the terpolymer) formed on cotton fabric in a broad range of curing temperatures. Consequently, it is transformed from a trifunctional acid to tetrafunctional one with the formation of an ester linkage with the homopolymer or the terpolymer. Thus, Cheng et al. (2004) proposed that, with sodium hypophosphite (NaH_2PO_2) as the catalyst, CA can interact with the cellulose in soy carbohydrates, and furthermore, CA in soy flour would create ester cross-linking within the straw complex so that CA would improve the bonding performance of soy flour for wheat straw particleboard preparation. They reported that the adhesive made from soy flour treated with 1.5 M urea, 0.4% N-(n-butyl) thiophosphoric triamide, 7% CA, 4% sodium hypophosphite (NaH_2PO_2) as a catalyst, 3% boric acid, and 1.85% NaOH, produced particleboard (PB) with the maximum mechanical strength and water resistance. Li et al. (2010) reported a simplified formulation of soy flour adhesive which was modified with urea, CA and boric acid. In this work, urea solution was prepared at 30°C and then soy flour was added and

Table 1. Selected studies of citric acid (CA) as adhesive additives for wood veneers and particleboards (PB).

Reference	Major adhesive component	CA content	CA function	Bonding subject
Imam et al., 1999; 2001	Corn starch and polyvinyl alcohol	0.9%	Cross-linking catalyst	Birch veneers
Cheng et al., 2004	Modified soy flour	5–9%	Cross-linking agent	Wheat straw PB
Li et al., 2010	Soy flour	9%	Cross-linking agent	Poplar veneers
Ciannamea et al., 2012	Soybean protein concentrate	7%	Cross-linking agent	Rice husk PB
Nordqvist et al., 2013	Soy protein, wheat gluten	0. 05 and 1 M	Dispersing agent	Beech veneers
Sridach et al., 2013	Tapioca starch/polyvinyl alcohol	0.24%	Cross-linking catalyst	Unspecified wood type
Khosravi et al., 2014; 2015	Wheat gluten	0.05 M	Dispersing agent	Pine and beech veneers
Liu et al., 2015	Sodium silicate/polyvinyl alcohol	0.9%	Cross-linking catalyst	Unspecified wood type
He et al., 2016b	Washed cottonseed meal	1%	Dispersing agent	Maple veneers

stirred at 30°C for 2 h. CA solution was added and stirred for another 0.5 h and then boric acid solution was added and heated at 30°C for a further 0.5 h. The resulting adhesive was used to bond poplar veneers. The optimum formulation was 100 g of soy flour powder treated with 9 g of CA at 30°C for 0.5 h in the presence of NaH_2PO_4. Whereas NaH_2PO_4 has been recognized as the most efficient catalyst in the esterification reaction of cellulosic fabrics with CA, Feng et al. (2014) reported that treatments with CA in the presence of NaH_2PO_4 provided wood properties comparable to wood treated with CA alone. Feng et al. (2014) questioned the application of NaH_2PO_4 due to the high cost and the environmentally harmful property of NaH_2PO_4. Thus, it may be interesting to study further the efficiency and ecomomic impact of NaH_2PO_4 as a catalyst in CA-related adhesive preparation.

To get a better insight into the potential changes of soy protein adhesives introduced by CA and boric acid modifications, Ciannamea et al. (2012) compared the attenuated total reflectance—Fourier transform infrared (ATR–FTIR) spectra of their soy protein adhesives. The ATR-FTIR analysis confirmed the changes in the soy protein secondary structures induced by CA and boric acid. The new peaks at 1,730 and 1,155 cm^{-1} (i.e., C=O and C–O ester stretching), along with the concomitant decrease in the absorption band at 1,064 cm^{-1} (i.e., C–O stretching of carbohydrate fraction) appeared in the spectrum of CA-modified adhesive, giving evidence of the esterification reaction between CA with hydroxyl groups of carbohydrate complex in soy protein concentrate. Similar peaks, but to a lesser extent, also occurred in boric acid-modified adhesive spectrum, confirming the capacity of boric acid also to react with OH from side-chain groups or carbohydrates in soy protein concentrate. Cross-linking reactions between the adhesive and the lignocellulosic substrate was shown by the slightly higher density of CA- and boric acid-glued boards. They also found that CA reduced the apparent viscosity of the modified-soy protein adhesive due to protein hydrolysis and formation of soluble aggregates. However, Ciannamea et al. (2012) reported that the addition of CA into soy protein concentrate adhesive had an insignificant effect on the inter bond value of rice husk particleboards, compared with untreated soy protein concentrate. In contrast, the addition of boric acid doubled the internal bond value of the glued rice husk particleboards. Ciannamea et al. (2012) attributed this observation to a result of the combination of many factors including the low viscosity value of CA-modified adhesive dispersion, the formation of protein aggregates at pH near that of the pI (isoelectric point), and the presence of salt (i.e., NaH_2PO_4). Even though, Ciannamea et al. (2012) concluded that their data were in accordance with the tensile strength data of 7% CA-treated soy flour–wheat straw particleboard reported by Cheng et al. (2004).

Serving as a Dispersing Agent

The isoelectric points (i.e., pI value) of wheat gluten and soy protein isolate are at pH 7.3 and 4.5, respectively. As pI will influence protein's solubility and how well a protein disperse in water solutions, Nordqvist et al. (2013) compared the bonding performance of the two types of protein dispersed in either CA or NaOH. They found that wheat gluten can be dispersed at 0.05 M CA while the CA concentration needs to be increased to 1 M to disperse soy protein isolate. They also advised that it is beneficial to keep the CA concentration as low as possible since a higher concentration might have a negative effect on the water resistance of the bond. In contrast, both protein were well dispersed in 0.1 M NaOH. Compared to the low value (2,700 mPa s) of the water dispersion, the viscosities of 11.5% soy protein isolate were 37,000 and 29,300 mPas s in NaOH and CA solutions, respectively. Similarily, the viscosities of 23% wheat gluten were 20,400 and 10,900 mPa s in NaOH and CA solutions, respectively. The beech wood panels were bonded together at a press temperature of 110°C and a press time of 15 min with these dispersions. The shear strength values of the beech veneer pairs bonded with the CA dispersions of both proteins are slightly lower than those of the veneers bonded with NaOH and H_2O protein dispersions, especially for the water soaked veneer pairs. The authors hypothesized that the relatively higher amount of a small molecule (i.e., 0.5 or 1 M CA vs. 0.1 M NaOH) in the CA protein dispersion may reduce hydrogen bonding between protein and wood molecules, thus lowering both cohesive and adhesive interactions of the corresponding wood veneer pairs.

Khosravi et al. (2014; 2015) also prepared wheat gluten adhesive with NaOH (0.1 M, pH 13) or CA (0.05 M, pH 2.2) as the dispersing agent and polyamidoamine-epichlorohydrin (PAAE) and trimethylolpropane triacetoacetate (AATMP) as cross-linkers. Their results show an enhanced performance of wheat gluten using PAAE as cross-linker. However, when CA was utilized as the dispersing agent it resulted in poor board properties, compared with NaOH as the dispersing agent. Thus, Khosravi et al. (2015) studied the wetting and film formation of wheat gluten dispersions applied to wood substrates to get the insights of the bonding mechanism and effect of dispersing agents. In this work, wheat gluten dispersions (12%, 16% and 20%) were brushed onto the pine or beech veneers, while the 24% dispersion was applied using a spatula due to higher viscosity. Without bonding together, these adhesive-applied veneers were kept in a conditioned room (temperature 20°C and relative humidity 65%) before they were analyzed for penetration and film formation. The optical microscopic images of these samples showed that the layers of protein on the surface of the veneers did not vary significantly, indicating the same extent of penetration of the adhesive into the veneer in all samples. In other words, the utilization of PAAE or CA did affect the initial film formation.

Washed cottonseed meal (WCSM) has been shown to be a promising biobased wood adhesive (He and Cheng, 2016; He et al., 2014a; 2014b). In these studies, the adhesive slurries were prepared in a low solid content (about 11%) applied at press temperature 100°C or higher. Recently, He et al. (2016c) prepared WCSM in a pilot scale for promoting its industrial application. However, some industrial applications, such as some furniture and small utensils, require high solid contents and/or lower press temperatures. Thus, He et al. (2016b) tested the adhesive strength of the adhesive preparation with high solid contents (up to 30%) and low press temperatures (40 and 60°C) using the pilot scale-produced WCSM in the presence and absence of 1% CA. Data in Fig. 2 shows that the solid content did not impact the adhesive strength much. On the other hand, the adhesive strength of WCSM greatly increased when the press temperature during veneer bonding was raised from 40 to 60, to 100°C. Similar to the observations of Nordqvist et al. (2013) and Khosravi et al. (2014), the addition of CA indeed lowered the adhesive strength slightly under most tested conditions. This effect might be mainly due to the effect

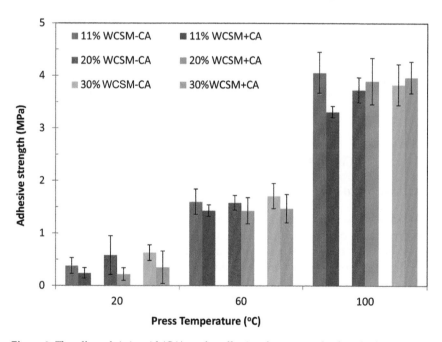

Figure 2. The effect of citric acid (CA) on the adhesive shear strength of washed cottonseed meal (WCSM) on bonding 2-ply maple veneers. Maple veneer bonding conditions: 11, 20, and 30% of WCSM in the absence (–CA) and presence (+CA) of 1% of CA with pressing temperature at 40, 60, and 100°C and pressing time of 20 min. Adapted from He et al. (2016b).

of CA lowering the pH of adhesive slurries (Table 2). Previous work has shown the lower adhesive strength of WCSM slurries prepared at pH 4.5 rather than at pH 6.0 (He et al., 2016a). It is worth pointing out that the maximal adhesive strength is not always required for bonding furniture and small domestic wood utensils. For example, an internal bond of 0.35 MPa would meet the strength requirement for particle boards used for furniture per European Standard (ECN, 2004). On the other hand, some other operational parameters, such as high solid content, appropriate pH and flowability, might be critical for practical applications of WCSM in the domestic furniture field. Addition of 1% of CA clearly improved the viscosity and stability of WCSM adhesive slurries, especially at the high solid content (Table 2).

Table 2. Apparent viscosity (mPa.s at shear rate of 10 s^{-1}) of the adhesive slurries of washed cottonseed meal (WCSM) in the absence (–CA) and presence (+CA) of 1% citric acid at day 0 and 7. Adapted from He et al. (2016b).

WCSM (% solid content)	pH	Viscosity	
		Day 0	Day 7
–CA			
11	6.60	29	18
20	6.51	460	7581
30	6.41	14590	9319
+CA			
11	3.97	36	28
20	4.57	166	3498
30	4.95	5953	10817

Citric Acid as Major Bonding Component

Application to Wood-based Molding

CA can also be used as a main component of adhesives for wood-based molding, particleboard and fiberboard manufacture (Table 3). When a wood-based molding is fabricated, a mixture of lignocellulose and CA powder is hot-pressed in a metal mold (Umemura et al., 2012a). A catalyst is not required in this process. Wood-based molding has been prepared by this process using dried *Acacia mangium* bark and dried CA powder in a 2:1 ratio with hot-pressing at 180°C and 4 MPa for 10 min. The prepared wood-based molding is shown in Fig. 3. In this case, the molding was black and had a uniform plastic-like surface with the densiy of 1.2 g/cm^3. The physical properties of the molding were improved significantly by CA addition of over 10 wt%. In particular, the molding with 20 wt%

Table 3. Recent studies of wood-based materials using only citric acid (CA) as an adhesive.

Reference	Wood-based material	Resin content	Pressing temperature	Pressing time	Bonding subject
Umemura et al., 2012a	Wood-based molding	0~40%	180°C	10 min	*Acacia mangium* Bark powder
Umemura et al., 2012b	Wood-based molding	0~40%	200°C	10 min	*Acacia mangium* Wood powder
Umemura et al., 2012c	Wood-based molding	20%	140~200°C	10 min	*Acacia mangium* Wood powder
Widyorini et al., 2014a	Particleboard	0~20%	200, 220°C	10, 15 min	Bamboo particle
Widyorini et al., 2014b	Particleboard	0~40%	180°C	10 min	Bamboo particle
Indrayani et al., 2015	Fiberboard	20%	200°C	10 min	Pineapple leaf fiber
Widyorini et al., 2016	Particleboard	0~30%	180°C	10 min	Bamboo particle
Kusumaha et al., 2016	Particleboard	0~30%	200°C	10 min	Sweet sorghum bagasse particle

Figure 3. Wood based molding composed of *Acacia mangium* bark and citric acid.

CA content showed the good specific bending properties which were the specific MOR of 18.1 MPa and the specific MOE of 4.9 GPa. The wood-based molding composed of *A. mangium* bark and CA did not decompose during repeated boiling treatment, indicating that the molding had high water resistance. Wood powder can also be used as a raw material. In studies on wood-based molding prepared using *A. mangium* wood powder as a raw material, the effects of the CA content and hot-pressing temperature on the physical properties were investigated (Umemura et al., 2012b,c). For CA content from 0 to 40 wt% with hot-pressing at 200°C and 4 MPa for 10 min, a CA content of 20 wt% provided the best mechanical properties (MOR 35.8 MPa, impact strength 0.94 kJ/m²). The wood-based molding also showed high resistance against boiling water. The effect of the hot-pressing temperature (140–220°C) on the physical properties was studied for molding with a CA content of 20 wt% and hot-pressed at 4 MPa for 10 min. The bending properties were affected by the hot-pressing temperature, and the molding fabricated at 180°C showed an MOR of 39.1 MPa. The average impact strength of the molding at 180°C was 1.1 kJ/m². Water resistance increased with the hot-pressing temperature, and good resistance against boiling water was achieved with a hot-pressing temperature of 180°C or more. Therefore, the optimum CA content and hot-pressing temperature were 20 wt% and 180°C, respectively. Both wood and bark powders of *Cryptomeria japonica* were also able to be used as raw materials for preparing wood-based molding with good physical properties.

As mentioned above, wood-based molding can be fabricated by using various lignocellulose powders. This indicates that some reaction of

CA contributes to the fabrication of wood-based molding, irrespective of the type of lignocellulose. To investigate this reaction, FT-IR analysis was performed on wood-based molding after repeated boiling treatment (Fig. 4.) (Umemura et al., 2012a,b,c). This treatment meant the effect of water-soluble substances could be excluded, and that the results would be a reflection of the inherent chemical structure. A distinct absorption peak at around 1734 cm^{-1} derived from ester linkages was observed in the wood-based molding with CA. This indicates CA reacted with wood components having hydroxyl groups such as cellulose and hemicellulose. Feng et al. (2014) revealed that the formation of ester linkages brings improvements in anti-swelling efficiency and modulus of elasticity and compression strength of wood itself. Ou et al. (2014) reported that esterification of wood flour with CA improved the processibility of wood flour/high density polyethylene thermoplastic composites via extrusion/injection molding processing. On the other hand, it is known that CA decomposes at temperatures >175°C and is converted to other chemical compounds such as aconitic acid (Windholtz, 1983). Considering that the wood-based molding fabricated at over 180°C showed excellent physical properties, it is possible that the decomposition of CA contributed to the development of adhesiveness. In any case, this research confirmed that CA could be used as a wood adhesive directly and without the requirement for any other chemical compounds such as catalysts. In addition, the adhesion was found to occur through the formation of ester linkages between CA and lignocellulose during hot-pressing.

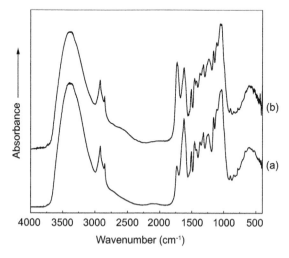

Figure 4. Infrared spectra of (a) wood-based molding with bark powder, (b) wood-based molding with bark powder and citric acid of 20 wt% after repeated boiling treatment. Adapted from Umemura et al. (2012a).

Application to Particleboard Production

Surface area of wood particles used for particleboard is generally smaller than that in wood powder used for wood-based molding. Therefore, investigation of some kind of additive to enhance bonding properties would be needed. Considering the environmental problems and potential shortage of fossil resources, the utilization of natural substances derived from bioresources is preferable. Umemura et al. (2013) tried to use sucrose because of the low cost, high safety and high water solubility. For particleboard manufacture, the adhesive needs to be in the soluble formulation, and aqueous solutions containing CA and sucrose have been used as adhesives. Figure 5 shows a particleboard bonded with CA and sucrose. The color of the board was a dark brown due to caramelization of sucrose and decomposition of citric acid. The effects of the weight ratio of CA and sucrose, resin content, board density, and hot-pressing temperature on the physical properties of the particleboard were evaluated (Umemura et al., 2013; 2015). The weight ratio of CA to sucrose of 100/0 to 0/100 and resin content from 5 to 40 wt% were investigated with hot-pressing at 200°C for 10 min. The mechanical properties and water resistance increased with increasing sucrose and resin content. The optimum weight ratio of CA to sucrose was 25/75 and the optimum resin content was 30 wt%. Particleboard bonded with only CA could be manufactured, but its physical properties were not as good as those of the board with the optimum ratio. Therefore, sucrose was a beneficial additive for particleboard

Figure 5. Particleboard bonded with citric acid and sucrose.

manufacture. Board densities between 0.4 and 1.0 g/cm^3 and hot-pressing temperatures between 140 and 240°C were investigated for particleboard manufactured with a CA to sucrose weight ratio of 25/75 and resin content of 30 wt%. The physical properties of the particleboard improved with increasing board density up to 0.8 g/cm^3. In addition, the particleboard manufactured at 200°C showed much better physical properties than the particleboard manufactured at other temperatures. The physical properties of the particleboard manufactured under the optimum conditions (CA to sucrose weight ratio of 25/75, resin content of 30 wt%, board density of 0.8 g/cm^3 and hot-pressing temperature of 200°C) were comparable to the requirements of the 18 type of JIS (Japanese Industrial Standards) A 5908 (2003). Therefore, particleboard with excellent physical properties could be manufactured using only CA and sucrose as an adhesive. According to the FT-IR analysis of the particleboard after a 4 h boiling water treatment, a distinct peak at around 1723 cm^{-1} was observed. This indicates that the formation of ester linkages would develop adhesiveness and contribute to the good physical properties of the particleboard.

Above studies suggest that saccharide is important for obtaining good bonding properties in the adhesion system using CA. Compared with wood lignocellulose, non-wood lignocellulose contains more hemicellulose, which contains saccharide. Therefore, CA will be useful as an adhesive for wood-based material manufactured using non-wood lignocellulose. Particleboard has been prepared using bamboo, and the effects of heating temperature, CA content, particle size, and layer composition on the physical properties were investigated (Widyorini et al., 2014a,b; 2016). Compared with particleboard prepared without CA (binderless board), the physical properties were greatly improved by the addition of CA. In particular, the board had very low thickness swelling (Table 4). FT-IR analysis confirmed the formation of ester linkages between CA and the bamboo, indicating that the adhesion through chemical bonds contributed to the good physical properties of the board. Sweet sorghum bagasse particleboard bonded with CA (Kusumaha et al., 2016) and low density sugarcane bagasse particleboard bonded with CA and sucrose (Liao et al., 2016) have also been manufactured, and showed good physical properties.

Table 4. Effects of citric acid content on water absorption and thickness swelling of bamboo particleboard. Adapted from Widyorini et al. (2016).

Water absorption (%)			Thickness swelling (%)		
0wt%	15wt%	30wt%	0wt%	15wt%	30wt%
121	32	13	62	5	2

Raw materials: Coarse particles of Petung (*D. asper*).

Conclusions

CA is produced on a large scale by fermentation with worldwide annual production at approximately 1.6×10^6 Mg. Anhydride formed via dehydration of carboxylic groups of CA can react with hydroxyl groups of the polysaccharide or protein to form an ester linkage. This makes CA usable as a wood adhesive additive or component. CA can function as a cross-linking catalyst, a cross-linking agent, or a dispersing agent to improve the adhesive strength and/or operational properties (such as viscosity) toward wood bonding. The direct ester linkages between CA and lignocellulose molecules (raw materials and/or sucrose) make CA workable, with or without sucrose, as a major adhesive component in manufacturing fiberboards and particleboards with a variety of lignocellulose raw materials. These laboratory research accomplishments demonstrated that application of CA as a renewable natural adhesive for wood and lignocellulosic composites is possible. Further effort is on optimizing practical manufacture conditions for specific types of wood-based molding and medium or low density lignocellulosic particleboards using CA as an adhesive.

Keywords: bagasse, citric acid, cross-linking, cottonseed, dispersing, molasses, molding, sucrose, viscosity

References

Angumeenal, A.R. and D. Venkappayya. 2005. Effect of transition metal ions on the metabolism of *Aspergillus niger* in the production of citric acid with molasses as substrate. J. Sci. Ind. Res. 64: 125–128.

Angumeenal, A.R. and D. Venkappayya. 2013. An overview of citric acid production. LWT-Food Sci. Technol. 50: 367–370.

Azeredo, H.M.C., C. Kontou-Vrettou, G.K. Moates, N. Wellner, K. Cross, P.H.F. Pereira and K.W. Waldron. 2015. Wheat straw hemicellulose films as affected by citric acid. Food Hydrocolloids 50: 1–6.

Berovic, M. and M. Legisa. 2007. Citric acid production. Biotechnol. Annual Rev. 13: 303–343.

Cheng, E., X. Sun and G.S. Karr. 2004. Adhesive properties of modified soybean flour in wheat straw particleboard. Composites Part A 35: 297–302.

Chiou, B.S., H. Jafri, T. Cao, G.H. Robertson, K.S. Gregorski, S.H. Imam, G.M. Glenn and W.J. Orts. 2013. Modification of wheat gluten with citric acid to produce superabsorbent materials. J. Appl. Poly. Sci. 129: 3192–3197.

Ciannamea, E., J. Martucci, P. Stefani and R. Ruseckaite. 2012. Bonding quality of chemically-modified soybean protein concentrate-based adhesives in particleboards from rice husks. J. Am. Oil Chem. Soc. 89: 1733–1741.

ECN. 2004. European Committee for Standardization (ECN) Particleboards Specifications. Standard EN 312.

Feng, X., Z. Xiao, S. Sui, Q. Wang and Y. Xie. 2014. Esterification of wood with citric acid: The catalytic effects of sodium hypophosphite (SHP). Holzforschung. 68: 427–433.

He, Z. and H.N. Cheng. 2017. Preparation and utilization of water washed cottonseed meal as wood adhesives. pp. 156–178. *In*: Z. He (ed.). Bio-based Wood Adhesives-Preparation, Characterization, and Testing. CRC Press/Taylor & Francis Group. This volume.

He, Z., D.C. Chapital and H.N. Cheng. 2016a. Effects of pH and storage time on the adhesive and rheological properties of cottonseed meal-based products. J. Appl. Polymer Sci. 13: 43637. doi: 10.1002/APP.43637.

He, Z., D.C. Chapital and H.N. Cheng. 2016b. Adhesive performance of washed cottonseed meal at high solid contents and low temperatures. Proceedings of the 70th Forest Products Society (FPS) International Convention, June 27–29, 2016, Portland, OR. Track 2.7.1. 4 pp.

He, Z., H.N. Cheng, D.C. Chapital and M.K. Dowd. 2014a. Sequential fractionation of cottonseed meal to improve its wood adhesive properties. J. Am. Oil Chem. Soc. 91: 151–158.

He, Z., D.C. Chapital, H.N. Cheng and M.K. Dowd. 2014b. Comparison of adhesive properties of water- and phosphate buffer-washed cottonseed meals with cottonseed protein isolate on maple and poplar veneers. Int. J. Adhes. Adhes. 50: 102–106.

He, Z., K.T. Klasson, D. Wang, N. Li, H. Zhang, D. Zhang and T.C. Wedegaertner. 2016c. Pilot-scale production of washed cottonseed meal and co-products. Modern Appl. Sci. 10(2): 25–33.

Imam, S.H., L. Mao, L. Chen and R.V. Greene. 1999. Wood adhesive from crosslinked poly (vinyl alcohol) and partially gelatinized starch: preparation and properties. Starch 51: 225–229.

Imam, S.H., S.H. Gordon, L. Mao and L. Chen. 2001. Environmentally friendly wood adhesive from a renewable plant polymer: characteristics and optimization. Polym. Degrad. Stability 73: 529–533.

Indrayani, Y., D. Setyawati, T. Yoshimura and K. Umemura. 2015. Decay resistance of medium density fibreboard (MDF) made from pineapple leaf fiber. J. Mathematical Fundamental Sci. 47(1): 76–83.

Japanese Industrial Standard (JIS) 2003. JIS A 5908-2003 Particleboards. Japanese Standard Association. Tokyo, Japan.

Khosravi, S., F. Khabbaz, P. Nordqvist and M. Johansson. 2014. Wheat-gluten-based adhesives for particle boards: Effect of crosslinking agents. Macromol. Mater. Engineer. 299: 116–124.

Khosravi, S., P. Nordqvist, F. Khabbaz, C. Ohman, I. Bjurhager and M. Johansson. 2015. Wetting and film formation of wheat gluten dispersions applied to wood substrates as particle board adhesives. Eur. Polymer J. 67: 476–482.

Kusumaha, S., K. Umemura, K. Yoshioka, H. Miyafuji and K. Kanayama. 2016. Utilization of sweet sorghum bagasse and citric acid for manufacturing of particleboard I: Effects of pre-drying treatment and citric acid content on the board properties. Ind. Crop. Prod. 84: 34–42.

Li, F., X.P. Li and W.H. Wang. 2010. Soy flour adhesive modified with urea, citric acid and boric acid. Pigment Resin Technol. 39: 223–227.

Liao, R., J. Xu and K. Umemura. 2016. Low density sugarcane bagasse particleboard bonded with citric acid and sucrose: Effect of board density and additive content. BioResources 11(1): 2174–2185.

Liu, X., Y. Wu, X. Zhang and Y. Zuo. 2015. Study on the effect of organic additives and inorganic fillers on properties of sodium silicate wood adhesive modified by polyvinyl alcohol. BioResources 10: 1528–1542.

Max, B., J.M. Salgado, N. Rodriguez, S. Cortes, A. Converti and J.M. Dominguez. 2010. Biotechnological production of citric acid. Brazilian J. Microbiol. 41: 862–875.

Newson, W.R., R. Kuktaite, M.S. Hedenqvist, M. Gallstedt and E. Johansson. 2014. Effect of additives on the tensile performance and protein solubility of industrial oilseed residual based plastics. J. Agric. Food Chem. 62: 6707–6715.

Nordqvist, P., N. Nordgren, F. Khabbaz and E. Malmstrom. 2013. Plant proteins as wood adhesives: Bonding performance at the macro- and nanoscale. Ind. Crop. Prod. 44: 246–252.

Ou, R., Q. Wang, M.P. Wolcott, S. Sui and Y. Xie. 2014. Rheological behavior and mechanical properties of wood flour/high density polyethylene blends: Effects of esterification of wood with citric acid. Polym. Compos. 37: 553–560.

Penniston, K.L., S.Y. Nakada, R.P. Holmes and D.G. Assimos. 2008. Quantitative assessment of citric acid in lemon juice, lime juice, and commercially-available fruit juice products. J. Endourol. 22: 567–570.

Reddy, N., Y. Li and Y.K. Yang. 2008. Wet cross-linking gliadin fibers with citric acid and a quantitative relationship between cross-linking conditions and mechanical properties. J. Agr. Food Chem. 57: 90–98.

Reddy, N., Q. Jiang and Y.K. Yang. 2012. Preparation and properties of peanut protein films crosslinked with citric acid. Ind. Crop. Prod. 39: 26–30.

Shi, R., J. Bi, Z. Zhang, A. Zhu, D. Chen, X. Zhou, L. Zhang and W. Tian. 2008. The effect of citric acid on the structural properties and cytotoxicity of the polyvinyl alcohol/starch films when molding at high temperature. Carbohydr. Polymer. 74: 763–770.

Sridach, W., S. Jonjankiat and T. Wittaya. 2013. Effect of citric acid, PVOH, and starch ratio on the properties of cross-linked poly (vinyl alcohol)/starch adhesives. J. Adhes. Sci. Technol. 27: 1727–1738.

Umemura, K., T. Ueda, S.M. Sasa and S. Kawai. 2012a. Application of citric acid as natural adhesive for wood. J. Appl. Polym. Sci. 123(4): 1991–1996.

Umemura, K., T. Ueda and S. Kawai. 2012b. Characterization of wood-based molding with citric acid. J. Wood Sci. 58(1): 38–45.

Umemura, K., T. Ueda and S. Kawai. 2012c. Effects of molding temperature on the physical properties of wood-based molding bonded with citric acid. Forest. Prod. J. 62(1): 63–68.

Umemura, K., O. Sugihara and S. Kawai. 2013. Investigation of a new natural adhesive composed of citric acid and sucrose for particleboard. J. Wood Sci. 59(3): 203–208.

Umemura, K., O. Sugihara and S. Kawai. 2015. Investigation of a new natural adhesive composed of citric acid and sucrose for particleboard II. Effects of board density and pressing temperature. J. Wood Sci. 61(1): 40–44.

Vandenberghe, L.P.S., C.R. Soccol, A. Pandey and J.M. Lebeault. 1999. Review - microbial production of citric acid. B. Brazil. Arch. Biol. Technol. 42: 263–276.

Verhoff, F.H. and H. Bauweleers. 2014. Citric Acid. Ullmann's Encyclopedia of Industrial Chemistry. 1–11. DOI: 10.1002/14356007.a07_103.pub3.

Widyorini, R., A.P. Yudha, Y. Adifandi, K. Umemura and S. Kawai. 2014a. Characteristics of bamboo particleboards bonded using citric acid. Wood Res. J. 4(1): 31–35.

Widyorini, R., A.P. Yudha, R. Isnan, A. Awaluddin, T.A. Prayitno, A. Ngadianto and K. Umemura. 2014b. Improving the physico-mechanical properties of eco-friendly composite made from bamboo. Adv. Mat. Res. 896: 562–565.

Widyorini, R., K. Umemura, R. Isnan, D.R. Putra, A. Awaludin and T.A. Prayitno. 2016. Manufacture and properties of citric acid-bonded particleboard made from bamboo materials. Eur. J. Wood Prod. 76(1): 57–65.

Windholtz, M. (ed.). 1983. The Merck Index. Tenth edition, pp. 330–331. Rahway, NJ, Merck & CO.

Yang, C.Q., X. Wang and I.-S. Kang. 1997. Ester crosslinking of cotton fabric by polymeric carboxylic acids and citric acid. Text. Res. J. 67: 334–342.

Zayat, M. and D. Levy. 2000. Blue $CoAl_2O_4$ particles prepared by the sol-gel and citrate-gel methods. Chem. Mater. 12: 2763–2769.

10

Synthesis of Polymers from Liquefied Biomass and Their Utilization in Wood Bonding

Hui Wan,[1] Zhongqi He,[2], An Mao[3] and Xiaomei Liu[1]*

ABSTRACT

As sustainable manufacturing becomes a mandatory requirement, more and more researchers are devoted to converting biomass as components for polymers, or as a substitution for a part of petroleum based polymers for different applications. Agricultural and forestry lignocellulosic biomass materials are mainly composed of cellulose, hemicellulose, and lignin which typically contain two or more hydroxyl groups per molecule. Biomass liquefaction is a unique thermochemical conversion process for biomass utilizations. In this chapter, we first review and discuss the biomass liquefaction and its major parameters (solvent, catalyst, and heating mode). Then, in the second part, we discuss the utilization of these liquefied biomass-based polymers in wood bonding. Literature data indicate that the liquefied biomass products have been tested as an adhesive, or blended with synthetic polymers and chemicals to form four types of adhesives or resins from biomass and liquidation materials. These liquefied biomass-based adhesives have been applied in bonding plywood, or making fiber and particleboards or other composite materials to reduce the cost and

[1] Department of Sustainable Bioproducts, Mississippi State University, 201 Locksley Way, Starkville, MS. 39759, USA.

[2] Southern Regional Research Center, USDA Agricultural Research Service, 1100 Robert E. Lee Blvd., New Orleans, LA 70124, USA.

[3] College of Forestry, Shandong Agricultural University, 271018, Taian, China.

* Corresponding author: zhongqi.he@ars.usda.gov

formaldehyde emissions. Whereas liquefied biomass show promise as a wood bonding component, improvement of its adhesive strength and water resistance is needed to meet the requirements of high end users. In future works, it will make sense to expand the liquefied biomass and adhesive applications into more biomass species which are currently undervalued. More important, developing new products or new processes that are able to accommodate the characteristics of liquefied biomass-based resins will greatly promote resin sustainable manufacturing.

Introduction

As the sustainable manufacturing concept becomes a mandatory requirement, more and more researchers are devoted to converting biomass as components for polymers, or as a substitution for a part of petroleum based polymers for different applications. Overall, the definition of biomass here is a broad concept that contains lignin, hemicellulose, cellulose, protein and other chemicals that can be used for phenolic, epoxy and polyurethane resin applications. There are many different ways to convert biomass into useful chemicals. Liquefaction is one of them. Generally liquefaction is the process of making or becoming liquid. Liquefaction techniques that convert biomass materials into fuels and chemicals have been investigated for almost a century. In the earliest work, Fierz-David (1925) converted wood, cellulose, lignin and starch into tarry product by dry distillation using hydrogen-pressure at up to 1000 atmosphere pressure and high temperature 200–800°C in the presence of nickel hydroxide as catalyst (Fierz-David, 1925). Most of these techniques required high temperature and pressure and were high energy consuming. More recently, simplified processes for liquefaction of lignocellulosic materials have been developed by using reactive solvents, i.e., phenols, polyhydric alcohols and carbonates, in the presence of strong acid at medium temperatures, usually at 150°C, under atmospheric pressure (Ono and Sudo, 1997; Yamada et al., 2002). These processes are actually based on solvolysis of lignocellulosic materials (Ono et al., 2001; Yamada and Ono, 2001).

Through the development of liquefaction processes, it is obvious that liquefied products can be used for different applications such as phenol formaldehyde (PF) resins (Ono et al., 1996), moldings (Lee et al., 2002), epoxy resins (Kobayashi et al., 2001), and polyurethane (PU) resins (Kurimoto et al., 2001). These have become an excellent approach in addressing recent greenhouse gas emission issues and sustainable manufacturing because they may make it possible to replace fossil resources partly by biomass, i.e., carbon neutral resources that can be applied at room temperature. To comply with the theme of the book, this chapter will focus on the liquefaction of wood and other lignocellulosic biomass materials and the utilization of these liquefied biomass-based polymers in wood bonding.

Biomass Liquefaction Processes

Figure 1 is an illustration of the biomass liquefaction process. Various biomass materials have been used for the biomass liquefaction and applied as wood bonding agents (Table 1). There are two major wood liquefaction methods. The first one is the preparation in the presence of phenol, which resulted in liquefaction products rich with phenol units. The second liquefaction method is achieved in the presence of alcohols, especially polyhydric alcohols, and the gained products can be used as polyols. In this section, liquefying solvent, catalyst and heating mode are the operational parameters which are discussed below.

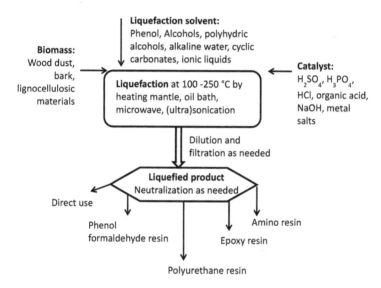

Figure 1. Illustration of the biomass liquefaction processes, major operational factors and adhesive applications.

Liquefaction Solvent

Phenol is a popular solvent in biomass liquefaction. The liquefaction method is called phenolysis, involving phenol with acids as catalysts, which results in liquefaction products rich in combined phenolic compounds. Most studies on biomass liquefaction have investigated the effect of liquefaction solvent to biomass ratio. This is because liquefaction solvent plays double functional roles during liquefaction reactions. The solvent reacts with the reactive sites on biomass components while it also dissolves reaction intermediates and final products, thus shifting the reaction to the liquefying

Table 1. Selected studies of polymers from liquefied biomass as wood adhesives.

Reference	Liquefied biomass	Liquefaction solvent	Adhesive blend/resin	Bonding usage
Alma and Basturk, 2006	Grapevine cane	Phenol	Resol PF	Black poplar veneers
Antonovic et al., 2010	Black poplar	Glycerol	UF, Novolac PF	Particleboard
Cuk et al., 2015	Spruce	Glycerol and diethylene glycol	MF	Spruce/beech particleboard
dos Santosa et al., 2016	Cork dust or granules	Diethylene glycol	PU	Cork
Esteves et al., 2015	Maritime pine	Ethylene glycol	UF, MUF	Beech veneers
Fidan et al., 2010	Cotton stalks	Phenol	Resol PF	Wood composite
Fu et al., 2006	Bamboo	Phenol	Resol PF	Plywood
Hassan et al., 2009	Pine	Phenol with H_2SO_4	Resol PF	Particleboard
Juhaida et al., 2010	Kenaf core	Glycerol and PEG 1000	Resol PF	Rubberwood blocks
Khan and Ashraf, 2005	Coffee bean shell (lignin)	Phenol	Resol PF	Teakwood
Kishi et al., 2006	German spruce	Resorcinol	Epoxy	Plywood plates
Kobayashi et al., 2001	Japanese cedar	PEG400/glycerol	Epoxy	Birch veneers
Kunaver et al., 2010	Poplar, oak, spruce, beech	glycerol – diethylene glycol	UF, ME, MUF	Particleboard
Lee and Liu, 2003	Taiwan acacia and China fir barks	Phenol	Resol PF	Particleboard
Lee and Lin, 2008	Taiwan acacia and China fir barks	Glycerol -polyethylene glycol	PU	Lauan veneers
Lee et al., 2014	Japanese cedar	Phenol	Resol PF	Molding plates
Lin et al., 2014	Chinese fir	25% NaOH	Resol PF	Eucalyptus veneers
Maldas et al., 1997	Birch meal	Phenol	Resol PF	Edge grained makamba veneers

Petric et al., 2015	Black poplar	Ethylene glycol	PU	Beech veneers
Roslan et al., 2014	Oil palm empty fruit bunch fiber	Phenol	Resol PF	Shorea sp. plywood
Sankar and Yan, 2014	Pine barks	Polyethylene glycol and glycerol	PU	Poplar strips
Tohmura et al., 2005	Sugi chips	poly(ethylene glycol) and glycerin	PU	Red meranti veneers
Ugovsek and Sernek, 2013a,b	Black poplar	Ethylene glycol	Direct use	Beech veneers
Wen et al., 2013	Larch bark	Phenol	Resol PF	Rice straw board
Wu and Lee, 2008	ma bamboo	phenol and polyhydric alcohol	Epoxy	Wood
Wu and Lee, 2011	Japanese cedar	Polyethylene glycol-glycerol	Epoxy	Formosan alder veneers

PF: phenol-formaldehyde; PU: polyurethane; UF: urea formaldehyde; MF: melamine formaldehyde; MUF: melamine-urea-formaldehyde direction and preventing the re-condensations of decomposed biomass components. Generally, an excessive amount of liquefaction solvent is necessary to achieve a satisfactory liquefaction (Pan, 2011). For example, Hassan et al. (2009) used phenol-liquefied southern pine to make PF resins for bonding particleboard. Their liquefaction experiments were carried out with three different weight charge ratios of phenol to wood (70/30, 65/35, and 60/40) on the basis of the oven-dried wood weights. Sulfuric acid was used as a catalyst at three weight percentages on the basis of the phenol weight in all of the experiments. The liquefaction temperature was slowly raised to 160–165°C over a period of 1 hour, and this temperature was maintained for 45 min to complete the liquefaction. The liquefied wood was cooled to room temperature. The acidity of the liquefied solution due to the catalyst sulfuric acid and thermal decomposed wood products including formic and levulinic acid was then neutralized by 50% sodium hydroxide. Whereas the un-reacted phenol after the liquefaction reaction can be removed by distillation under reduced pressure (Lin et al., 1994), a more practical approach is direct synthesis of PF-type polymers without removal of the extra phenol in the liquefied biomass. In such a case, free phenol (Alma and Basturk, 2006) or the formaldehyde reactivity (Hassan et al., 2009) of the liquefied wood should be determined for appropriate percentages and ratios of phenol and formaldehyde for the PF polymer synthesis. While the liquefied-wood products could be used directly to prepare resins in the reaction flask for particleboard preparation (Hassan et al., 2009), methanol dilution and filtration are needed to separate the liquefied biomass from the residues for better adhesive resin preparation (Alma and Basturk, 2006).

The second liquefaction method is achieved in the presence of alcohols, especially polyhydric alcohols, and the resulting products can be used as polyols for the preparation of polyurethane and epoxy products (Hajime et al., 2011; Niu et al., 2011). Lee and Lin (2008) reported the liquefaction of Taiwan acacia and China fir which were liquefied in polyethylene glycol (PEG)–glycerol cosolvent, with sulfuric acid as catalyst. The liquefaction was performed in the weight ratio of poly(hydric alcohol)/wood/sulfuric acid set at 3/1/0.09 and at 130–150°C heated by an electric heating mantle. Lee and Lin (2008) blended the liquefied woods with three types of isocyanate, such as poly-4,4-diphenylmethane diisocyanate (PMDI), Desmodur L (adduct of toluene diisocyanate with trimethylol propane), and Desmodur N (trimer of hexamethylene diisocyanate), to form PU resins for wood bonding test. More applications of polyhydric alcohols as liquefaction solvent are listed in Table 1. It is worth pointing out that the liquefaction of biomass could induce a high degree of modification to the biomass components. D'Souza and Yan (2013) used beetle killed mountain pine bark to synthesize polyols via a solvent liquefaction with polyethylene glycol (PEG)/glycerol as cosolvent. Their experiments were carried out at mild, medium, and high temperatures (90, 130, and 160°C) to study the impact of the liquefaction temperature on bark-based polyols. Polyethylene glycol, glycerol, sulfuric acid, and a co-solvent of water with xylene were the matrix for the bark liquefaction. NaOH-neutralized liquefied mixture was diluted and filtrated to obtained the bark-based polyols. Their work demonstrated that at the mildest liquefaction temperature of 90°C, the liquefaction mechanism was a hybridization of a solvent liquefaction with a hot-water extraction while the highest liquefaction temperature of 160°C produced polyols containing highly degraded bark-biopolymers and lignin. The medium liquefaction temperature of 130°C represented an intermediate condition with some levels of bark polymer degradation. Esteves et al. (2015) recently used liquefied pine sawdust prepared by the polyhydric alcohol process as a partial substitute of urea formaldehyde (UF) and melamine urea formaldehyde (MUF) resins in particleboard production. Their maximum liquefaction was obtained after 30 min at 180°C (about 80%) and 60 min at 160°C (about 70%). It was found that the ratio of wood/ethylene glycol influenced the percentage of liquefaction that ranged between 62% and 88% and the ratio which showed best results was 1:6.

In addition to the two major solvents of biomass liquefaction, the application of other polar solvents have also been reported. These solvents include cyclic carbonate such as ethylene carbonate (Mun et al., 2001), dioxane/water (Mun and Hassan, 2004), and ionic liquids (Lu et al., 2015; Xie et al., 2015). Recently, the biomass liquefaction in alkaline medium has been reported as an attractive way to obtain low molecular weight compounds for bio-oil (Yin et al., 2011). The alkaline treatment at elevated temperature can enhance the reactivity of the crystalline cellulose through decreasing degrees of polymerization and increasing accessibility of cellulose. Meanwhile, the lignin components depolymerize to form monomeric and oligomeric phenolic compounds. Lin et al. (2014) reported liquefaction of Chinese fir powder in this way. In their work, Chinese fir powder was liquefied in a sealed reactor using a stainless steel autoclave reactor (maximal pressure, 12 MPa). After the wood powder was loaded into the reactor, a concentration of 25% sodium hydroxide solution was gradually charged into the reactor with stirring. Under the sealing conditions, the mixture was kept in oil bath at 200°C for 15 min. The liquefied resultant was cooled and preserved for synthesis. The residues in the liquefied product accounted for only 0.65 wt% based on the total mass of the final liquefaction. Thus, no filtration was applied prior to its direct use in the adhesive synthesis. In order to enhance the bonding strength and boiling water resistance, formaldehyde and phenol were used to co-polymerize with alkaline Chinese fir liquid.

Catalyst

Most biomass liquefaction reactions are carried out with an acid catalyst which can substantially lower the reaction temperature and increase the extent of liquefaction compared to that without an acid catalyst (Pan, 2011). Catalysts such as sulfuric acid (Hassan et al., 2009), phosphoric acid (Zhang et al., 2005) and oxalic acid (Pan et al., 2008) are used in biomass liquefaction. Kishi et al. (2006) liquefied German spruce meal in resorcinol with and without a sulfuric acid catalyst at high temperature. The catalytic effect of sulfuric acid was not obvious based on the amount of insoluble residues generated. They found that about 1.5 to 2 h reaction time gave the least amount of insoluble residues for both noncatalyzed liquefaction and acid-catalyzed liquefaction. More than 2 h of heating resulted in increased insoluble residues as recondensation of components from degraded wood (polysaccharide and lignin) had occurred. Gel permeation chromatography of the resorcinol-liquefied wood without catalyst and with acid catalyst showed that much more resorcinol could react with the wood components during the liquefaction process in the case with acid catalyst. In other words, the acid-catalyst method in wood liquefaction allowed more wood components to react with resorcinol than the noncatalyst method.

Liquefaction of biomass with phenol could also be conducted under alkaline conditions (Maldas and Shiraishi, 1997; Maldas et al., 1997). In these studies, wood meal (birch, aspen or other biomass samples), phenol and NaOH aqueous solution were mixed and reacted in a closed pressure-proof tube at 250°C for 1 h. Characterization of the liquefied mixture showed that liquefaction under such conditions resulted in substantially higher un-reacted phenol than strong acid catalyzed liquefaction. These researchers observed that pH changed roughly four to five units after the liquefaction, and the final pH was always acidic generated from wood components, mostly cellulose and hemicelluloses derivatives.

There are also reports of the use of other catalysts in biomass liquefaction although such liquefied products may not be suitable for wood adhesives. Mun and Hassan (2004) and Mun et al. (2001) reported cellulose liquefaction using ethylene carbonate (EC) as a liquefying agent and organic sulfonic acids, such as methane sulfonic acid (MSA) or p-toluenesulfonic acid (PTSA), as a catalyst. Xie and Shi (2006) reported rapid wood liquefaction in imidazole-based ionic liquid 3,3'-ethane-1,2-diylbis(1-methyl-1*H*-imidazol-3-ium) without residue at 120°C for 25 min. In contrast, liquefaction in phenol/H_2SO_4 (liquid/wood ratio 20:1) left 25.9% residue. In the imidazole-based ionic liquid system, $AlCl_3$ plays a dual role as a catalyst and a liquefaction reagent.

Heating Mode

The biomass liquefaction is conducted at moderate temperatures (100–250°C). The conventional heading mode is direct heat transfer such as electric heating mantle or oil bath (Lee and Liu, 2003; Lin et al., 2014). Microwaves have sometimes been utilized as an alternative source of energy for chemical reactions (Biswas et al., 2015). Compared to conventional heating, microwave energy penetrates and produces a volumetrically distributed heat source, and heat is generated throughout the material and leads to faster heating rate and improved kinetics (Xie et al., 2015). Thus, microwave radiation has been explored as the heating source of biomass liquefaction (Dos Santos et al., 2015; Krzan and Zagar, 2009; Xie et al., 2016). Xie et al. (2015) optimized the microwave-assisted direct liquefaction of bamboo residues in glycerol/methanol mixtures. Their microwave-assisted liquefaction was carried out in a laboratory microwave oven. They recommended the optimal liquefaction conditions as the reaction temperature of 120°C and the reaction time of 7 min with microwave power of 300 W. Under such conditions with glycerol–methanol–bamboo ratio of 8/0/2 on weight basis, the maximum conversion yield was 96.7%. Dos Santos et al. (2015) reported microwave-assisted liquefaction of cork powder in 2-ethyl hexanol/DEG with p-toluene sulfuric acid as a catalyst. They applied pulsed microwaves for 5, 10, and 20 min under 150, 225, and 300 W as starting microwave power. They found that the alternative energy source led to a high conversion of cork powder into liquid bio-polyols. The efficiency of liquefaction increased with higher microwave power and shorter reaction time. Their study evidenced that cork powder can be liquefied via microwave.

High energy ultrasound has been used in liquefaction reactions with different wood waste materials. The frequency range from 20 kHz to 1 MHz is used in chemical processing while higher frequencies are used in medical and diagnostic applications. Kunaver et al. (2012) applied high energy ultrasound for the depolymerization and liquefaction of different lignocellulosic materials, wood wastes in particular. In their work, the reactor was charged with a range of diethylene glycol (DEG)/glycerol mixtures together with p-toluene sulfonic acid or concentrated sulfuric acid as the catalyst. The liquefaction of different wood waste materials was achieved for a DEG:glycerol = 1:4 mixture. The high frequency (24 kHz) power output was regulated by the adjustment of the amplitude from 20% to 100% of the nominal power of 400 W. The high frequency output was transferred through a titanium cylindrical horn, introduced into the reactor through the side neck and submerged 20 mm into the reaction mixture. They (Kunaver et al., 2012) found that the use of the ultrasound process inhibited the formation of the large molecular structures during the liquefaction from the degradation products by keeping the reactive segments apart and due to such a short reaction time being used. The authors concluded that the short reaction time and subsequent low energy consumption for the liquefaction reaction should lead to the creation of a new method for the transformation of the wood waste materials into valuable chemicals. Mateus et al. (Mateus et al., 2015) conducted a comparative study on conventional and ultrasound liquefaction of cork powder. Similar to Kunaver et al. (2012), Mateus et al. (2015) also utilized the output power of the high frequency (24 kHz) regulated with an amplitude from 60% to 100% of the nominal power of 400 W. Compared to the conventional heating, Mateus et al. (2015) achieved a rapid liquefaction of cork within 60 min with the presence of p-toluene sulfonic acid by using ultrasound as energy source. A pseudo-first order reaction model showed that ultrasound increased the speed of the reaction up to 4.5 times, when compared with that from the conventional method.

Dos Santos et al. (2015) further compared the yields and characteristics of the liquefied cork products produced by conventional heating, ultrasound-, and microwave-assisted methods. The microwave-assisted liquefaction was faster than the ones assisted by ultrasounds and conventional heating. They reported that the microwave-assisted liquefaction with the minimal residue content of 4.56% was obtained with a microwave irradiation power of 300 W for 5 minutes. In contrast, much longer reaction time (135 min) was needed to reach similar residue content (5.35%) with the conventional heating at the same temperature of 160°C. Thus, microwave heating greatly accelerated the depolymerization of cork. The authors proposed that a different temperature regime caused by microwave heating is the main contributing factor to the acceleration of the reaction, leading to a more rapid heating and a more uniform temperature distribution. The microwave-assisted liquefaction was also faster than that assisted by ultrasounds. However, ultrasounds provided higher yields for longer reactions since it was able to produce less decomposition and re-polymerized products. The microstructure morphology of residues that remained after the liquefaction processes showed that the ultrasounds and microwave systems led to smaller, but similar disintegrated and disrupted artifacts than those from the conventional heating. By using attenuated total reflectance (ATR) FTIR analysis, they (Dos Santos et al., 2015) revealed that CH stretching bands and C=O stretching band were higher in the spectra of liquefied products resulting from conventional and ultrasounds reactions, indicating the presence of a longer carbon chain and higher hydroxyl content. On the other hand, the C=O band was stronger in the spectrum of microwave-assisted liquefaction product than that from other procedures, indicating a higher content of carbonyl function that resulted from the oxidation of hydroxyl groups by microwave irradiation.

Liquefied Biomass Used for Wood Bonding

Independent Bonding Material

Theoretically, liquefied biomass is able to form a cross-linked structure as a condensation reaction occurs due to the reaction between the depolymerized cellulose and the aromatic derivatives of lignin or due to the nucleophilic displacement reaction of cellulose by phenoxide ions. Ugovsek and Sernek (2013b) examined the influence of pH value, press temperature, and pressing time on the shear strength of two-layer beech wood lamellas directly bonded with low solvent liquefied black poplar wood. In their work, the wood sawdust was liquefied at 180°C using ethylene glycol as the solvent and sulphuric acid as a catalyst. Ugovsek and Sernek (2013a) pointed out that liquefied wood bonding strength was a combination of physical and chemical phenomena and both were dependent on temperature. In their case, liquefied wood functioned as a binder with the initial elimination of water and the solvent used for liquefaction (ethylene glycol) followed by a chemical reaction at higher temperatures. So, Ugovsek and Sernek (2013b) applied the liquefied product with different pH values for the bonding of solid wood at 200°C for 15 min. Then, the liquefied product with an optimal pH value was used for bonding at different press temperatures for 15 min. They (Ugovsek and Sernek, 2013b) found that unmodified liquefied wood with acid pH value, a press temperature of 180°C, and a pressing time of 12 min were the optimal conditions for bonding the 5 mm thick wood lamellas. They also observed high wood failure (100%) attributed to the low pH value and high press temperature which caused damage of the surface of beach lamellas while the bonding strength was too low to attain standard requirements. Ugovsek et al. (2013) applied light microscopy, scanning electron microscopy, FT-IR micro-spectroscopy and elemental carbon, nitrogen and sulphur (CNS) analysis techniques to investigate the formation of the wood bond line with the liquefied wood as an adhesive. They found that the bonds were very untypical compared to bonds formed by synthetic wood adhesives. Their data suggested that the entire zone of bonding was formed by: (i) a central broad zone of partly carbonized wood cells, and (ii) a narrow zone of delignified wood cells in connection with the original undamaged wood tissue of the two strips. This delignified layer was assumed to be the weak boundary layer where cracks were generated. Such observations provided an explanation of the phenomenon of high wood failure with low shear strength of the wood lamellas bonded with the liquefied wood.

Resol Phenol-formaldehyde Resins

Acid-catalyzed liquefied biomass with phenol has been used to partially substitute phenol in PF resin synthesis. Both novolac and resol type phenolic resins have been synthesized from liquefied biomass. Novolacs are phenol-formaldehyde resins with formaldehyde to phenol molar ratio of less than one and synthesized under acidic conditions. Resol type phenolic resin is thermosets and synthesized with formaldehyde to phenol ratio of greater than one (usually around 1.5) (Fig. 2). Acid-catalyzed liquefied biomass with phenol is favorable for novolac resin synthesis as an extra step is not needed to neutralize the acidic liquefied biomass. Novolac resins are widely used for manufacturing moldings as they are thermoplastic prepolymers without hydroxymethyl group in its molecular structure. A crosslinking agent such as hexamine is needed for it to form a cured network structure under heating. On the other hand, resol-type of PF resin can be cured under heating and is widely used as an adhesive (Lee et al., 2010; Wan and Kim, 2008).

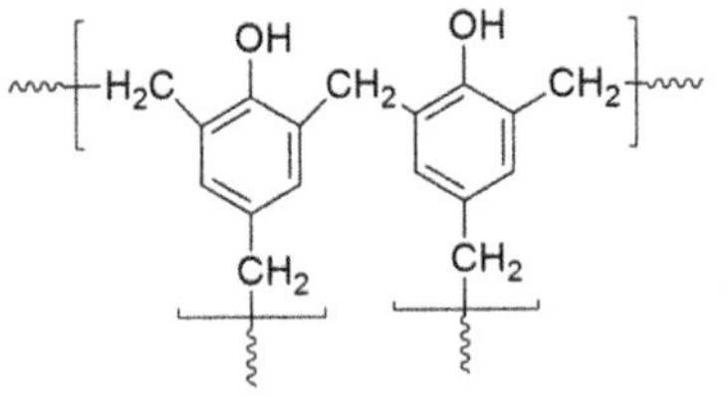

PF (resol) resin 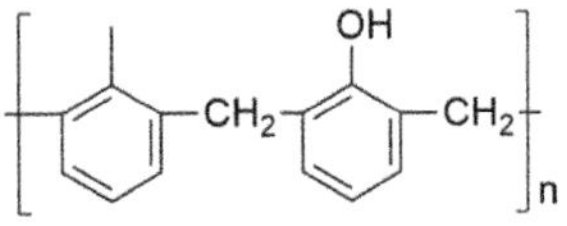   PF (Novolac) resin

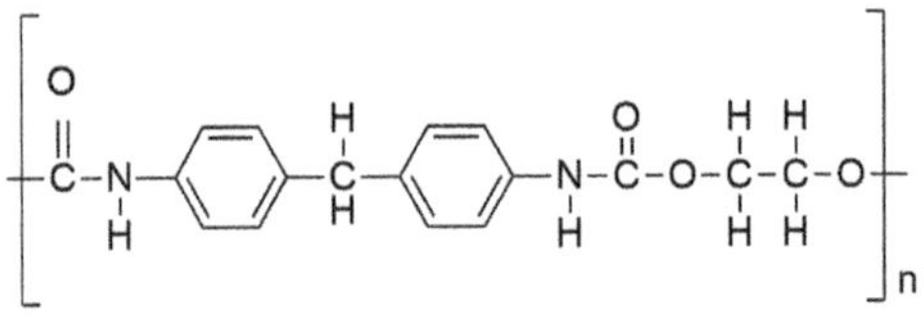

Polyurethane resin

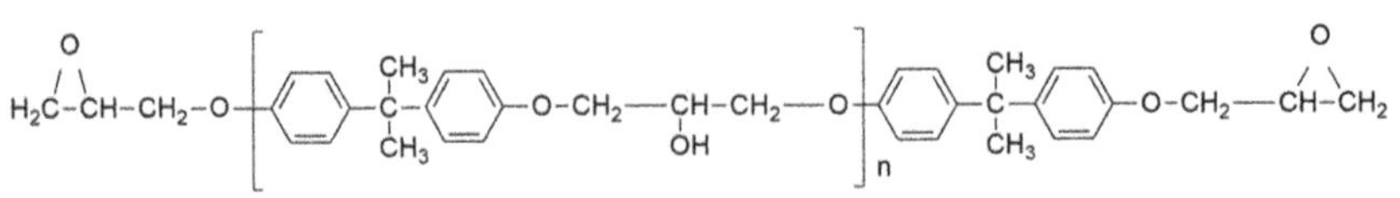

Epoxy resin

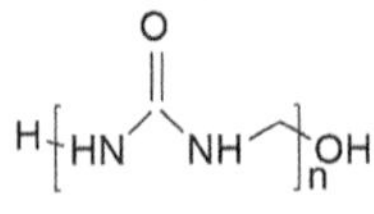

UF resin

Figure 2. contd....

Figure 2. contd.

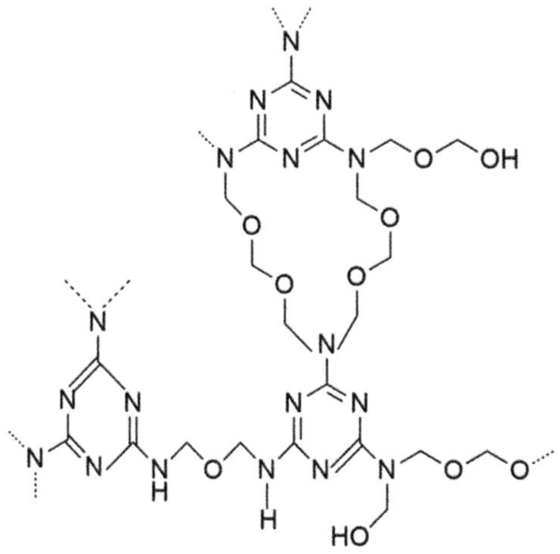

MF resin

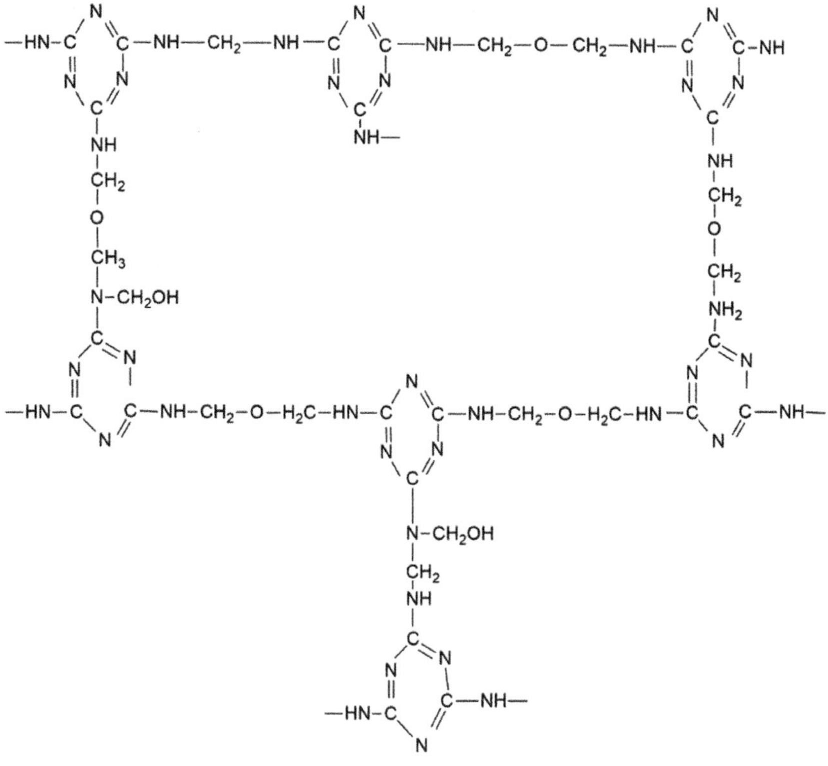

MUF resin

Figure 2. Structures of phenol-formaldehyde (PF), polyurethane (PU), epoxy, urea formaldehyde (UF), melamine formaldehyde (MF), and melamine-urea-formaldehyde (MUF) resins.

For this reason, most applications of the liquefied biomass with phenol for adhesive are resol type of PF resins whereas the novolac PF resin is rare (Table 1). Therefore, we are taking Hassan et al. (2009) as an example to discuss the resol type of PF resins made from phenol-liquefied biomass used in wood and composite bonding.

In their work, Hassan et al. (2009) liquefied southern pine wood in phenol in 30–40:70–60 weight ratios which resulted in homogeneous liquefied materials. The liquefied wood was cooled to room temperature and the liquefaction catalyst sulfuric acid was neutralized by 50% sodium hydroxide solution. Then, it was used directly to prepare resins in the liquefaction flask. For comparison, three different PF resins were prepared from the three liquefied-wood products at three different phenol/wood ratios (Table 2). The viscosity, free formaldehyde content, resin solid content, and alkalinity percentage of the prepared resins are also listed in Table 2. Hassan et al. (2009) also reported that these synthesized resins showed good physical and handling properties: low viscosity, stability for storage and transportation, and applicable by a common sprayer. Therefore, particleboard panels were manufactured with the three synthesized PF resins with a control panel bonded with a commercial UF resin under the same hot-pressing conditions and their bonding performance properties are listed in Table 3. Compared to UF resin, these PF resins with liquefied wood generally resulted in low panel density, internal bonding strength (IB), thickness swelling (TS), water absorption (WA) and free formaldehyde emission; higher modulus of rupture (MOR) and about the same modulus of elasticity (MOE). As the authors (Hassan et al., 2009) implied in their panel preparation section, the panel manufacturing method in their work was adopted from the one of manufacturing UF resin bonded panels for the comparison purpose. If the authors had used PF resin bonded panel manufacturing method, such as hot stacking after hot press, the PF resin bonded panels should have had better panel performance. As the authors (Hassan et al., 2009) showed, the free formaldehyde emission data indicated that increasing liquefied wood in PF resin reduced free formaldehyde

Table 2. Preparative parameters and selected physical properties of the phenolic liquefied-wood (LW) resol phenol formaldehyde (PF) resins. Compiled from Hassan et al. (2009).

	PF resin 1	PF resin 2	PF resin 3
Phenol/wood of LW	70/30	65/35	60/40
Formaldehyde/LW	104/100	86/100	88/100
Solid content (%)	48.14	44.26	46.24
Free HCHO (%)	0.53	1.02	0.25
Alkalinity (%)	4.18	4.46	4.68
Viscosity (cP)	300	220	210

Table 3. Panel performance with liquefied wood modified PF resins. Reprinted from Hassan et al. (2009).

	Control (UF resin)	PF resin 1	PF resin 2	PF resin 3
Density (kg/m³)	724	712	707	709
IB (MPa)	0.703	0.475	0.512	0.486
MOR (MPa)	14.81	15.47	17.24	15.94
MOE (MPa)	2392	2308	2558	2342
TS (%)	23.3	16.7	14.0	17.7
WA (%)	49.8	41.8	34.6	39.8
Free formaldehyde (ppm)	0.16	0.09	0.09	0.06

emission from the panels, indicating that liquefied wood would function as scavenger for formaldehyde. Overall, this work showed that the process of wood liquefaction with limited amounts of phenol as a solvent had the potential of providing practical, low-cost PF-type resins with very low formaldehyde emission potential.

Polyurethane Resins

Polyurethane (PU) resin (Fig. 2) that has the urethane linkage in its molecular structure is one of the most important synthetic resins used nowadays (Lee and Lin, 2008; Mao et al., 2014). It is composed of two major components— one is the compound containing active hydrogen and the other is the isocyanate containing the NCO group. Biomass is naturally rich in hydroxyl groups, and liquefied biomass with polyhydric alcohols has been studied as polyols for polyurethane production. Whereas most of the researches about the polyhydric alcohol liquefied biomass were focused on finding the suitable liquefaction condition and their utilization as PU foams and films, Lee and Lin (2008) prepared a liquefied wood polyurethane adhesive for wood laminate. In their work, polyhydric alcohol-liquefied Taiwan acacia and China fir were used to blend with various kinds of isocyanate to understand the potential of which one to be used as the cold-setting PU adhesive. Three types of isocyanate (i.e., PMDI-aromatic isocyanate polymer, Desmodur L-aromatic isocyanate monomer, and Desmodur N -aliphatic isocyanate monomer) were used to compare the influence of molecular structure of isocyanate on the gluing properties of PU resins. Their data (Lee and Lin, 2008) indicated that the PU resins prepared with Desmodur L had a suitable gel time for processing. As wood adhesives, blending liquefied Taiwan acacia with Desmodur L had better dry and wet bonding strength than with Desmodur N. At certain range, the bonding strength of these PU resins could be increased with a higher molar ratio of NCO/(OHCOOH).

In an earlier work, Tohmura et al. (2005) prepared the PU adhesive from polyhydric alcohol-liquefied wood blending with poly-4,40-diphenylmethane diisocyanate (PMDI) and used it in the manufacturing of plywood with hot-pressing. Whereas all the dry test results of the shear strength met the Japanese Agricultural Standard (JAS) criteria for plywood, only the plywood bonding with adhesives from the liquefied wood with a reaction time of 1.5 h satisfied the JAS criteria after a cyclic steaming treatment. On the other hand, with low emissions of formaldehyde and acetaldehyde, and bond durability, Tohmura et al. (2005) predicted that liquefied wood-based PU adhesives have the potential to become ideal wood adhesives. Recently, Petric et al. (2015) reported bonding and surface finishing of wood with liquefied wood-based PU resins. These researchers found that the pressing time of 12 minutes and pressing temperature at 180°C were the optimal parameters for the bonding of the 5 mm thick beech wood lamellas with their liquefied black poplar wood PU resin. As liquefied wood based products are of a dark brown or black color, these researcher also prepared PU coatings from bleached liquefied wood for an aesthetically acceptable light color. Petric et al. (2015) reported that the properties of the H_2O_2 bleached liquefied wood based PU films were all comparable to the properties obtained with non-bleached liquefied wood based films. They suggested that the bleaching pretreatment could be applied to liquefied wood before its use in the preparation of PU without any harmful effects on the properties of the cured liquefied woodbased PU coatings, which was an important new finding.

Epoxy Resins

Epoxy resin is one of the most important polymers due to the fact that this resin does not shrink during curing like PF resin (Fig. 2). While the most widely used epoxy resin is made by condensation of epichlorohydrin with bisphenol A or diphenylol propane, other hydroxyl containing compounds, including resorcinol, glycols, and glycerol, can replace bisphenol A (Pan, 2011). Xie and Chen (2005) reported liquefaction of sugarcane bagasse in ethylene carbonate and preparation of epoxy resin from the liquefied product with bisphenol A. Their resin was cured with triethylene tetramine at 100°C. These researchers reported that their resin possessed higher adhesive shear strength and better thermal stability than a commercial epoxy resin when single lap shear joints fabricated with stainless steel substrate were used to evaluate the adhesive tensile shear strength of the cured resin. Kobayashi et al. (2001) used liquefied wood to prepare liquefied wood/epoxy resin. They found liquefied wood could be incorporated into the resin to form a cross-linked copolymer network. The liquefied wood/epoxy resin had a shear bonding strength similar to commercial epoxy resin.

Kishi et al. (2006) synthesized epoxy resins from resorcinol-liquefied German spruce wood. By the glycidyl etherification, epoxy functionality was introduced to the liquefied wood. They reported that the epoxy functionality of the resins was controlled by the concentration of phenolic OH groups in the liquefied wood, which would be a dominant factor for crosslink density and properties of the cured epoxy resins. Their data indicated that the MOE and the strength of the cured woodbased epoxy resin were equivalent to those of the bisphenol A type epoxy resin at room temperature. Thus, these wood-based epoxy resins would be well suited for matrix resins of natural plant-fiber reinforced composites. Wu and Lee (2011) evaluated the epoxy resins blended with polyhydric alcohol-liquefied Japanese cedar. They demonstrated that blended epoxy resins could cure under room temperature with an exothermic reaction. Whereas blended epoxy resins had a good dry bonding strength for wood when cured at room temperature, curing with heat treatment improved the wet bonding strength of blended epoxy resins, especially for those prepared with PEG-400-liquefied wood. Using the same strategy, the same researchers (Wu and Lee, 2008; 2010a; 2010b) have also blended and characterized the liquefied bamboo with epoxy resin and used it to bond wood. It is worth pointing out that bio-oil from pyrolysis of wood can also be used to make epoxy blends for wood bonding, such as reported by Liu et al. (2014).

Amino Resins

Amino resins (Fig. 2) are polymeric condensation products of the reaction of aldehydes (most commonly formaldehyde) with compounds carrying aminic or amidic groups (urea, melamine) and can be prepared for different purposes by varying the preparation procedure (Cuk et al., 2015). Amino adhesives have been used in wood bonding, especially for the production of particleboards and medium density fiberboards. There are three common amino resins [i.e., urea–formaldehyde (UF), melamine–formaldehyde (MF), and urea–melamine–formaldehyde (MUF)]. MF and MUF resins tend to give lower formaldehyde emission compared to UF resins. However, MF resin adhesives are expensive so that addition of urea has been adopted to form relatively cheaper MUF adhesives (Pizzi and Mittal, 2003). Thus, liquefied biomass has been blended with the three common amino resins in order to reduce the formaldehyde emission with potentially low adhesive cost (Table 1).

Kunaver et al. (2010) studied liquefied spruce wood polymerized in the hot press with different UF and MUF resin precursors and used them as adhesives for wood particleboards. They (Kunaver et al., 2010) reported that the temperature of the press unit was lowered from 180°C to 160°C with no significant influence on the mechanical properties of these particleboards.

Furthermore, addition of 50% of the liquefied wood to such resin precursors allowed the product to meet the European standard requirement for particleboards. In addition, up to 40% reduction of the formaldehyde emission was achieved. Esteves et al. (2015) liquefied pine wood as a partial substitute of MUF and UF resins. Their work demonstrated that the IB of the particleboards decreased with the increase of the content of liquefied wood in the resins. The reduction in IB by 20% of liquefied wood addition was minimal, meeting the minimum standards requirement. However, inclusion of 70% of liquefied wood significantly decreased the IB. Thus, they concluded that it was possible to use a small amount of maritime pine sawdust liquefied wood as a partial substitute for UF and MUF resins in the particleboard production, thus decreasing the formaldehyde content. On the other hand, Cuk et al. (2015) reported that MF resin adhesive with 30% liquefied wood substitution could be used to produce particleboards with suitable mechanical properties and reduced formaldehyde release content up to 46%. The formaldehyde emission reduction could be due to the fact that the substitution of MF resin adhesive with liquefied wood led to less commercial resin that contains free formaldehyde, so it introduced less free formaldehyde into the particleboard. Additionally, the lignin-like substances in liquefied wood could react with the formaldehyde resulting in reduced content of formaldehyde in the board (Cuk et al., 2015).

Conclusions

Wood and other agricultural lignocellulosic biomass materials are natural resources available in huge quantity. The hydroxyl groups in biomass materials provide the sites for reaction in many chemical modification works involving the natural biomass-based polymers. Biomass liquefaction is a unique thermochemical conversion process for biomass utilizations. Liquefaction by phenol resulted in liquefaction products rich in combined phenolic compounds. Biomass can also be liquefied in the existence of alcohols, especially polyhydric alcohols, to result in liquefied products as polyols. From the biomass sources and liquefaction solvents, the liquefied products can be blended or cooked with synthetic chemical to make PF, PU, UF, MF, or MUF resins. These liquefied biomass-based resins have been applied in bonding plywood, making fiberboards, particleboards and other composite materials. As the biomass is a versatile raw material including hundreds of thousands of forestry and agricultural species, one aspect of future work should be on expanding the research into those species which are currently undervalued. Another aspect of future research is the improvement of the adhesive strength and water resistance of the liquefied biomass to meet the requirements of high end users. More important, developing new products or new processes that are able to accommodate the characteristics of liquefied biomass-based resins will greatly promote the resin sustainable manufacturing.

Keywords: Liquefaction, Liquefied biomass, Polyurethane, Phenolic, Resin, Bonding

References

Alma, M.H. and M.A. Basturk. 2006. Liquefaction of grapevine cane (*Vitis vinisera* L.) waste and its application to phenol-formaldehyde type adhesive. Ind. Crop. Prod. 24: 171–176.

Antonovic, A., B. Jambrekovic, J. Kljak, N. Spanic and S. Medved. 2010. Influence of urea-formaldehyde resin modification with liquefied wood on particleboard properties. Drvna Industrija. 61: 5–14.

Biswas, A., S. Kim, Z. He and H.N. Cheng. 2015. Microwave-assisted synthesis and characterization of polyurethanes from TDI and starch. Int. J. Polymer Anal. Charact. 20: 1–9.

Cuk, N., M. Kunaver, I. Poljansek, A. Ugovsek, M. Sernek and S. Medved. 2015. Properties of liquefied wood modified melamine-formaldehyde (MF) resin adhesive and its application for bonding particleboards. J. Adhes. Sci. Technol. 29: 1553–1562.

Dos Santos, R., J. Bordado and M. Mateus. 2015. Microwave-assisted liquefaction of cork-From an industrial waste to sustainable chemicals. Ind. Eng. Manag. 4: 173.

dos Santos, R.G., R. Carvalho, E.R. Silva, J.C. Bordado, A.C. Cardoso, M. do Rosario Costa and M.M. Mateus. 2016. Natural polymeric water-based adhesive from cork liquefaction. Ind. Crop. Prod. 84: 314–319.

D'Souza, J. and N. Yan. 2013. Producing bark-based polyols through liquefaction: Effect of liquefaction temperature. Sustainable Chem. Eng. 1: 534–540.

Esteves, B., J. Martins, J. Martins, L. Cruz-Lopes, J. Vicente and I. Domingos. 2015. Liquefied wood as a partial substitute of melamine-urea-formaldehyde and urea-formaldehyde resins. Maderas. Ciencia y tecnologla. 17: 277–284.

Fidan, M.S., M.H. Alma and I. Bektas. 2010. Liquefaction of cotton stalks (*Gossypium hirsutum* L.) with phenol. Wood Res. 55: 71–80.

Fierz-David, H.E. 1925. The liquefaction of wood and cellulose and some general remarks on the liquefaction of coal. Chem. Ind. Rev. 44: 942–944.

Fu, S., L. Ma, W. Li and S. Cheng. 2006. Liquefaction of bamboo, preparation of liquefied bamboo adhesives, and properties of the adhesives. Front. Forest. China. 1: 219–224.

Hajime, K., A. Yuki, N. Masayuki, F. Akira, M. Satoshi and N. Hirofumi. 2011. Synthesis of epoxy resins from alcohol-liquefied wood and the mechanical properties of the cured resins. J. Appl. Polymer Sci. 120: 745–751.

Hassan, E.B., M. Kim and H. Wan. 2009. Phenol-formaldehyde-type resins made from phenol-liquefied wood for the bonding of particleboard. J. Appl. Polymer Sci. 112: 1436–1443.

Juhaida, M., M. Paridah, M.M. Hilmi, Z. Sarani, H. Jalaluddin and A.M. Zaki. 2010. Liquefaction of kenaf (*Hibiscus cannabinus* L.) core for wood laminating adhesive. Bioresour. Technol. 101: 1355–1360.

Khan, M. and S. Ashraf. 2005. Development and characterization of a lignin-phenol-formaldehyde wood adhesive using coffee bean shell. J. Adhes. Sci. Technol. 19: 493–509.

Kishi, H., A. Fujita, H. Miyazaki, S. Matsuda and A. Murakami. 2006. Synthesis of wood-based epoxy resins and their mechanical and adhesive properties. J. Appl. Polymer Sci. 102: 2285–2292.

Kobayashi, M., Y. Hatano and B. Tomita. 2001. Viscoelastic properties of liquefied wood/epoxy resin and its bond strength. Holzforschung. 55: 667–671.

Krzan, A. and E. Zagar. 2009. Microwave driven wood liquefaction with glycols. Bioresour. Technol. 100: 3143–3146.

Kunaver, M., E. Jasiukaityte and N. Cuk. 2012. Ultrasonically assisted liquefaction of lignocellulosic materials. Bioresour. Technol. 103: 360–366.

Kunaver, M., S. Medved, N. ÄŒuk, E. Jasiukaityte, I. Poljansek and T. Strnad. 2010. Application of liquefied wood as a new particle board adhesive system. Bioresour. Technol. 101: 1361–1368.

Kurimoto, K., M. Takeda, S. Doi and H. Ono. 2001. Network structure and thermal properties of polyurethane films prepared from liquefied wood. Bioresour. Technol. 77: 33–40.

Lee, S.-H., Y. Teramoto and N. Shiraishi. 2002. Acid-catalyzed liquefaction of waste paper in the presence of phenol and its application to novolak-type phenolic resin. J. Appl. Polymer Sci. 83: 1473–1481.

Lee, W.J., Y.H. Wu, C.C. Wu, K.C. Chang and I.M. Tseng. 2010. Properties of compressio-molded plates made from wood powders impregnated with liquefied wood-based novolak-type phenol-formaldehyde resins. J. Appl. Polymer Sci. 118: 3471–3476.

Lee, W.-J. and C.-T. Liu. 2003. Preparation of liquefied bark-based resol resin and its application to particle board. J. Appl. Polymer Sci. 87: 1837–1841.

Lee, W.-J. and M.-S. Lin. 2008. Preparation and application of polyurethane adhesives made from polyhydric alcohol liquefied Taiwan acacia and China fir. J. Appl. Polymer Sci. 109: 23–31.

Lee, Y.-Y., W.-J. Lee, L.-Y. Hsu and H.-M. Hsieh. 2014. Properties of molding plates made with various matrices impregnated with PF and liquefied wood-based PF resins. Holzforschung. 68: 37–43.

Lin, L., M. Yoshioka, Y. Yao and N. Shiraishi. 1994. Liquefaction of wood in the presence of phenol using phosphoric acid as a catalyst and the flow properties of the liquefied wood. J. Appl. Polymer Sci. 52: 1629–1636.

Lin, R., J. Sun, C. Yue, X. Wang, D. Tu and Z. Gao. 2014. Study on preparation and properties of phenol-formaldehyde-chinese fir liquefaction copolymer resin. Maderas. Ciencia y Tecnologla. 16: 159–174.

Liu, Y., J. Gao, H. Guo, Y. Pan, C. Zhou, Q. Cheng and B.K. Via. 2014. Interfacial properties of loblolly pine bonded with epoxy/wood pyrolysis bio-oil blended system. BioResources 10: 638–646.

Lu, Z., L. Fan, Z. Wu, H. Zhang, Y. Liao, D. Zheng and S. Wang. 2015. Efficient liquefaction of woody biomass in polyhydric alcohol with acidic ionic liquid as a green catalyst. Biomass Bioener. 81: 154–161.

Maldas, D. and N. Shiraishi. 1997. Liquefaction of biomass in the presence of phenol and H_2O using alkalies and salts as the catalyst. Biomass Bioener. 12: 273–279.

Maldas, D., N. Shiraishi and Y. Harada. 1997. Phenolic resol resin adhesives prepared from alkali-catalyzed liquefied phenolated wood and used to bond hardwood. J. Adhes. Sci. Technol. 11: 305–316.

Mao, A., R. Shmulsky, Q. Li and H. Wan. 2014. Recycling polyurethane materials: a comparison of polyol from glycolysis with micronized polyurethane powder in particleboard applications. BioResource 9: 4253–4265.

Mateus, M.M., N.F. Acero, J.C. Bordado and R.G. dos Santos. 2015. Sonication as a foremost tool to improve cork liquefaction. Ind. Crop. Prod. 74: 9–13.

Mun, S.P. and E.B.M. Hassan. 2004. Liquefaction of lignocellulosic biomass with dioxane/polar solvent mixtures in the presence of an acid catalyst. J. Ind. Eng. Chem. 10: 473–477.

Mun, S.P., E.B. Hassan and T.H. Yoon. 2001. Evaluation of organic sulfonic acids as catalyst during cellulose liquefaction using ethylene carbonate. J. Ind. Eng. Chem. 7: 430–434.

Niu, M., G. Zhao and A. Hakki. 2011. Thermogravimetric studies on condensed wood residues in polyhydric alcohols liquefaction. Bioresour. 6: 615–630.

Ono, H. and K. Sudo. 1997. The manufacturing method of the resin raw material. Japan Patent JP2611166.

Ono, H., Y.-C. Zhang and T. Yamada. 2001. Dissolving behavior and fate of cellulose in phenol liquefaction. Trans. Mat. Res. Soc. Japan 26: 807–810.

Ono, H., T. Yamada, Y. Hatano and K. Motohashi. 1996. Adhesive from waste paper by means of phenolation. J. Adhes. 59: 135–145.

Pan, H. 2011. Synthesis of polymers from organic solvent liquefied biomass: A review. Renew. Sustain. Energy Rev. 15: 3454–3463.

Pan, H., T.F. Shupe and C.Y. Hse. 2008. Synthesis and cure kinetics of liquefied wood/phenol/ formaldehyde resins. J. Appl. Polymer Sci. 108: 1837–1844.

Petric, M., A. Ugovsek and M. Sernek. 2015. Bonding and surface finishing of wood with liquefied wood. Pro Ligno 11.

Pizzi, A. and K.L. Mittal. 2003. Handbook of Adhesive Technology. Marcel Dekker, New York.

Roslan, R., S. Zakaria, C.H. Chia, R. Boehm and M.-P. Laborie. 2014. Physico-mechanical properties of resol phenolic adhesives derived from liquefaction of oil palm empty fruit bunch fibres. Ind. Crop. Prod. 62: 119–124.

Sankar, G. and N. Yan. 2014. Bio-based two component (2 K) polyurethane adhesive derived from liquefied infested lodgepole pine barks. J. Biobased Mater. Bioenergy. 8: 457–464.

Tohmura, S.-I., G.-Y. Li and T.-F. Qin. 2005. Preparation and characterization of wood polyalcohol-based isocyanate adhesives. J. Appl. Polymer Sci. 98: 791–795.

Ugovsek, A. and M. Sernek. 2013a. Characterisation of the curing of liquefied wood by rheometry, DEA and DSC. Wood Sci. Technol. 47: 1099–1111.

Ugovsek, A. and M. Sernek. 2013b. Effect of pressing parameters on the shear strength of beech specimens bonded with low solvent liquefied wood. J. Adhes. Sci. Technol. 27: 182–195.

Ugovsek, A., A. Skapin, M. Humar and M. Sernek. 2013. Microscopic analysis of the wood bond line using liquefied wood as adhesive. J. Adhes. Sci. Technol. 27: 1247–1258.

Wan, H. and M.G. Kim. 2008. Distribution of phenol-formaldehyde resin in impregnated southern pine and effects on stabilization. Wood Fiber Sci. 40(2): 181–189.

Wen, M.-Y., J.-Y. Shi and H.-J. Park. 2013. Dynamic wettability and curing characteristics of liquefied bark-modified phenol formaldehyde resin (BPF) on rice straw surfaces. J. Wood Sci. 59: 262–268.

Wu, C.C. and W.J. Lee. 2008. Gluability of polyblend resins prepared from polyhydric alcohol liquefied ma bamboo with epoxy resin. Forest Prod. Ind. 27: 31–40.

Wu, C.-C. and W.-J. Lee. 2010a. Curing and thermal properties of copolymer epoxy resins prepared by copolymerized bisphenol-A and epichlorohydrin with liquefied Dendrocalamus latiflorus. Polymer J. 42: 711–715.

Wu, C.-C. and W.-J. Lee. 2010b. Synthesis and properties of copolymer epoxy resins prepared from copolymerization of bisphenol A, epichlorohydrin, and liquefied Dendrocalamus latiflorus. J. Appl. Polymer Sci. 116: 2065–2073.

Wu, C.-C. and W.-J. Lee. 2011. Curing behavior and adhesion properties of epoxy resin blended with polyhydric alcohol-liquefied Cryptomeria japonica wood. Wood Sci. Technol. 45: 559–571.

Xie, H. and T. Shi. 2006. Wood liquefaction by ionic liquids. Holzforschung. 60: 509–512.

Xie, J., J. Qi, C. Hse and T.F. Shupe. 2015. Optimization for microwave-assisted direct liquefaction of bamboo residue in glycerol/methanol mixtures. J. Forestry Res. 26: 261–265.

Xie, J., C.Y. Hse, T.F. Shupe, H. Pan and T. Hu. 2016. Extraction and characterization of holocellulose fibers by microwave-assisted selective liquefaction of bamboo. J. Appl. Polymer Sci. 133.

Xie, T. and F. Chen. 2005. Fast liquefaction of bagasse in ethylene carbonate and preparation of epoxy resin from the liquefied product. J. Appl. Polymer Sci. 98: 1961–1968.

Yamada, T. and H. Ono. 2001. Characterization of the products resulting from ethylene glycol liquefaction of cellulose. J. Wood Sci. 47: 458–464.

Yamada, T., S. Toyoda, K. Shimize and H. Ono. 2002. The manufacturing method of the resin raw material from the woody system matter with annular carbonic acid. Japan Patent JP3343564.

Yin, S., A.K. Mehrotra and Z. Tan. 2011. Alkaline hydrothermal conversion of cellulose to bio-oil: Influence of alkalinity on reaction pathway change. Bioresour. Technol. 102: 6605–6610.

Zhang, Q., G. Zhao and S. Jie. 2005. Liquefaction and product identification of main chemical compositions of wood in phenol. For. Stud. China 7: 31–37.

11

Preparation, Properties, and Bonding Utilization of Pyrolysis Bio-oil

An Mao,[1] Zhongqi He,[2] Hui Wan[3] and Qi Li[1],*

ABSTRACT

The rapid increase in energy consumption, limited fossil fuel resources, and environmental concerns have stimulated research need for biomass-derived fuels and chemicals. Pyrolysis is a thermal degradation process of biomass in the absence of oxygen. The liquid product from pyrolysis is known as bio-oil. Currently, it can be used as an energy source as well as feedstock for chemical production. In this chapter, we first discuss the pyrolysis process used in the preparation of bio-oil and its chemical components and properties. The bio-oil obtained from pyrolysis process is a multi-component mixture of molecules with different molecular weights derived from depolymerization of cellulose, hemicellulose, and lignin. Then, we review the current research progress of bio-oil in wood bonding utilization. Bio-oil can be blended or synthesized with commercial chemicals to prepare phenol formaldehyde (PF), polymeric diphenylmethane diisocyanate (pMDI), epoxy, and starch resin/adhesive. These bio-oil modified/blended adhesives have been applied in plywood, oriented strand board (OSB), particleboard, and flakeboard manufacturing.

[1] College of Forestry, Shandong Agricultural University, 271018, Taian, China.
[2] Southern Regional Research Center, USDA Agricultural Research Service, 1100 Robert E. Lee Blvd., New Orleans, LA 70124, USA.
[3] Department of Sustainable Bioproducts, Mississippi State University, 201 Locksley Way, Starkville, MS. 39759, USA.
* Corresponding author: wonderfulliqi2364@126.com

Finally, we conclude with discussions and future research directions. Efforts still have to be made to meet the challenges of making quality comparable products at a reasonable cost. Future research should focus on improving resin properties, developing new phenol separation/oil processing technology, and optimizing panel press parameters. Moreover, the bio-oil production process should also correspond with its target application to make better use of its potential as renewable resources.

Introduction

Increasing renewable energy use has stimulated the need for biomass-derived fuels and chemicals. Here, the word biomass refers to the plants or plant-based materials that can be used as the source of energy and chemicals. The scope of biomass includes, but is not limited to, woody and herbaceous materials, energy crops, bark, and forest/agricultural residues. Biomass can either be used directly via combustion to produce heat (Uchmiya and He, 2012), or indirectly by converting it into various forms of biofuel. Conversion of biomass to biofuel can be achieved with thermochemical or biochemical methods (Guo et al., 2012; Shen et al., 2015) and this chapter deals with a method called pyrolysis (Bridgewater et al., 1999). Different from liquefaction (Wan et al., 2016), pyrolysis is the thermal degradation of biomass in the absence of oxygen, resulting in production of solid, liquid, and gaseous products (Fig. 1) (He et al., 2016; Bridgewater et al., 1999). The solid product from pyrolysis of biomass is biochar (Guo et al., 2016) while the gaseous products are composed of CO_2, CO, CH_4, and small amounts of H_2 and O_2 (syngas) (He et al., 2016). The liquid product is known as pyrolysis oil or bio-oil (Mohan et al., 2006).

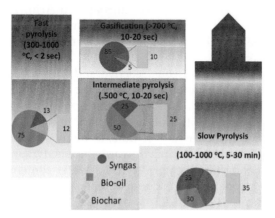

Figure 1. Thermochemical technologies used in conversion of biomass to biochar, bio-oil, and syngas. General range of pyrolysis temperature and time are indicated in the parenthesis after each individual process. Reprinted from He et al. (2016).

Bio-oil comes from renewable biomass resources. Currently, it can be used as an energy source as well as a feedstock for chemical production (Demirbas, 2009; Sukhbaatar et al., 2014). Through the development of pyrolysis processes, bio-oil has found its application in many industries, such as in the formulation of PF resin (Nakos et al., 2001; Amen-Chen et al., 2002; Yi et al., 2012; Choi et al., 2015) and starch-based adhesive (Zhang et al., 2014a), as part of pMDI binder system (Gagnon et al., 2004; Mao et al., 2011), and in epoxy blends (Liu et al., 2014) in resin industry. In this chapter, we will discuss the preparation and properties of bio-oil derived from pyrolysis of lignocellulosic biomass, and their utilization in wood bonding.

Preparation and Properties of Bio-oil

Pyrolysis is an efficient and cost-effective way of converting lignocellulosic biomass into liquids (bio-oil). Depending on the operating parameters, such as temperature, heating rate and retention time, pyrolysis can be divided into three subclasses: conventional (slow or intermediate) pyrolysis, fast pyrolysis, and flash pyrolysis (Demirbas and Arin, 2002; He et al., 2016). Fast pyrolysis is a commonly used technology for obtaining bio-oil with high yield (Fig. 1) (He et al., 2016; Zhang et al., 2007). The required operational parameters for fast pyrolysis are rapid application of temperature of 400 to 650°C with a heating rate from 600°C/sec to 10,000°C/sec, short retention time (<2 sec), and the absence of oxygen (Diebold, 2000). This temperature range results in the decomposition of the biomass cell structure into their molecular components. During pyrolysis, the hemicelluloses are broken down first at temperatures of 197 to 257°C, cellulose follows in the temperature from 237 to 347°C, and lignin is the last one to be pyrolyzed at temperatures of 277 to 497°C (Demirbas and Arin, 2002). At these temperatures biomass is converted to approximately 10%–15% gases by pyrolysis and the majority of biomass is converted to bio-oil (60%–75%) or charcoal (15%).

Bio-oil is a dark, acidic, viscous, reactive, and thermally unstable liquid comprising of substantial amounts of carboxylic acids which lead to a low pH value of 2–3 (Zhang et al., 2007). Depending on biomass feedstocks and pyrolysis processes, viscosities of bio-oils vary in a large range. During storage, the viscosity increases over time, as many chemical compounds comprising the bio-oil react slowly towards thermodynamic equilibrium (Scahill et al., 1997). Bio-oil typically has a water content of 15%–30%, which is derived from the original moisture in the biomass feedstock and the dehydration product during the pyrolysis (Zhang et al., 2007). The oxygen content of bio-oil is usually 35%–40%, distributed in more than 300 compounds depending on the source of feedstock and pyrolysis operating conditions (temperature, retention time, heating rate, etc.) (Oasmaa and Czernik, 1999; Scholze and Meier, 2001; Monreal and Schintzer, 2012).

The composition and properties of bio-oil depend on the source of biomass feedstock, reaction system, operating conditions, and the type of catalyst (Mohan, 2006). Thus, to have a clear picture of the production and properties of pyrolysis bio-oil, we are taking the work of Li et al. (2013a) as a case study. In this work, loblolly pinewood was used as the feedstock for pyrolysis. The trees were debarked and reduced to paper chip size (20–35 mm). The wood chips were then grounded and sieved to a particle size of 0.6–3 mm for efficient heat transfer. The grounded pinewood particles were air-dried for 1–2 weeks to moisture content of 8%–10%. The compositional and elemental analyses results of pinewood are shown in Table 1. The fast pyrolysis reaction was conducted in a 7 kg/h auger-fed proprietary pyrolysis reactor at Mississippi State University (MSU) bio-oil laboratory. Pinewood particles were pyrolyzed at 450°C with a retention time of 1–2 seconds. After biomass residuals were removed by a charcoal filter, the bio-oil was obtained with a water content of 18% and a viscosity of 368 mPa·s at 25°C. The yields of liquid, gaseous, and solid products are summarized in Table 2. The chemical compounds of bio-oil were determined by Gas Chromatography-Mass Spectrometer (GC-MS) (Table 3). Data in Table 3 shows the presence of numerous phenols and polyols in the bio-oil. Those compounds are the active ingredients for wood bonding application

Table 1. Compositional and elemental analyses of loblolly pinewood. Adapted from Li et al. (2013a).

Composition analyses	Value (oven dry %)
Ash	0.25
Extractives (by benzene/ethanol mixture)	6.03
Cellulose	39.46
Hemicellulose	19.77
Klason lignin	27.84
Elemental analysis	**Value (%)**
Carbon	46.81
Hydrogen	6.52
Nitrogen	0.18
Oxygen[a]	46.49
Sodium (ppm)	5.8
Potassium (ppm)	38.9
Calcium (ppm)	71.2
Magnesium (ppm)	20.3
Total alkali metals (ppm)	136.1

[a] Oxygen content was calculated by subtraction

Table 2. Percentage yields (feedstock weight based) of bio-oil, char, and non-condensable gases of loblolly pinewood by fast pyrolysis at 450°C. Average ± standard deviation (n = 3). Adapted from Li et al. (2013a).

Product	Yield (%)
Bio-oil	58.0 ± 1.5
Char	25.0 ± 2.9
Gases	10.4 ± 0.7

Table 3. Chemical composition of bio-oil produced from fast pyrolysis of loblolly pinewood at 450°C in a stainless steel proprietary auger reactor. Adapted from Li et al. (2013a).

Compounds	Concentration in bio-oil (%)
Acetic acid	5.59
Furfural	0.10
2-Methyl-2-cyclopentenone	0.06
3-Methyl-2-cyclopentenone	0.04
Phenol	0.06
3-methyl-1,2-cyclopentaned	0.38
2-Methylphenol	0.03
3-Methylphenol	0.06
2-Methoxyphenol	0.45
2,6-Dimethylphenol	0.00
2,4-Dimethylphenol	0.02
3-Ethylphenol	0.01
2,3-Dimethylphenol	0.00
1,2-benzenediol	0.30
5-(Hydroxymethyl)-2-furanc	0.41
3-Methyl-1,2-benzenediol	0.09
4-Ethyl-2-methoxy-phenol	0.24
4-methyl-1,2-benzenediol	0.28
2,6-dimethoxyphenol	0.00
Eugenol	0.18
2-Methoxy-4-propylphenol	0.08
Vanillin	0.33
cis-Isoeugenol	0.73
3,4-dimethylbenzoic acid	0.00
4-ethylcatechol	0.00
Levoglucosan	4.70
Oleic Acid	0.69

(Mao et al., 2014; Wan et al., 2016). In addition, selective extraction may be further applied as needed for enrichment of certain compounds (Li et al., 2013b).

Separation of Phenol-rich Bio-oil or Fraction

Phenolic compounds found in bio-oil come in two forms: monomeric phenol with sites substituted by alcohol, aldehyde, or carboxylic acid groups; oligmeric polyphenols with various number of phenolic structure units (Kim, 2015). Both of these compounds come mainly from pyrolyzed lignin (Yang et al., 2007; Bu et al., 2011). To obtain a higher percentage of reactive phenolic compounds or lignin fraction in bio-oil, separation is usually carried out prior to use of bio-oil (Kim, 2015). Other than that, biomass feedstock (Yoshikawa et al., 2013), pretreatment (Lu et al., 2013; Zhang et al., 2014b), operating parameters (Murwanashyaka et al., 2001; Kim et al., 2010), type of additive/catalyst (Butt 2006; Bu et al., 2012; 2013; Peng et al., 2014) may also influence the availability of phenolic compounds in bio-oil. Thus, for a high-quality phenol-rich bio-oil or its fraction, additional separation or extraction procedures are applied after the biomass pyrolysis (Fig. 2).

1. Solvent Extraction

Commonly used separation methods are solvent extraction and distillation (Garcia-Perez et al., 2007; Effendi et al., 2008; Zilnik and Jazbinsek, 2012). In solvent extraction, the separation of lignin fraction is based on the lignin's good solubility in organic solvents and poor solubility in water (Chum and Black, 1990). Ethyl acetate is a common organic solvent. The lignin and neutral fractions of bio-oil are soluble in ethyl acetate, while the other components derived from cellulose and hemicellulose are much more soluble in water (Sukhbaatar et al., 2009). As described in a patent (Chum and Kreibich, 1992), by mixing bio-oil and acetate, a fraction with lignin and neutral was obtained. This ethyl acetate-soluble fraction was separated and washed with water and sodium bicarbonate solutions to remove organic acids. A lignin fraction was obtained after the evaporation of ethyl acetate. Dobele et al. (2010) used a simple solvent, water, to extract pyrolytic lignin from bio-oil. The bio-oil was obtained from fast pyrolysis of alder, ash-tree, and aspen mixture in a laboratory scale reactor. When water was added into bio-oil, precipitation occurred. The water-insoluble fraction was separated and dried to obtain pyrolytic lignin fraction. The content of phenolic compounds in this fraction was between 75%–83%. The authors (Dobele et al., 2010) also proposed another procedure to separate monomeric phenol from bio-oil using methyl tert-butyl ether (MTBE) as organic solvent (Fig. 2).

In addition to the conventional solvents, other carriers have also been introduced and applied, such as supercritical CO_2 (Zhang et al., 2013).

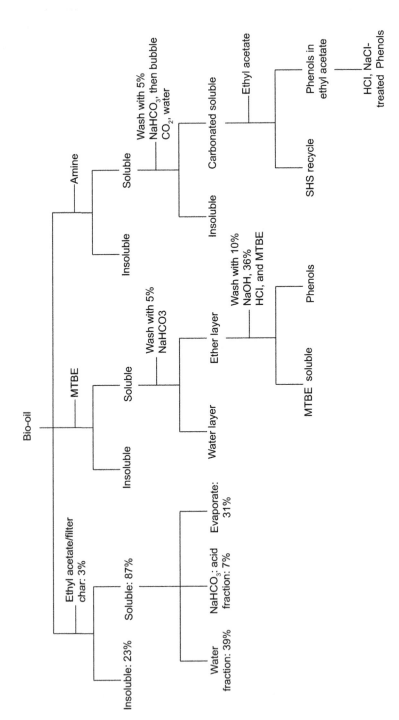

Figure 2. Three solvent extraction procedures used in bio-oil separation and fractionation. SHS - switchable hydrophilicity solvent. MTBE - methyl tert-butyl ether. Compiled per Chum and Kreibich (1992), Dobele (2010), and Fu et al. (2014).

Patel et al. (2011) reported extraction of phenol-rich fraction from bio-oil using CO_2 as the supercritical fluid. The biomass feedstock (cashew net shells and sugarcane bagasse) was vacuum pyrolyzed under the pressure range of 120–300 bar, at temperature range of 30–60°C and the mass flow rate range of 0.7–1.2 kg/h. A phenol concentration of about 72% was obtained and a total yield of 15% of the bio-oil was extracted. More recently, Fu et al. (2014) introduced a procedure using switchable hydrophilicity solvents (SHS) to extract phenolic compounds from lignin microwave-pyrolysis oil. With the extraction procedure illustrated in Fig. 2, three fractions were obtained. 96% of the bio-oil was recovered in its three fractions, 72% of guaiacol and 70% of 4-methylguaiacol were extracted, and 91% of the solvent SHS was recovered.

2. Distillation

Because of their complex compositions, bio-oils show a wide range of boiling point temperature (Zhang et al., 2013). Distillation separates bio-oil components according to their different boiling points. Atmospheric pressure, vacuum, steam, and molecular distillation can be applied to bio-oil separation (Kim, 2015). Czernik and Bridgwater (2004) reported that bio-oils started boiling below 100°C, while stopping at 250–280°C, leaving 35%–50% as solid residues. Compared to atmospheric pressure, vacuum pressure provides some advantages for its lower temperature operation, because during the distillation, heat could induce the polymerization of some reactive components in bio-oil (Zhang et al., 2013). However, the energy consumption of vacuum distillation would be much higher.

A piece of molecular distillation equipment was used by Guo et al. (2010a) to fractionate woody bio-oil obtained from fluidized-bed flash pyrolysis into three fractions: light fraction (LF), middle fraction (MF), and heavy fraction (HF). The LF was mainly composed of water and acids. MF and HF contained more phenols derived from lignin pyrolysis. Another work by Guo et al. (2010b) combined conventional vacuum distillation and molecular distillation. Vacuum distillation was first applied to remove most of the water from bio-oil. Then two distillation processes based on molecular weight of chemicals were conducted. Two fractions were finally obtained under each combination of the operating conditions (80°C, 1600 Pa and 80°C, 340 Pa). Each condition combination resulted in a distilled fraction and a residual fraction. In each distilled fraction, phenolic compounds are the most abundant chemicals.

Other Methods to Obtain a Phenol-rich Bio-oil

To obtain a higher concentration of phenolic compounds without extensive post-processing of bio-oil, the choice of feedstock could turn to lignin-rich materials instead of common woody biomass (Yoshikawa et al., 2013). Other than that, pretreatment of feedstock, type of pyrolysis, operating conditions, type of additive/catalyst could also influence phenol concentration in bio-oil and these studies have been reported (Amen-Chen et al., 2001; Murwanashyaka et al., 2001; Butt, 2006; Kim et al., 2010; Bu et al., 2012; 2013; Lu et al., 2013; Peng et al., 2014; Zhang et al., 2014b).

Bonding Application of Bio-oil

As discussed in the previous part of this chapter, bio-oil is a multi-component mixture of molecules with different molecular weights derived from depolymerization of cellulose, hemicellulose, and lignin (Zhang et al., 2007). With water excluded, over 99% of bio-oil is composed of acids, alcohols, aldehydes, esters, ketones, sugars, phenols, guaiacols, syringols, furans, lignin derived phenols, and extractible terpene with multi-functional groups (Guo et al., 2001). Therefore, it has been considered as a potential source for various industrial applications (Radlein, 1999). In particular, phenolic compounds in bio-oil have various uses in forest products industry, for example, as part of bonding materials for composite wood products (Amen-Chen et al., 2002; Effendi et al., 2008; Yi et al., 2012; Chaouch et al., 2014). Furthermore, some compounds in bio-oil with hydroxyl groups may also be reactive with groups in isocyanate resins and might form new adhesive systems (Gagnon et al., 2003; Mao and Shi, 2012). Applications of bio-oil also have been found in epoxy resin and starch adhesive (Liu et al., 2014; Zhang et al., 2014a). Selected studies are summarized and listed in Table 4. In this part, we will review the studies in utilization of bio-oil as wood binders in recent years. Specifically, we classify the type of usable bio-oil into two categories: the whole bio-oil and the phenol-rich bio-oil/fraction. We will discuss their applications separately.

Use of Whole Bio-oil in Wood Bonding

1. *Phenolic Resins*

   Phenolic resins are synthesized by reacting formaldehyde with phenols in the presence of alkaline catalyst. As the first commercial synthetic resins, they have been widely used for bonding hot-pressed wood panels, such as plywood, OSB, and laminated veneer lumber (LVL) for their good bonding performance and high exterior durability (Gardziella et al., 2000). Phenol is a petroleum-derived chemical, which means it cannot be obtained directly from biomass in a commercial scale now and its price and availability are closely related to those of fossil fuels. This contradicts with current trends of reducing dependence on nonrenewable resources and promoting the utilization of sustainable bio-based products.

Table 4. Selected studies of utilization of bio-oil in wood bonding.

Reference	Biomass feedstock	Resin/adhesive blend	Bio-oil type	Bonding usage
Nakos et al., 2001	Unreported	Resol PF	Whole bio-oil	OSB and Plywood
Amen-Chen et al., 2002	Spruce bark, fir bark with some hardwood bark	Resol PF	Whole bio-oil	OSB
Gagnon et al., 2004	Spruce bark, fir bark with some hardwood bark	pMDI	Whole bio-oil	Particleboard
Sukhbaatar et al., 2009	Loblolly pine	Resol PF	Phenol-rich oil/ fraction	OSB
Mao et al., 2011	Pinewood	pMDI	Whole bio-oil	Flakeboard
Yi et al., 2012	Poplar, larch, and bamboo	Resol PF	Whole bio-oil	Plywood
Chaouch et al., 2014	Trembling aspen and white spruce	Resol PF	Whole bio-oil	Maple blocks
Li et al., 2014	Larch sawdust	UF	Whole bio-oil	Poplar plywood
Liu et al., 2014	A common American hardwood	Epoxy	Whole bio-oil	Loblolly pine veneers
Zhang et al., 2014a	Unreported	Starch	Whole bio-oil	Plywood
Aslan et al., 2015	Scotch pine	Resol PF	Phenol-rich oil/ fraction	Scotch pine lamellas
Choi et al., 2015	Palm kernel shell	Resol PF	Whole bio-oil	Maple blocks
Ozbay and Ayrilmis, 2015	Scotch pine	Scotch pine	Phenol-rich oil/ fraction	Beech lamellas

Therefore, developing technologies for using alternative raw materials, such as biomass-derived phenols is necessary.

Nakos et al. (2001) studied the use of bio-oil to replace part of the phenol in the formulation of PF resin for plywood and OSB. They (Nakos et al., 2001) reported that PF resin with phenol substitution of up to 50% by bio-oil could be produced with modified synthesis procedure. The reactivity and bonding performance of bio-oil-PF resin were comparable to those of conventional PF. Amen-Chen et al. (2002) synthesized PF resins using bark (mixture of spruce, fir, and some hardwood barks) vacuum pyrolysis oil. The effects of bio-oil substitution percentage, formaldehyde/ phenol molar ratio, and NaOH/phenol molar ratio on properties and performances of PF resins were investigated. Homogeneous OSB panels bonded with bio-oil-PF resins (25% and 50% bio-oil substitutions, respectively) have mechanical properties similar to those bonded with commercial PF. Three-layer OSB panels with core layer bonded with resin of 25% bio-oil substitution and face layers bonded with resin of 50% bio-oil substitution met the requirements specified by Canadian Standards CSA O437.0-93 for OSB products. Yi et al. (2012) prepared three types of bio-oil-PF resins using bio-oil obtained from pyrolyzed poplar, larch, and bamboo, respectively, at substitution percentage of 30%. These resins were tested as plywood binders. The results showed that all the plywood specimens achieved the standard requirements of GB/T 9846-2004. The bonding performances of these resins followed the order: larch bio-oil-PF > poplar bio-oil-PF > bamboo bio-oil-PF, the same order of phenolics concentrations in bio-oil (Phenolics in larch, poplar, and bamboo bio-oils were 14.8%, 13.9%, and 10.4%, respectively).

Bio-oil derived from fast pyrolysis of larch sawdust was also used to modify urea-formaldehyde (UF) resin (Bio-oil amounts varied from 5%–15%, based on the weight of urea) (Li et al., 2014). The bio-oil was found to be able to co-condense with urea and formaldehyde to prepare urea-phenol-formaldehyde type co-condensed resin with improved thermal stability. The fabricated three-layer poplar plywood displayed remarkably reduced formaldehyde emission. Although the wet strength of plywood decreased with increase of bio-oil percentage in the modified resin, it still met the Chinese standard requirement set by GB/T 9846-2004. Chaouch et al. (2014) pyrolyzed trembling aspen and white spruce wood into bio-oil and used it to substitute 25%, 50%, and 75% of phenols in synthesis of bio-oil-PF resins, respectively. The obtained resin with 50% substitution showed good storage stability and thermal stability comparable to the control PF resin. The shear strength of bonded maple blocks was improved. In another study (Choi et al., 2015), palm kernel shells were fast pyrolyzed in a fluidized bed reactor to obtain a bio-oil. The contents of phenolic compounds in bio-oil varied from 17.9% to 24.8% depending on different pyrolysis temperatures. The phenol content was 8.1%, relatively high compared to that in bio-oil obtained from woody materials. PF resins with acceptable performances were synthesized by substituting phenol with up to 25% bio-oil.

2. *pMDI*

Isocyanate resin, such as pMDI is another resin for bonding durable wood composite panels. Since they were first introduced to the German particleboard market in the early 1970s, the use of isocyanate binders in composite boards has grown significantly. Although pMDI resin can exhibit some enhanced performance over PF resin, they are a lot more costly than PF and their high reactivity may cause an adhesion to press platen (Gagnon et al., 2003).

Bio-oil has the potential to be a suitable component in pMDI-bio-oil binder system for wood composite boards, because bio-oil components such as sugars, carboxylic acids, phenols, alcohols, water, tannin/lignin derivatives may react with pMDI (Gagnon et al., 2003) to form polyurethane type of adhesive for wood bonding purposes. Gagnon et al. (2003) developed a pMDI-softwood bark bio-oil hybrid adhesive system. The reaction between these two components was examined by differential scanning calorimetry (DSC) and rheology. Results showed that chemical reactions occurred between pMDI and bio-oil at room temperature. The bio-oil was used to replace 30%–40% of pMDI in a 4% resin loading particleboard (Gagnon et al., 2004). The mechanical properties and thickness swelling of the boards exceeded the minimum requirements set by ANSI A208.1-1993 and ASTM 1037-96a standards. The study indicated that bio-oil could be mixed at weight ratios as high as 40% with pMDI and give acceptable interior grade particleboard properties. It was also found that bio-oil could help solve the adhesion to press platen problem.

Mao et al. (2011) investigated the feasibility of using pinewood bio-oil as part of a pMDI binder system for flakeboard. The bio-oil was obtained from MSU bio-oil laboratory (Li et al., 2013a,b). The pMDI was mechanically mixed with 25%, 50%, and 75% of bio-oil to make adhesive systems. Acetone was added into these systems to maintain suitable viscosity for easy spraying. The effects of resin loading rate and the pMDI/bio-oil weight ratio on the physical and mechanical properties of the flakeboard were examined. It was found that adding acetone lowered the viscosities of the adhesives (Table 5) and improved resin spraying efficiency; at resin loading rate of 7%, the bonding properties of flakeboard made with an adhesive containing 25% bio-oil was comparable to that made with a pure pMDI (Table 6). The authors (Mao et al., 2011) also

Table 5. Viscosity of the adhesive systems with different mixing formulations. Adapted from Mao et al. (2011).

Resin code[a]	Resin composition (%)			Viscosity (mPa·s)
	pMDI	Bio-oil	Acetone	
1	100	0	0	226
2	75	25	10[b]	188
3	50	50	10[b]	146
4	25	75	10[b]	127
5	0	100	10[b]	121
6	0	100	0	368

[a] The resin codes represent the indicated resin compositions.
[b] Based on bio-oil solid weight.

Table 6. Mechanical and physical properties of flakeboard bonded with different resin formulations. Adapted from Mao et al. (2011).

Resin code[a]	Resin loading rate (%)	Board density (g/cm³)	Internal bond (MPa)	Modulus of rupture (MPa)	Modulus of elasticity (MPa)	Thickness swelling (%)	Water absorption (%)
1	3	0.79	0.65	46.0	9,182	20.0	25.1
1	5	0.84	1.21	56.4	9,737	14.9	19.0
1	7	0.88	1.44	63.1	10,536	13.1	14.2
2	3	0.85	0.58	43.6	7,908	26.5	26.4
2	5	0.85	1.01	49.5	8,523	20.5	22.6
2	7	0.85	1.40	51.8	9,005	14.5	13.0
3	3	0.79	0.50	43.9	6,721	27.1	28.8
3	5	0.81	0.69	46.5	6,826	21.9	24.4
3	7	0.80	0.74	48.4	8,525	18.9	18.5
4	3	0.76	0.25	33.6	6,399	45.7	40.7
4	5	0.84	0.52	35.7	6,585	34.7	34.1
4	7	0.81	0.55	46.7	8,177	29.8	32.6

[a] Resin codes are defined in Table 5.

attempted to make flakeboard with whole bio-oil as the sole binder, but without success. In the successive study, Mao and Shi (2012) tested the thermal properties and curing behavior of these adhesive systems using dynamic mechanical analysis (DMA). The results indicated that less time was needed to reach the maximum storage modulus of each adhesive system with an increase in bio-oil content. The adhesive containing 75% bio-oil presented the fastest curing rate but the lowest storage modulus value. The DMA results of this study (Mao and Shi, 2012) correlated well with the mechanical properties of flakeboards made (Mao et al., 2011).

3. *Epoxy Resin*

Here, epoxy resins refer to low molecular weight pre-polymers that normally contain at least two epoxide groups. Epoxy resins create strong bonds between epoxy and wood. Therefore, they can be used in structure wood products applications (Vick et al., 1995). Liu et al. (2014) blended epoxy resin with a hardwood pyrolysis bio-oil at 10%, 30%, and 50% weight percentages, respectively. The bio-oil was produced in a circulating fluidized bed. The viscosities of the blended resins increased with the content of bio-oil in blends. The shear strength of loblolly pine veneers bonded with epoxy resin blends decreased slowly with an increase of the bio-oil content (up to 50%) in blends. Chemical reactions between wood and epoxy resin blends were confirmed by Attenuated Total Reflectance-Fourier Transform Infrared spectroscopy (ATR-FTIR). DSC, scanning electron microscope (SEM), and optical microscope analyses results also proved an improved resin polymerization crosslinking process and enhanced moisture resistance of resin blends bond line.

4. *Starch Adhesive*

Derived from natural materials, starch adhesives are normally for interior uses. Their applications in wood composite panels are limited by their low water resistance, susceptible to microbial invasion, and low bonding strength. Zhang et al. (2014a) synthesized modified starch adhesives using starch and bio-oil as the primary components with other agents added as well. Prior to use, the modified adhesive was blended with a certain amount of flour powders to reach a target solid content of 40% and then applied onto birch veneers. Plywood was made at pressure of 1.0 MPa, temperature of 120°C, and press time of 1.0 min/mm. The bonding strength met the requirements specified by Chinese standard GB/T 9846-2001 for interior-grade plywood.

Use of Phenol-rich Bio-oil/Fraction in Wood Bonding

Although the separation of phenol-rich fraction from bio-oil has been extensively studied, its application in wood bonding is limitedly reported. Compared to the procedures discussed above, a lower cost procedure of separating lignin was introduced by Sukhbaatar et al. (2009). The authors extracted phenol-rich oil (pyrolytic lignin fraction) from pinewood bio-oil and used it in formulating PF resin as OSB binders. The separation procedure employed water and methanol in two steps to obtain a water-insoluble pyrolytic lignin fraction. The procedure is illustrated in Fig. 3 and described as follows: First, water was added to bio-oil to obtain a water-rich layer and a water-insoluble fraction. Then the water-rich layer was removed. The water-insoluble fraction left behind, was mixed with methanol in a 1:1 weight ratio to give a clear solution. A certain amount of water was gradually added again to give a precipitate, which was a viscous liquid with dark color, and was considered as lignin fraction in bio-oil. This procedure yielded a lignin fraction of 25% based on dry weight of bio-oil and may be more cost-effective than the ethyl acetate procedure. The lignin fraction was used to replace 30%, 40%, and 50% of phenol in PF resin production. OSB panels were made with these modified PF as core layer binders. The test results showed that resin made with 30% lignin substitution could be comparable to control PF resin in bonding core-layers of OSB, and 40% lignin substitution gave somewhat lower values but might be improved by some adjustments of resin synthesis procedures and/or board hot-pressing parameters.

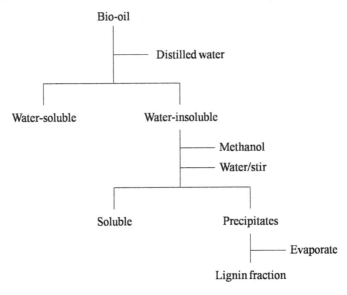

Figure 3. Separation of phenol-rich oil (pyrolytic lignin fraction) from pinewood bio-oil. Drawn as per the procedure described in Sukhbaatar et al. (2009).

In a recent study (Aslan et al., 2015), scotch pine wood sawdust was pyrolyzed at 500°C within a fixed-bed type reactor under a nitrogen environment. The obtained bio-oil was extracted with diethyl ether in two stages to obtain an aqueous fraction and an organic fraction. The organic fraction was evaporated to recover the diethyl ether. The extracted bio-oil was obtained and analyzed by GC-MS. It was composed of aldehydes, ketones, phenols, acids, benzenes, and alcohols, but the primary components were phenols. This phenolic-rich fraction was mechanically mixed with available commercial PF resin at different ratios (up to 40%). FT-IR analysis showed that the phenolic-rich fraction had similar chemical structure to commercial PF resin. The viscosities of modified PF increased with the content of phenolic-rich fraction. Single lap-joint wood specimens were prepared to test bond performance of these adhesives. The results showed that PF resin mixed with 10% of phenolic-rich fraction could give good bonds. However, further increment in the fraction content decreased the bond strength. The reason may be the lower reactivity of the high molecular weight phenolic compounds in this fraction as compared to the commercial phenols (Effendi et al., 2008). Other reasons might be the increased viscosity of the mixed adhesives resulting in an insufficient penetration of resin into wood and the acidity of the adhesive mixtures with increased organic acids content in the bio-oil. In another study (Ozbay and Ayrilmis, 2015), the same bio-oil and extraction methods were used, but the lap-joint test specimens were different. The authors concluded that the bonding performances of the PF resins improved with increasing bio-oil content up to 20% under dry condition.

In addition, Shahid et al. (2014; 2015) reported the characterization and bonding performance of phenol-formaldehyde resins modified with crude bio-oil prepared from Ziziphus mauritiana endocarps. However, in their work, the shell stone powder (50 g) was first mixed with 50% aqueous ethanol (500 g). The reactor was then pressured to 2.0 MPa with nitrogen and with the reaction temperature raised to 300°C (10°C/min) and kept at it for 15 min. Although the liquid product is rich in phenols, it may be more suitable in the category of liquefied biomass product (Wan et al., 2016).

Conclusions

This chapter covers the preparation and properties of bio-oil from pyrolysis of biomass, and its applications in wood bonding. Bio-oil can be blended or synthesized with available commercial chemicals to prepare PF, pMDI, epoxy, and starch resin/adhesives. These bio-oil modified/blended resins/adhesives have been applied in plywood, OSB, particleboard, and flakeboard panel manufacturing. As we discussed in the second part, to use an untreated whole bio-oil in resin synthesis process or simply mechanically blend with other resins seems to be the most economical way of applying bio-oil. However, it is worth pointing out that the chemical components in bio-oil are complex and the percentage of real reactive components in bio-oil is relatively low, which might result in panels with inconsistent quality. Other than that, from our experience, strong smoky odor of bio-oil could continuously emanate from panels even after the resin is fully cured, which could create a concern about air pollution. On the other hand, the separation/processing technologies for phenol-rich bio-oil/fraction have been introduced for years but still not commercialized, probably due to the high cost of bio-oil separation and the complexity of the procedure. Using reactive phenol-rich bio-oil/fraction should obtain better or consistent resin performance, but with the possible sacrifice of economy. However, limited reports have been found on the comparison of structures, properties, application performances, and costs of resins made/blended by whole bio-oil and phenol-rich bio-oil/fraction, respectively. This kind of study would be interesting and meaningful.

Generally, considerable progress has been made towards reaching the goal of developing new products from renewable sources. However, efforts still have to be made to meet the challenges of making competitive quality products at reasonable cost. From this point of view, future research should focus on improving resin properties, developing new phenol separation/oil processing technology, and optimizing panel press parameters. Moreover, since the composition of bio-oil is complicated and varied by biomass feedstock, pyrolysis operating parameters, catalyst/additive type, and pretreatments, the bio-oil production process should also correspond with its target application to make better use of its potential as renewable resources.

Keywords: Pyrolysis, Bio-oil, PF resin, pMDI, Biomass, Bonding

References

Amen-Chen, C., H. Pakdel and C. Roy. 2001. Production of monomeric phenols by thermochemical conversion of biomass: a review. Bioresour. Technol. 79: 277–299.

Amen-Chen, C., B. Riedl and C. Roy. 2002. Softwood bark pyrolysis oil-PF resols. Part 1. Resin synthesis and OSB mechanical properties. Holzforschung. 56: 167–175.

Aslan, M., G. Ozbay and N. Ayrilmis. 2015. Adhesives characteristics and bonding performance of phenol formaldehyde modified with phenol-rich fraction of crude bio-oil. J. Adhes. Sci. Technol. 457(24): 2679–2691.

Bridgewater, A.V., D. Meier and D. Radlein. 1999. An overview of fast pyrolysis of biomass. Org. Geochem. 30: 1479–1493.

Bu, Q., H. Lei, S. Ren, L. Wang, J. Holladay, Q. Zhang, J. Tang and R. Ruan. 2011. Phenol and phenolics from lignocellulosic biomass by catalytic microwave pyrolysis. Bioresour. Technol. 102: 7004–7007.

Bu, Q., H. Lei, S. Ren, L. Wang, Q. Zhang, J. Tang and R. Ruan. 2012. Production of phenols and biofuels by catalytic microwave pyrolysis of lignocellulosic biomass. Bioresour. Technol. 108: 274–279.

Bu, Q., H. Lei, L. Wang, Y. Wei, L. Zhu, Y. Liu, J. Liang and J. Tang. 2013. Renewable phenols production by catalytic microwave pyrolysis of Douglas fir sawdust pellets with activated carbon catalysts. Bioresour. Technol. 142: 546–552.

Butt, D.A.E. 2006. Formation of phenols from the low-temperature fast pyrolysis of Radiata pine (Pinus radiata) Part I. Influence of molecular oxygen. J. Anal. Appl. Pyrol. 76: 38–47.

Choi, G., S. Oh, S. Lee and J. Kim. 2015. Production of bio-based phenolic resin and activated carbon from bio-oil and biochar derived from fast pyrolysis of palm kernel shells. Bioresour. Technol. 178: 99–107.

Chum, H.L. and S.K. Black. 1990. Process for fractionating fast pyrolysis oils, and products derived therefrom. U.S. Patent 4,942,269.

Chum, H.L. and R.E. Kreibich. 1992. Process for preparing phenol-formaldehyde resin products derived from fractionated fast-pyrolysis oil. U.S. Patent 5,091,499.

Chaouch, M., P.N. Diouf, A. Laghdir and S. Yin. 2014. Bio-oil from whole-tree feedstock in resol-type phenolic resins. J. Appl. Polymer Sci. 131: 596–602.

Czernik, S. and A.V. Bridgwater. 2004. Overview of applications of biomass fast pyrolysis oil. Energy Fuels. 18: 590–598.

Demirbas, A. 2009. Chemical valorization of wood for bio-fuels and bio-chemicals. Energy Sources. 32: 1–9.

Demirbas, A. and G. Arin. 2002. An overview of biomass pyrolysis. Energy Sources. 24: 471–482.

Diebold, J.P. 2000. A review of the chemical and physical mechanisms of the storage stability of fast pyrolysis bio-oils. Rep. NREL/SR-570-27613, National Renewable Energy Laboratory. Golden, CO.

Dobele, G., T. Dizhbite, J. Ponomarenko, I. Urbanovich, J. Kreicberga and V. Kampars. 2010. Isolation and characterization of the phenolic fractions of wood pyrolytic oil. Holzforschung 65: 503–510.

Effendi, A., H. Gerharser and A.V. Bridgwater. 2008. Production of renewable phenolic resins by thermochemical conversion of biomass: A review. Renewable Sustainable Energy Reviews. 2008: 2092–2116.

Fu, D., S. Farag, J. Chaouki and P.G. Jessop. 2014. Extraction of phenols from lignin microwave-pyrolysis oil using a switchable hydrophilicity solvent. Bioresour. Technol. 154: 101–108.

Gagnon, M., C. Roy and B. Riedl. 2004. Adhesives made from isocyanates and pyrolysis oils for wood composites. Holzforschung. 58: 400–407.

Gagnon, M., C. Roy, D. Rodrigue and B. Riedl. 2003. Calorimetric and rheological study of isocyanate-pyrolysis oil blends. J. Appl. Polymer Sci. 89: 1362–1370.

Garcia-Perez, A. Chaala, H. Pakdel, D. Kretschmer and C. Roy. 2007. Characterization of bio-oils in chemical families. Biomass and Bioenergy. 31: 222–242.

Gardziella, A., A. Knop and L.A. Pilato. 2000. Phenolic Resins: Chemistry, Application, Safety and Ecology. Springer-Verlag, Berlin Heidelberg.

Guo, Y., Y. Wang and F. Wei. 2001. Research progress in biomass flash pyrolysis technology for liquids production. Chemical Industry Engineering Progress. 8: 13–17.

Guo, X., S. Wang, Z. Guo, Q. Liu, Z. Luo and K. Cen. 2010a. Pyrolysis characteristics of bio-oil fractions separated by molecular distillation. Applied Energy. 87: 2892–2898.

Guo, Z., S. Wang, Y. Gu, G. Xu, X. Li and Z. Luo. 2010b. Separation characteristics of biomass pyrolysis oil in molecular distillation. Separation and Purification Technology. 76: 52–57.

Guo, M., Y. Shen and Z. He. 2012. Poultry litter-based biochar: preparation, characterization, and utilization. pp. 171–102. In: Z. He (ed.). Applied Research of Animal Manure: Challenges and Opportunities beyond the Adverse Environmental Concerns. Nova Science Publishers, Inc., New York.

Guo, M., Z. He and S.M. Uchimiya (eds.). 2016. Agricultural and Environmental Applications of Biochar: Advances and Barriers, pp. 1–504. Soil Science Society of America, Inc., Madison, WI.

He, Z., S.M. Uchimiya and M. Guo. 2016. Production and characterization of biochar from agricultural by-products: Overview and use of cotton biomass residues. pp. 63–86. In: M. Guo et al. (eds.). Agricultural and Environmental Applications of Biochar: Advances and Barriers. Soil Science Society of America, Inc., Madison, WI.

Kim, J.S. 2015. Production, separation and applications of phenolic-rich bio-oil–A review. Bioresour. Technol. 178: 90–98.

Kim, S.J., S.H. Jung and J.S. Kim. 2010. Fast pyrolysis of palm kernel shells: Influence of operation parameters on the bio-oil yield and the yield of phenol and phenolic compounds. Bioresour. Technol. 101: 9294–9300.

Li, B., J.-Z. Zhang, X.-Y. Ren, J.-m. Chang and J.-s. Gou. 2014. Preparation and characterization of bio-oil modified urea-formaldehyde wood adhesives. BioResources 9: 5125–5133.

Li, Q., P.H. Steele, F. Yu, B. Mitchell and E.B.M. Hassan. 2013a. Pyrolytic spray increases levoglucosan production during fast pyrolysis. J. Anal. Appl. Pyrolysis 100: 33–40.

Li, Q., P.H. Steele, B.K. Mitchell, L.L. Ingram and F. Yu. 2013b. The addition of water to extract maximum levoglucosan from the bio-oil produced via fast pyrolysis of pretreated loblolly pinewood. BioResour. 8: 1868–1880.

Liu, Y., J. Gao, H. Guo, Y. Pan, C. Zhou, Q. Cheng and B.K. Via. 2014. Interfacial properties of loblolly pine bonded with epoxy/wood pyrolysis bio-oil blended system. BioResour. 10: 638–646.

Lu, Q., Z. Zhang, X. Yang, C. Dong and X. Zhu. 2013. Catalytic fast pyrolysis of biomass impregnated with K_3PO_4 to produce phenolic compounds: Analytical Py-GC/MS study. J. Anal. App. Pyrolysis. 104: 139–145.

Mao, A., S.Q. Shi and P.H. Steele. 2011. Flakeboard bonded with polymeric diphenylmethane diisocyanate/bio-oil adhesive systems. Forest Products Journal. 61: 240–245.

Mao, A. and S.Q. Shi. 2012. Dynamic mechanical properties of polymeric diphenylmethane diisocyanate/bio-oil adhesive system. Forest Products Journal. 62: 201–206.

Mao, A., R. Shmulsky, Q. Li and H. Wan. 2014. Recycling polyurethane materials: a comparison of polyol from glycolysis with micronized polyurethane powder in particleboard applications. BioResources 9: 4253–4265.

Mohan, D., C.U. Pittman and P.H. Steele. 2006. Pyrolysis of wood/biomass for bio-oil: A critical review. Energy & Fuels 20: 848–889.

Monreal, C.M. and M.I. Schnitzer. 2012. Bio-oil production from poultry litter: Potential energy, environmental and chemcial opportunities. pp. 83–138. *In*: Z. He (ed.). Applied Research of Animal Manure-Challenges and Opportunites beyond the Adverse Environmental Concerns. Nova Science Publishers, New York.

Murwanashyaka, J.N., H. Pakdel and C. Roy. 2001. Step-wise and one-step vacuum pyrolysis of birch-derived biomass to monitor the evolution of phenols. Journal of Analytical and Applied Pyrolysis 60: 219–231.

Nakos, P., S. Tsiantzi and A. Eleftheria. 2001. Wood adhesives made with pyrolysis oils. pp. 1–8. *In*: Proceedings of 3rd European Wood-based Panel Symposium, 2001 Sep 12–14; European Panel Federation & Wilhelm Klauditz Institute, Hannover.

Oasmaa, A. and S. Czernik. 1999. Fuel oil quality of biomass pyrolysis oils: state of the art for the end users. Energy & Fuels 13: 914–921.

Ozbay, G. and N. Ayrilmis. 2015. Bonding performance of wood bonded with adhesive mixtures composed of phenol-formaldehyde and bio-oil. Industrial Crops and Products. 66: 68–72.

Patel, R.N., S. Bandyopadhyay and A. Ganesh. 2011. Extraction of cardanol and phenol from bio-oils obtained through vacuum pyrolysis of biomass using supercritical fluid extraction. Energy 36: 1535–1542.

Peng, C., G. Zhang, J. Yue and G. Xu. 2014. Pyrolysis of lignin for phenols with alkaline additive. Fuel Processing Technology. 124: 212–221.

Radlein, D. 1999. The production of chemicals from fast pyrolysis bio-oil. pp. 164–188. *In*: Bridgwater et al. (eds.). Fat Pyrolysis of Biomass: A Handbook. Vol. 1. CPL Press, Newbury, UK.

Scahill, J.W., J.P. Diebold and C. Feik. 1997. Removal of residual char fines from pyrolysis vapors by hot gas filtration. pp. 256–266. *In*: A.V. Bridgwater and D.G.B. Boocock (eds.). Developments in Thermochemical Biomass Conversion, Blackie Academic and Professional, London.

Scholze, B. and D. Meier. 2001. Characterization of the water-insoluble fraction from pyrolysis oil (pyrolytic lignin). Part 1. PY-GC/MS, FTIR, and functional groups. Journal of Analytic and Applied Pyrolysis 60: 41–54.

Shahid, S.A., M. Ali and Z.I. Zafar. 2014. Characterization of phenol-formaldehyde resins modified with crude bio-oil prepared from Ziziphus mauritiana endocarps. BioResour. 9: 5362–5384.

Shahid, S.A., M. Ali and Z.I. Zafar. 2015. Cure kinetics, bonding performance, thermal degradation, and biocidal studies of phenol-formaldehyde resins modified with crude bio-oil prepared from Ziziphus mauritiana Endocarps. BioResour. 10: 105–122.

Shen, Y., L. Jarboe, R. Brown and Z. Wen. 2015. A thermochemical-biochemical hybrid processing of lignocellulosic biomass for producing fuels and chemicals. Biotechnol. Adv. 33: 1799–1813.

Sukhbaatar, B., P.H. Steele and M.G. Kim. 2009. Use of lignin separated from bio-oil in oriented strand board binder phenol-formaldehyde resins. Bioresources 4: 789–804.

Sukhbaatar, B., Q. Li, C. Wan, F. Yu, E.-B. Hassan and P. Steele. 2014. Inhibitors removal from bio-oil aqueous fraction for increased ethanol production. Bioresour. Technol. 161: 379–384.

Uchimiya, M. and Z. He. 2012. Calorific values and combustion chemistry of animal manure. pp. 45–62. *In*: Z. He (ed.). Applied Research of Animal Manure: Challenges and Opportunities beyond the Adverse Environmental Concerns. Nova Science Publishers, New York.

Vick, C.B., K. Richter, B.H. River and A.R. Fried. 1995. Hydroxymethylated resorcinol coupling agent for enhanced durability of bisphenol-A epoxy bonds to sitka spruce. Wood Fiber Sci. 27: 2–12.

Wan, H., Z. He, A. Mao and X. Liu. 2017. Synthesis of polymers from liquefied biomass and their utilization in wood bonding, pp. 239–259. *In*: Z. He (ed.). Bio-based Wood Adhesives: Preparation, Characterization, and Testing. CRC Press, Boca Raton, FL.

Yang, H.R. Yan, H. Chen, D.H. Lee and C. Zheng. 2007. Characteristics of hemicellulose, cellulose and lignin pyrolysis. Fuel. 86: 1781–1788.

Yi, J.P., J.Z. Zhang and S.X. Yao. 2012. Preparation of bio-oil-phenol-formaldehyde resins from biomass pyrolysis oil. Applied Mechanics and Materials 174-177: 1429–1432.

Yoshikawa, T., T. Yagi, S. Shinohara, T. Fukunaga, Y. Nakasaka, T. Tago and T. Masuda. 2013. Production of phenols from lignin via depolymerization and catalytic cracking. Fuel Processing Technology 108: 69–75.

Zhang, L., R. Liu, R. Yin and Y. Mei. 2013. Upgrading of bio-oil from biomass fast pyrolysis in China: A review. Renewable and Sustainable Energy Reviews 24: 66–72.

Zhang, Q., J. Chang and J. Zhang. 2014a. Synthesis technology of bio-oil starch adhesive for interior plywood. Journal of Nanjing Forestry University 38: 121–124. (In Chinese)

Zhang, Q., J. Chang, T. Wang and Y. Xu. 2007. Review of biomass pyrolysis oil properties and upgrading research. Energy Conversion and Management 48: 87–92.

Zhang, Z., Q. Lu, X. Ye, L. Xiao, C. Dong and Y. Liu. 2014b. Selective production of phenolic-rich bio-oil from catalytic fast pyrolysis of biomass: comparison of K_3PO_4, K_2HPO_4, and KH_2PO_4. Bioresour. 9: 4050–4062.

Zilnik, L.F. and A. Jazbinsek. 2012. Recovery of renewable phenolic fraction from pyrolysis oil. Separ. Purif. Technol. 86: 157–170.

12

Application of the Rosin from White Pitch (Protium heptaphyllum) for use as Wood Adhesive

Raimundo Kennedy Vieira[1,*] *Adalena Kennedy Vieira*[1] and *Anil Narayan Netravali*[2]

ABSTRACT

The use of adhesives for wood obtained from natural renewable sources comes from the beginning of human technological development. Nonetheless, due to some of its disadvantages such as low durability and low water resistance, sustainable adhesives rapidly lost ground. This situation has resulted in introduction of synthetic resins in the market. These resins captured the market, because they have better mechanical properties and high resistance to humidity. However, synthetic adhesives are based on non-renewable resources of petroleum and natural gas. They not only generate a large uncertainties in terms of the source availability, in the long-term but in some cases could represent a risk to the human health. Thus, in this chapter, suggested another source of raw material for adhesive wood production has been: the rosin obtained directly from trees. Since rosin can be obtained from a large number of tree species in almost all the continents, are a source of the rosin, as an alternative source for commercial adhesive production.

[1] Faculty of Technology, Federal University of Amazonas, Manaus, AM 69077-000, Brazil.
[2] Department of Fiber Science and Apparel Design, Cornell University, Ithaca, NY 14853-4401, USA.
* Corresponding author: maneiro01@ig.com.br

Introduction

The use of adhesives for wood obtained from natural renewable sources comes from the beginning of human technological development. Among these natural sources, one can mention casein and soybean protein as the prominent ones (Umemura et al., 2003). Another class of natural adhesives is starch based glue. These have been widely used for general purposes and are made using animal or plant derived material as a main component (Amico et al., 2010).

The discovery of petroleum and the benefits of using its derivatives led to the development of synthetic adhesives. Being inexpensive, these adhesives experienced fast growth due to their low cost and ease of application (Kaboorani and Riedl, 2011). During the same period, sustainable natural adhesives rapidly lost ground, mainly due to their low cost and ease of disadvantages such as low durability and low water resistance (Umemura et al., 2003). In comparison, synthetic resins have better mechanical properties and high resistance to humidity (Amico et al., 2010).

Currently, the resins used in the production of wood composites formaldehyde (Li et al., 2014) and include phenol-formaldehyde (PF), urea-formaldehyde (UF) (Cui et al., 2015; Liu, 2005; Wang and Pizzi, 1997), melamine (Fan et al., 2011) and isocyanates (Jang et al., 2011). However, these compounds are derived from petroleum and emit formaldehyde, as volatile organic compound (VOC), which is a known carcinogen (Liu, 2005).

Due to the fact that the synthetic adhesives are based on non-renewable resources of petroleum and natural gas, there are uncertainties in terms of their long-term application. Furthermore, these oil products can also cause damage to the environment and human health (International Agency for Research on Cancer, 2004). In 2007, to regulate the use of this type of material the California Air Resources Board (CARB) created a law limiting the emission of formaldehyde from wood-based products that are marketed in California (Cui et al., 2015; Jang et al., 2011). Alson, in 2007, the President of the United States of America signed the Formaldehyde Standards for Composite Wood Products Act into law. This law determines the limits of formaldehyde used in composite wood. Several studies on using renewable materials have been conducted to assess and solve the problems associated with synthetic adhesives. These renewable resources-based adhesives without formaldehyde include, but are not limited to, lignin (Ibrahim et al., 2013; Mansouri et al., 2006; Moubarik et al., 2013a; Ping et al., 2012), tannin (Cui et al., 2015; Moubarik et al., 2013b; Saad et al., 2014), protein (Chen et al., 2015; He et al., 2014; Liu and Li, 2007), starch (Li et al., 2014; 2015; Zhang et al., 2015) and chitosan (Umemura et al., 2003).

Each of the raw materials that studied in this research has advantages and disadvantages. However, adhesive-related characteristics have been improved through the development of new treatments or technologies.

Among these techniques, the approaches of the use of environmentally friendly additives, the use of non-petroleum based cross-linkers and incorporation of nanoparticles have been presented in other chapters of this book. In this chapter, we will discuss another source of raw material, rosin from trees, as adhesive wood production.

Vegetable Rosin

The Amazon Forest is one of the main reservoirs of flora biodiversity throughout the world. In the Amazon forest, there are more species of animals and plants than anywhere else in the world, both with respect to species inhabiting the region (gamma diversity) or coexisting in the same place (alpha diversity). This variety of species of plants and animals that live here, can sometimes hide some of its greatest treasures, the so-called green chemicals (Maia, 2011). In fact, numerous green compounds were produced from plants in the Amazon Forest and used by the Brazilian indigenous population, to cure diseases even before America was discovered. The Amazon biodiversity of compounds has been increasingly valued recently. Currently, plants constitute as the greatest variety of Amazon chemicals (Maia and Andrade, 2009).

The Amazonian flora comprises approximately 30,000 species, about 10% of plants from all over the planet. There are about 5,000 species of trees (greater than 15 cm in diameter) and between 40 and 300 different species can exist per hectare. This biodiversity provides trees from which rosin can be obtained and used for various purposes (Myers, 1988; Steege et al., 2013). Among them, there are dozens of tree species within the Burseraceae family that produce different qualities of "Pitch" or "white or black pitch." The Burseraceae family consists of 18 genera with 700 species separated into three tribes: Canarieae (eight genera) and Bursereae (seven genera), Protieae (three genera). The Protium genus (Tribe Protieae) is the primary family member with 150 species, and it is the only species that produces a rosin known as "white pitch", which is the most aromatic and has greater amount of medicinal properties (Marques et al., 2010). Tribe Protieae can be found in the America, Africa, and Indo-Asian tropics, with the highest diversity found in the Southern Hemisphere (Weeks et al., 2005). These species have widely proliferated in all Brazilian territories, mainly in the Amazon Region, where the genus's protium makes up 80% of the Burseraceae (Fine et al., 2014; Siani et al., 2004).

White Pitch

The white-pitch is a sweet-smelling rosin that comes from the heart of the tree trunk (Protium heptaphyllum). Figure 1 shows the typical tree, secreted

Figure 1. Protium heptaphyllum cycle in nature. Source: Google.

rosin and the hardened rosin. It has several other names as well: pitch, pitch-black, pitch-blend, almecega-brava, almecega-true, copal. The White pitch is classified as oil rosin, because it shows the dissolution of property in nonpolar solvents, due to a highly hydrophobic constitution. The rosin is partly soluble in alcohol and insoluble in water (Zoghbi et al., 1993).

The rosin appears as a mass of soft consistency, light yellow, acquiring a resinous appearance after exposure to air. The tree rosin naturally secreted by it from the tree as a form of self-protection when it is injured or bitten by an animal in the forest (Siani et al., 2004).

The white pitch can be obtained from over a large area, due to which, there is a great variety of products with different chemical compositions are available. The pitch name "white" or "black" is directly related to the color of the rosin. It may vary from tree to tree and according to the species. The white pitch can be found throughout the Amazon region and other species of the same genus in the Brazilian rainforest. For this reason, several studies have been conducted to characterize the rosin obtained from different sources.

Some studies have identified the compounds present in the rosin Heptaphyllum Protium. The main components in this species are: pentacyclic triterpene and amyrins (an isomeric mixture). This composition is important because it permits the use of these two compounds as marker components in the recognition and certification of raw materials regarded as White pitch (Bandeira et al., 2002; Lima-Júnior et al., 2006; Maia et al., 2000; Pinto et al., 2008).

Applications

Several researchers have used White Pitch as a source for research for its medical properties. Most of these properties are anti-inflammatory (Pontes et al., 2007; Marques et al., 2010; Siani et al., 2004), gastroprotective effect

Figure 2. The incineration of the rosin. Source: Google.

(Oliveira et al., 2004), antimicrobial and antioxidant activities (Bandeira et al., 2006).

The rosin has been widely used by people in folk medicine, particularly in the treatment of wounds as an analgesic and antiseptic. The smoke from the incineration of the rosin is used as an insect repellent. Besides the use in traditional medicine, the rosin is burned as incense in religious rituals (Fig. 2).

The rosin may also be used for fixing of ornamental stones, marble, and glass. The paper industry uses the rosin as an additive to control water absorption. The volatile fraction finds use in perfumery and disinfection of environments. The medicinal study of the mixture of alpha- and beta-amyrin (AMY) (Fig. 3), with pentacyclic triterpenes isolated from the stem bark rosin of Protium heptaphyllum, evidenced the sedative and anxiolytic properties of AMY. It involves an action on benzodiazepine-type receptors of mice, producing an antidepressant effect. The mixture of AMY could show a good indication of the animal's emotional state (Aragão et al., 2006).

The white pitch has its own pleasant smell. It is partially soluble in cold alcohol and insoluble in water. This property gives the white pitch an important role within the perfume industry. The paint industry, especially art, uses the power of adhesive rosins as varnishes and as a fixative (Zoghbi et al., 1993). Another use of white pitch is in the paint industry and caulking boats (Correia, 1984). A video on Youtube (Duarte, 2015) shows the processing of this material as an adhesive. In this video, a fisherman presents the adhesive preparation procedure used for caulking wood canoe.

Notwithstanding all the studies about the use of White pitch, there are few articles related to its application as wood adhesive. A concern is the vast chemical composition variation of this White pitch. The variation

(A) **(B)**

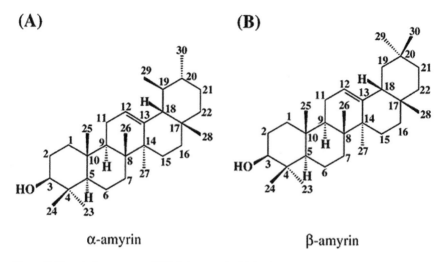

α-amyrin   β-amyrin

Figure 3. Chemical structures of (A) alpha-amyrin (3β-hydroxyurs-12-eno) and (B) beta-amyrin (3β-hydroxyolean-12-eno). Source: Aragão et al. (2006).

in the White pitch composition could be due to its locations, the weather conditions in the region where the plants grow, this concern was addressed by Vieira et al. (2014). In their work, the same researchers found a way to characterize whether a commercial available. The white pitch sample could be utilized independent of where it was purchased, because the identification of its main compounds will confirm its origin. Thus, part of this work is presented in the following section of this chapter.

Chemical Characterization of White Pitch

Purification

In the experimental study of Vieira et al. (2014), the rosin obtained from the Ver-o-peso market hall in Belem, Brazil was originally obtained from the trunk wood of Protium heptaphyllum. Figure 4 shows the White pitch

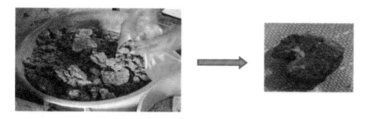

Figure 4. White Pitch. Source: Authors.

sample. Before any characterization, the purchased white pitch was purified. The purification process was carried out by extraction with ethyl acetate at 70°C. It was then filtered using a paper filter, and dried at room temperature. Its chemical composition was examined by FTIR and GC/MS analyses. The results obtained were comparable to the study of Bertan (2003).

FTIR Spectroscopy Analysis

FTIR spectra of the white pitch samples were collected in absorbance mode using the Perkin Elmer Magna-IR 560 spectrometer in the range of 4000–400 cm^{-1}. Each spectrum was an average of 130 scans with a resolution of 4 cm^{-1}.

FTIR analysis was used to confirm the results obtained through the GC/MS analyses, and to compare the changes in the composition of White pitch before and after purification (Fig. 4). The most notable changes, after purification, are at bands of 1010, 1074, 1143, 1346, 2923 cm^{-1} and in the range from 3200 to 3500 cm^{-1} wavenumbers. These bands are characteristic of cellulose, according to the literature (Liu et al., 2016). It indicates that extraction of the white pitch by ethyl acetate was able to separate the polar organic phase of the wood residual. Whereas the reduction of the band intensities around 1010 and 3340 cm^{-1} in FT-IR spectrum of purified white pitch was mainly due to the removal of cellulose with compensation by a relative increase of peak intensities of ethyl acetate, amyrins, and keto-ester, which are related to bonds: CO, CH$_2$ and CH as illustrated in Fig. 5. Specifically, 3303 cm^{-1} belonged to the

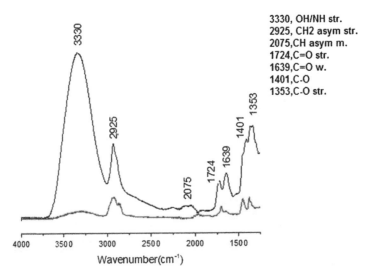

3330, OH/NH str.
2925, CH2 asym str.
2075,CH asym m.
1724,C=O str.
1639,C=O w.
1401,C-O
1353,C-O str.

Figure 5. FTIR spectra of White pitch without treatment (red lower curve) and purified (black upper curve). Adapted from: Vieira et al., 2014.

OH of the amyrins (Fig. 3); peak 1970 cm^{-1} was assigned to the absorption by CO and peak 1708 cm^{-1} was assigned to the absorption by CO relative to α,β unsaturated ketone. These assignments confirmed the structure of keto-ester found in GC/MS analysis (Pinto et al., 2008).

GC/MS Analysis

Purified White pitch (15 mg) was dissolved in 15 mL of ethyl acetate. The sample was then analyzed by an Agilent 6890N Network GC system equipped with an Agilent 5973 Network mass selective detector and Agilent 7683 series injector. The GC/MS conditions were as follows: a 30 cm × 0.25 mm (i.d.) fused silica capillary column with 0.25 μm film thickness (HP19091S-433) and a carrier gas of helium (10.50 psi) were used. The initial temperature was 150°C for 5 min, increasing at 5°C min^{-1} to 290°C, then kept for 30 min; injector port temperature was 200°C; detector temperature was 290°C.

Analysis by gas chromatography coupled to mass spectrometry showed the presence of three main components in the white pitch (Fig. 6), and the mixture of alpha and beta-amyrin constitute 76% of the total blend. Table 1 presents a comparison with the data submitted by Bertan (2003), which verified that the correlation between the components and the retention time is true in the current analysis. These results (i.e., the dominant presence of alpha and beta-amyrin and confirmation by comparison with literature data) certified the material used for this study as authentic White pitch.

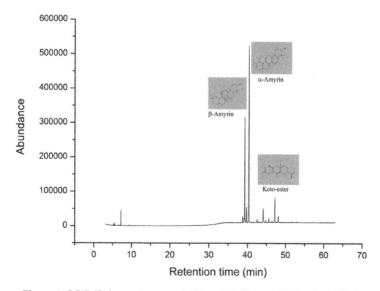

Figure 6. GC/MS chromatogram of white pitch. Source: Vieira et al., 2014.

Table 1. Retention time and constituent percentages of the fixed fraction of white pitch based on relative areas of peaks corresponding to this study compared with the data from Bertan (2003).

Retention time (min)		Compounds (Molecular weight)	mix (%)	
Present data	Bertan (2003) data		Present data	Bertan (2003) data
39.26	38.03	β amyrin (426)	26.02	22.0
40.33	39.07	α amyrin	50.04	43.1
-	41,19	steroidal acid	-	10.8
-	41.69	steroidal acid	-	14.1
-	44.20	steroidal acid	-	9.85
47.18	-	Keto-ester	9.10	-

TG Analysis

Thermo-gravimetric analysis of white pitch rosin was carried out using TA Instruments, Thermo-Gravimetric Analyzer (TGA), model 2050. The specimens were scanned in a nitrogen atmosphere from 30°C to 400°C at a ramp rate of 15°C min^{-1}.

A typical thermo-gravimetric curve (thermogram) of the white pitch sample is shown in Fig. 7. Initially, a slight weight loss, about 9%, was observed below 200°C, which was ascribed to the release of absorbed water and residual solvent. After that, a rapid weight loss percent appeared between 200°C and 325°C, which was caused by the decomposition of the sample. This second step could be mainly ascribed to evaporation of amyrins. This change started at a higher temperature because these compounds have OH groups, which implies stronger intermolecular bonding (Marques et al., 2010).

Potential Application of White Pitch as Wood Adhesives

Vieira et al. (2014) tested the adhesion properties of the purified White pitch by measuring both shear and tensile strengths of bonded maple wood strips. For shear strength measurement, two 15-mm wide maple wood strips were placed in parallel direction with bonded area of 15×15 mm at the end of the strips, and two short pieces of the same wood strip were attached to the other ends to avoid undergoing torsion in the grip during the shear test. To obtain adhesive strength in tension mode, one wood strip was placed in a perpendicular direction on top of another wood strip with the 15×15 mm interface area between the top and bottom wood strips. The tensile adhesive strength was measured using a simple small-scale test method developed by Kim and Netravali (2013).

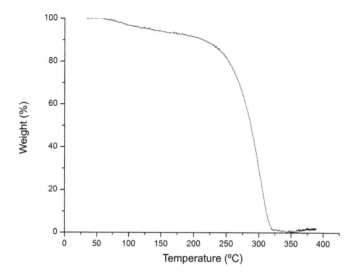

Figure 7. TGA thermogram for the white pitch. Source: Vieira et al., 2014.

The shear strength of the rosin-bonded maple pair was found to be 1.45 MPa which is comparable to that of soy protein concentrate-based adhesive (1.2 MPa). The adhesive tensile strength of the rosin with maple wood specimens ranged between 0.10 and 0.18 MPa with average of 0.14 MPa, lower than that of soy protein concentrate-based adhesive (0.34 MPa). Furthermore, the values of both shear and tensile adhesive strengths of the rosin were lower than those (6.0 and 1.78 MPa, respectively) of commercial Titebond-II wood glue in bonding maple specimens. Thus, there are many areas to improve the bonding ability of white pitch rosen in addition to the purification.

General Considerations

The study of Vieira et al. (2014) provided the possibility of using White pitch rosin for adhesives application. In addition, it was possible to identify the origin of commercial white pitch samples based on the presence of Amyris in their composition.

There are a large number of tree families which produce the type of rosin (White pitch) described in this chapter. Thus, it is reasonable to find a sustainable source (tree) for this rosin. With the appropriate source and considering its use as a boat caulker, it would be natural to consider this sort of material as a new alternative to synthetic glues for using as a wood adhesive—at the least, it is an excellent material for further study and commercialization.

keywords: Rosin, White Pitch, Wood Adhesive, Application

References

Amico, S.D., M. Hrabalova, U. Müller and E. Berghofer. 2010. Bonding of spruce wood with wheat flour glue—Effect of press temperature on the adhesive bond strength. 31: 255–260.

Aragão, G.F., L.M.V. Carneiro, A.P.F. Junior, L.C. Vieira, P.N. Bandeira, T.L.G. Lemos and G.S. de B. Viana. 2006. A possible mechanism for anxiolytic and antidepressant effects of alpha- and beta-amyrin from Protium heptaphyllum (Aubl.) March. Pharmacology Biochemistry and Behavior 85(4): 827–834.

Bandeira, P.N., O. Deusdênia, L. Pessoa, M. Teresa, S. Trevisan and L. Gomes. 2002. Metabólitos secundários de Protium heptaphyllum march Quim. Química Nova. 25(6): 1078–1080.

Bandeira, P.N., A.M. Fonseca, S.M.O. Costa, M.U.D.S. Lins, O.D.L. Pessoa, F.J.Q. Monte, N.A.P. Nogueira and T.L.G. Lemosa. 2006. Chemical composition: Antimicrobial and antioxidant activities of the essential oil from resin of Protium heptaphyllum. Natural Product Communications 1(2): 117–120.

Bertan, L.C. 2003. Development and characterization of simple films and compounds gelatine base, fatty acids and white pitch (in Portuguese).

Chen, M., Y. Chen, X. Zhou, B. Lu, M. He, S. Sun and X. Ling. 2015. Improving water resistance of soy-protein wood adhesive by using hydrophilic additives. 10(1): 41–54.

Correia, M.P. 1984. Dicionário de Plantas Úteis do Brasil. Rio de Janeiro: IBDF.

Cui, J., X. Lu, X. Zhou, L. Chrusciel, Y. Deng and H. Zhou. 2015. Enhancement of mechanical strength of particleboard using environmentally friendly pine (Pinus pinaster L.) tannin adhesives with cellulose nanofibers. 27–32. http://doi.org/10.1007/s13595-014-0392-2.

Duarte, J. 2015. A fisherman of 84 years shows how to caulk a wooden boat using resin in Pirapora mg (in Portuguese). Brazil: Youtube.

Fan, D.B., T.F. Qin and F.X. Chu. 2011. A soy flour-based adhesive reinforced by low addition of MUF resin. Journal of Adhesion Science and Technology 25(1-3): 323–333.

Fine, P.V.A., F. Zapata and D.C. Daly. 2014. Investigating processes of neotropical rain forest tree diversification by examining the evolution and historical biogeography of the protieae (Burseraceae). Evolution 68(7): 1988–2004.

He, Z., H.N. Cheng, D.C. Chapital and M.K. Dowd. 2014. Sequential fractionation of cottonseed meal to improve its wood adhesive properties. Journal of the American Oil Chemists' Society 91(1): 151–158. http://doi.org/10.1007/s11746-013-2349-2.

Ibrahim, V., G. Mamo, P.J. Gustafsson and R. Hatti-Kaul. 2013. Production and properties of adhesives formulated from laccase modified Kraft lignin. Industrial Crops and Products 45: 343–348. http://doi.org/10.1016/j.indcrop.2012.12.051.

International Agency for Research on Cancer. 2004. IARC classifies formaldehyde as carcinogenic to humans. International Agency for Research on Cancer Press Release No. 153.

Jang, Y., J. Huang and K. Li. 2011. A new formaldehyde-free wood adhesive from renewable materials. International Journal of Adhesion and Adhesives 31(7): 754–759.

Kaboorani, A. and B. Riedl. 2011. Effects of adding nano-clay on performance of polyvinyl acetate (PVA) as a wood adhesive. Composites Part A: Applied Science and Manufacturing 42(8): 1031–1039.

Kim, J.T. and A.N. Netravali. 2013. Performance of protein-based wood bioadhesives and development of small-scale test method for characterizing properties of adhesive-bonded wood specimens. J. Adhes. Sci. Technol. 27: 2083–2093.

Li, Z., J. Wang, L. Cheng, Z. Gu, Y. Hong and A. Kowalczyk. 2014. Improving the performance of starch-based wood adhesive by using sodium dodecyl sulfate. Carbohydrate Polymers 99: 579–583.

Li, Z., J. Wang, C. Li, Z. Gu, L. Cheng and Y. Hong. 2015. Effects of montmorillonite addition on the performance of starch-based wood adhesive. Carbohydrate Polymers 115: 394–400.

Lima-Júnior, R.C.P., F.A. Oliveira, L.A. Gurgel, I.J. Cavalcante, K.A. Santos, D.A. Campos, C. A. Vale, R.M. Silva, M.H. Chaves and F.A. Santos. 2006. Attenuation of visceral nociception by α- and β-amyrin, a triterpenoid mixture isolated from the resin of Protium heptaphyllum, in mice. Planta Medica. 72(1): 34–39.

Liu, Y. 2005. Formaldehyde-free wood adhesives from soybean protein and lignin development and characterization. Retrieved from https://ir.library.oregonstate.edu/xmlui/bitstream/handle/1957/20040/LiuYuan2006.pdf?sequence=1.

Liu, Y. and K. Li. 2007. Development and characterization of adhesives from soy protein for bonding wood. International Journal of Adhesion and Adhesives 27(1): 59–67.

Liu, Y., Z. He, M. Shankle and H. Tewolde. 2016. Compositional features of cotton plant biomass fractions characterized by attenuated total reflection Fourier transform infrared spectroscopy. Ind. Crop. Prod. 79: 283–286.

Maia, R.M., P.R. Barbosa, F.G. Cruz, N.F. Roque and M. Fascio. 2000. Triterpenos da resina de protium heptaphyllum march (Bourseraceae): Caracterização em misturas binárias. New Chemistry = Química Nova. 23(5): 623–626.

Maia, J.G.S. and E.H.A. Andrade. 2009. Database of the Amazon aromatic plants and their essential oils. New Chemistry = Quimica Nova. 32(3): 595–622.

Maia, V.C. 2011. Characterization of insect galls, gall makers, and associated fauna of Platô Bacaba (Porto de Trombetas, Pará, Brazil). Biota Neotropica. 11(4): 37–53.

Mansouri, N-N.El., A. Pizzi and J. Salvado. 2006. Lignin-based polycondensation resins for wood adhesives. Journal of Applied Polymer Science 103(3): 1690–1699.

Marques, D.D., R.A. Sartori, T.L.G. Lemosi, L.L. Machado, J.S.N. Souza and F.J.Q. Monte. 2010. Chemical composition of the essential oils from two subspecies of Protium heptaphyllum. Acta Amazonica. 40(1): 227–230.

Moubarik, A., N. Grimi, N. Boussetta and A. Pizzi. 2013a. Isolation and characterization of lignin from Moroccan sugar cane bagasse: Production of lignin-phenol-formaldehyde wood adhesive. Industrial Crops and Products 45: 296–302.

Moubarik, A., H.R. Mansouri, A. Pizzi, F. Charrier, A. Allal and B. Charrier. 2013b. Corn flour-mimosa tannin-based adhesives without formaldehyde for interior particleboard production. Wood Science and Technology 47(4): 675–683.

Myers, N. 1988. Threatened biotas: "hot spots" in tropical forests. The Environmentalist 8(3): 187–208.

Oliveira, F.A., G.M. Vieira-Júnior, M.H. Chaves, F.R.C. Almeida, K.A. Santos, F.S. Martins, R.M. Silva, F.A. Santos and V.S.N. Rao. 2004. Gastroprotective effect of the mixture of α- and β-amyrin from Protium heptaphyllum: role of capsaicin-sensitive primary afferent neurons. Planta Medica. 70: 780.

Ping, L., F. Gambier, A. Pizzi, Z.D. Guo and N. Brosse. 2012. Wood adhesives from agricultural by-products: Lignins and Tannins for the Elaboration of Particleboards 46(7-8): 457–462.

Pinto, S.A.H., L.M.S. Pinto, G.M.A. Cunha, M.H. Chaves, F.A. Santos and V.S. Rao. 2008. Inflammopharmacology short communication anti-inflammatory effect of a, b-Amyrin, a pentacyclic triterpene from Protium heptaphyllum in rat model of acute periodontitis. Inflammopharmacology 16(01): 48–52.

Pontes, W.J.T., J.C.G. Oliveira, C.A.G. Câmara, A.C.H. Lopes, M.G.C. Gondim Júnior, J.V. Oliveira, R. Barros and M.O.E. Schwartz. 2007. Chemical composition and acaricidal activity of the leaf and fruit essential oils of Protium heptaphyllum (Aubl.) Marchand (Burseraceae). Acta Amazonica. 37(1): 103–110.

Saad, H., A. Khoukh, N. Ayed, B. Charrier and F.C.-E. Bouhtoury. 2014. Characterization of Tunisian Aleppo pine tannins for a potential use in wood adhesive formulation. Industrial Crops and Products 61: 517–525.

Siani, A.C., I.S. Garrido, S.S. Moteiro, E.S. Carvalho and M.F.S. Ramos. 2004. Protium icicariba as a source of volatile essences. Biochemical Systematics and Ecology 32(5): 477–489.

Steege, H.T., Nigel C.A. Pitman, Daniel Sabatier, Christopher Baraloto, Rafael P. Salomão, Juan Ernesto Guevara, Oliver L. Phillips, Carolina V. Castilho, William E. Magnusson, Jean-François Molino, Abel Monteagudo, Percy Núñez Vargas, Juan Carlos Montero, Ted R.

Feldpausch, Eurídice N. Honorio Coronado, Tim J. Killeen, Bonifacio Mostacedo, Rodolfo Vasquez, Rafael L. Assis, John Terborgh, Florian Wittmann, Ana Andrade, William F. Laurance, Susan G.W. Laurance, Beatriz S. Marimon, Ben-Hur Marimon, Jr., Ima Célia Guimarães Vieira, Iêda Leão Amaral, Roel Brienen, Hernán Castellanos, Dairon Cárdenas López, Joost F. Duivenvoorden, Hugo F. Mogollón, Francisca Dionízia de Almeida Matos, Nállarett Dávila, Roosevelt García-Villacorta, Pablo Roberto Stevenson Diaz, Flávia Costa, Thaise Emilio, Carolina Levis, Juliana Schietti, Priscila Souza, Alfonso Alonso, Francisco Dallmeier, Alvaro Javier Duque Montoya, Maria Teresa Fernandez Piedade, Alejandro Araujo-Murakami, Luzmila Arroyo, Rogerio Gribel, Paul V.A. Fine, Carlos A. Peres, Marisol Toledo, Gerardo A. Aymard C., Tim R. Baker, Carlos Cerón, Julien Engel, Terry W. Henkel, Paul Maas, Pascal Petronelli, Juliana Stropp, Charles Eugene Zartman, Doug Daly, David Neill, Marcos Silveira, Marcos Ríos Paredes, Jerome Chave, Diógenes de Andrade Lima Filho, Peter Møller Jørgensen, Alfredo Fuentes, Jochen Schöngart, Fernando Cornejo Valverde, Anthony Di Fiore, Eliana M. Jimenez, Maria Cristina Peñuela Mora, Juan Fernando Phillips, Gonzalo Rivas, Tinde R. van Andel, Patricio von Hildebrand, Bruce Hoffman, Eglée L. Zent, Yadvinder Malhi, Adriana Prieto, Agustín Rudas, Ademir R. Ruschell, Natalino Silva, Vincent Vos, Stanford Zent, Alexandre A. Oliveira, Angela Cano Schutz, Therany Gonzales, Marcelo Trindade Nascimento, Hirma Ramirez-Angulo, Rodrigo Sierra, Milton Tirado, María Natalia Umaña Medina, Geertje van der Heijden, César I.A. Vela, Emilio Vilanova Torre, Corine Vriesendorp, Ophelia Wang, Kenneth R. Young, Claudia Baider, Henrik Balslev, Cid Ferreira, Italo Mesones, Armando Torres-Lezama, Ligia Estela Urrego Giraldo, Roderick Zagt, Miguel N. Alexiades, Lionel Hernandez, Isau Huamantupa-Chuquimaco, William Milliken, Walter Palacios Cuenca, Daniela Pauletto, Elvis Valderrama Sandoval, Luis Valenzuela Gamarra, Kyle G. Dexter, Ken Feeley, Gabriela Lopez-Gonzalez and Miles R. Silman. 2013. Hyperdominance in the amazonian tree flora. Science 342(6156): 1243092.

Umemura, K., A. Inoue and S. Kawai. 2003. Development of new natural polymer-based wood adhesives I: dry bond strength and water resistance of konjac glucomannan, chitosan, and their composites. Journal of Wood Science 49: 221–226.

Vieira, R.K., A.K. Vieira, J.T. Kim and A.N. Netravali. 2014. Characterization of amazonic white pitch (*Protium heptaphyllum*) for potential use as "green" adhesive. Journal of Adhesion Science and Technology 28(10): 963–974.

Wang, S. and A. Pizzi. 1997. Waste nylon fibre hardeners for improved UF wood adhesives water resistance. 55: 91–95.

Weeks, A., D.C. Daly and B.B. Simpson. 2005. The phylogenetic history and biogeography of the frankincense and myrrh family (Burseraceae) based on nuclear and chloroplast sequence data. Molecular Phylogenetics and Evolution 35(1): 85–101.

Zhang, Y., L. Ding, J. Gu, H. Tan and L. Zhu. 2015. Preparation and properties of a starch-based wood adhesive with high bonding strength and water resistance. Carbohydrate Polymers 115: 32–7.

Zoghbi, M.G.B., E.V. Cunha and W. Wolter Filho. 1993. Essential oil of protium unifoliolatum. Acta Amazonica. 23(1): 15–16.

13

Effects of Rheology and Viscosity of Bio-based Adhesives on Bonding Performance

Alejandro Bacigalupe,[1] *Zhongqi He*[2] and
Mariano M. Escobar[1,*]

ABSTRACT

Rheology is the science of deformation and flow of the matter due to the application of a force. Most rheological tests involve applying a force to a material and measuring its flow or change in shape. Rheological characterization is useful to study the flowability and viscoelastic properties of adhesive materials. This chapter reviews and discusses the rheological behavior of protein-based adhesives and the effect on their bonding properties. Rheological characterization allows the analysis of the internal structure of the adhesives when they are modified by both physical and chemical processes. Most of the discussion focuses on soy-based and cottonseed-based adhesives modified by heat treatments, varying pH, enzymatic treatment and blending with commercial latex. Rheological characterization may also provide operational parameters for the preparation of wood composites. As an example, adhesives with a high flow point value will not flow easily on the surface of wood particles, hindering the dispersion of the adhesive and wetting of such particles.

[1] Rubber Center, National Institute of Industrial Technology (INTI), Avenida General Paz 5445, B1650WAB, San Martín, Buenos Aires, Argentina.
[2] Southern Regional Research Center, USDA Agricultural Research Service, 1100 Robert E. Lee Blvd., New Orleans, LA 70124, USA.
* Corresponding author: mescobar@inti.gob.ar

Introduction

Formaldehyde-based resins are common wood adhesives due to their high adhesion strength and low cost. One of their major disadvantages is that they are derived from non-renewable and limited fossil sources. It was recognized that most petroleum-based adhesives are not environmentally friendly, and in the case of formaldehyde based resins, are hazardous to human health because of the emission of free formaldehyde. Therefore, there is a renewed interest in developing adhesives based on biopolymers which are environmentally friendly and originated from renewable resources. Plant proteins (e.g., wheat gluten or soybean proteins) are such attractive biopolymers. Nevertheless, the strength and water resistance of plant protein based adhesives have to be improved to fulfill today's requirements for wood adhesives application (Nordqvist et al., 2012).

In order to achieve desirable mechanical properties of biobased adhesives, it is essential to adjust their preparation and application parameters. Within this context, rheology has become an important and useful technique to study the properties of viscoelastic materials. Using an appropriate rheometer, various rheological parameters [viscosity, storage (G') and loss (G'') modulus] can be determined (Ugovsek and Sernek, 2013; Bacigalupe et al., 2015; Zhu et al., 2016).

Related to the elastic behavior, G' is a measure of the deformation energy stored in the sample during the shear process. G'' is a measure of the deformation energy used in the sample during the shear process and then lost, reflecting the viscous behavior of a sample. Therefore, information about viscosity, G' and G'' would also allow us to analyze the internal structure of the adhesive, which will be affected by chemical phenomena (such as interactions due to weak intermolecular forces) and physical phenomena (such as entanglement and coiling of the polymer chains). For example, Ugovsek and Sernek (2013) were able to characterize the wood bonding mechanism with liquefied wood-based adhesives through a rheological oscillatory test. Their approach provided a curing profile of the tested material by means of its rheological response to an oscillating load generated by a stress control rheometer. Information about the curing material was gained in terms of the storage modulus (G') and the loss modulus (G'') or complex viscosity (η^*).

Luo et al. (2015) studied the interaction of soy protein with lignin resin (LR). The viscosity varied with the addition of the resin, which was explained as follows: LR contains sodium hydroxide and denatured the protein. The excess of alkali unfolded and increased the friction among protein molecules, increasing the adhesive viscosity (Luo et al., 2015). Further addition of LR decreased wet shear strength due to the high excess of alkali that decomposed the protein molecules into small pieces. The presence of short protein chains reduced the viscosity of the adhesive, which caused an over-penetration into wood. Moubarik et al. (2010) evaluated the rheological properties of cornstarch-tannin adhesives with hexamine as hardener. Their studies compared the viscosity of alkali modified cornstarch adhesive with cornstarch-tannin-hexamine natural adhesive. The addition of tannin and hexamine to the formulation produced a significant decrease in the viscosity and improved application conditions. More recently, Moubarik (2015) studied the rheology of sugar cane bagasse lignin-phenol formaldehyde (PF) adhesive. Lignin is a by-product and has similar structure to the PF resin but its cost is considerably lower than that of PF (Turunen et al., 2003).

According to Moubarik (2015), the adhesives were prepared at a 30/70 weight ratio of lignin/PF, and characterized through oscillatory rheology. Time sweep measurements of lignin-PF and control PF resins showed that both adhesives had predominant elastic behavior, which indicates an excellent structural stability of the resins. Moreover, frequency sweep test showed that G' was larger than G", which was associated with a typical behavior of a weak gel. The structure was stabilized by weak bonds such as hydrogen bonds (Moubarik, 2015). The authors have also characterized the rheological properties at industrial polymerization temperature (125°C). G' and G" increased with time due to cross-linking reaction, which improved the cohesion and resulted in higher adhesion values. Lignin-PF resins reach a plateau region at 7.5 minutes, which implies that the adhesive is fully crosslinked. In this chapter, we have reviewed and discussed the rheological behavior and modification of some biobased (mainly protein) adhesive materials and their implication in adhesive or operative properties. As case studies and mostly per our published data, we presented more details on the effect of pH on rheology and viscosity of soy protein adhesives and the viscosity of cottonseed-base adhesive slurries.

Bonding Performance and Rheological Behavior

There is no single theory describing the interaction between adhesives and substrates. Several adhesion mechanisms have been proposed including mechanical interlocking, electron transfer, boundary layers and interfaces, adsorption, diffusion, and chemical bonding (Schultz and Nardin, 1994; Gardner et al., 2014). Adhesion theory between protein polymers and wood substrates is mainly attributed to a combination of three major mechanisms: mechanical bonding, physical adsorption and chemical bonding (Sun, 2005). To achieve a proper adhesion, the adhesive must have the capacity to flow, wet and penetrate into the roughened surface of the substrate, and work as an anchor. Adhesive penetration is generally believed to have a strong influence on bonding mechanical performance. Adequate penetration provides a substantial interphase that promotes interaction, perhaps reaction, and also mechanical interlocking. On the other hand, excessive penetration could lead to a "starved" bond-line having poor performance.

From a rheological point of view, adhesives behave as non-Newtonian fluids. Rheological measurement has become an important and useful technique to study the properties of viscous fluids (Asghari et al., 2016; Irani et al., 2016; Zhang et al., 2016). Figure 1 presents the viscosity behavior of washed cottonseed meal and protein isolate adhesive slurries as a function of shear rate. The apparent viscosities of all slurries were shear-rate dependent, indicating shear thinning behavior for all of them. Due to the shear thinning nature, the viscosity measurement of biobased adhesives typically is performed at shear rate range from 10 to 250 s[-1] (Wang et al., 2009; Gao et al., 2013; Qi et al., 2013).

Generally, adhesives with high viscosity and high flow point value do not flow easily on the surface, resulting in undesirable bond strength, while adhesives with very low viscosity lead to excessive and undesirable water absorption and decreasing bonding properties (Bacigalupe et al., 2015). The viscosity of the adhesive is partly dependent on the solid content, which influences the balance between viscosity and energy (and time) spent to

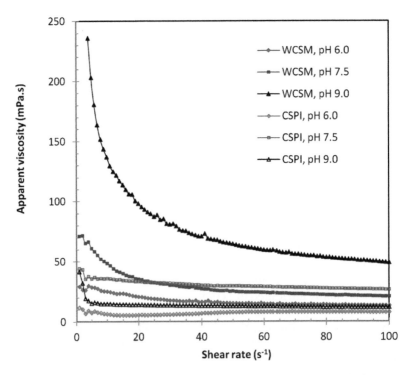

Figure 1. Effect of shear rate on the viscosity of the fresh adhesive slurries of washed cottonseed meal (WCSM) and cottonseed protein isolate (CSPI) with preparative pH of 6.0, 7.5 and 9.0. Reprinted from He et al. (2016).

evaporate the water. It would be desirable to maximize the solid content of the adhesives, minimizing the increase of viscosity adhesive and thereby improving its ability to properly wet, flow, and penetrate into the wood substrate. In the case of soybean glues, the operating viscosity limits range from 500 to 75,000 cP depending upon the application and the nature of the materials to be glued. A viscosity of 500–5,000 cP is needed for gluing materials which are highly absorbing such as paper, soft board and dried wood aggregates, 5,000–25,000 cP for most wood laminating purposes (both cold or hot press) and over 50,000 cP for mastic consistency wood laminating operations (Kumar et al., 2002). A less viscous adhesive is preferable as it is easier to produce and to apply.

Typically, bio-based adhesives have low solid contents due to viscosity constraints. However, low solid content is a major problem for composites, such as particleboard and fiber-board. The excess water generates steam in the hot press that can cause internal voids called blows when the pressing pressure is released (Frihart and Satori, 2013). Thus, the balance between viscosity and solids content is a key issue in order to get high bond strength and spread easiness of the adhesive. Bacigalupe and collaborators studied the effect of solid content on the viscosity of dry blood powder (DBP) based adhesives which is an inexpensive secondary product of the food industry (Lin and Gunasekaran, 2010). DBP is rich in proteins, complex macromolecules and contain a number of chemically linked amino acid monomers, which form polypeptide chains and constitute the primary structure (Bacigalupe et al., 2013). Table 1 presents viscosity values and bonding strength of DBP adhesives as a dependence of the solid content. At low shear rates all the slurries show high viscosity as a result of Brownian motion (Lin and Gunasekaran, 2010; Bacigalupe et al., 2013). However, viscosity decayed by the increased shear rate and reached desirable rheological behavior for application with spray. Moreover, the increment of solid content not only improved the hot-press curing time, but also enhanced the bonding performance of the DBP-based adhesive (Table 1).

Table 1. Effect of solid content on dry adhesive strength (MPa) and viscosity (mPa.s) of dry blood powder (DBP) based adhesive at different shear rates (1.0E − 02; 1.0E + 00 and 1.0E + 02 s^{-1}). Compiled per Bacigalupe et al. (2013).

DBP Percent	Viscosity			Strength
	1.0E−02	1.0E + 00	1.0E + 02	
20	44,900	1,030	18.5	2.894
25	717,000	16,000	127	3.066
30	241,000	5,430	70.6	3.253
35	731,000	16,400	212	3.420
40	62,000	1,990	629	4.055

Nordqvist et al. (2010) compared dry and wet bond strength of alkali-modified soy protein isolate (SPI) and wheat gluten (WG). Different dispersion methods were applied to overcome the problem of differences in the viscosity of each system. The results indicated that the buffer capacity was slightly different for WG and SPI, although the dispersing agent was the same, apparently due to the difference in the amino acid composition between WG and SPI. Thus, the solids content was adjusted depending on the source of protein to obtain similar viscosity values. The authors argued that the viscosity of a suspension with 23% of WG was similar to that containing 11.5% of SPI, both being easy to apply. Decreasing the WG content from 23% to 11.5% would result in very low viscosity dispersions that either drained off or penetrated too deep into the wood substrate, leaving an adhesive layer on the surface of wood substrate that is too thin. On the other hand, the increase of SPI content (from 11.5 to 23%) resulted in too high viscosity dispersions, hindering its workability and decreasing its capacity for wetting, flowing, and penetrating into the substrate (Nordqvist et al., 2013). Kalapathy et al. (1996) studied the effect of salts and disulfide bond cleavage on adhesion and viscosity of alkali-modified soy proteins. They reported that both viscosity and adhesive strength decreased with increasing concentrations of salts. At a concentration of 0.1 M, salts reduced the viscosity of soy proteins with no significant adverse effects on adhesive strength and water resistance. The addition of 0.1 M NaCl, Na_2SO_4, or Na_2SO_3 slightly reduced bond strength from 1230 N to 1120, 1060, or 1013 N, respectively. The viscosity of protein isolate modified at pH 10.0 and 50°C in the absence of salts was >30,000 cP. Treatment with NaCl or Na_2SO_4 resulted in viscosity reductions to 6000 or 1050 cP, respectively. Treatment with Na_2SO_3 led to adhesives with the lowest viscosity without modifying wet and dry mechanical performance.

Kim and Sun (2015) studied the correlation between the physical properties and bonding performance of trypsin-modified soy protein-based adhesives. It was reported that the adhesive viscosity increased with the concentration of glutaraldehyde, regardless of the concentration of trypsin. The highest viscosity was reach after an incubation time of 9.7 h for different glutaraldehyde contents. The increase of viscosity could be attributed to the hydrolysis mechanism of trypsin on soy protein, which enhanced the surface area of soy protein, leading to intermolecular interactions at certain solid contents, thus increasing viscosity. The subsequent decrease in viscosity after 9.77 h may be due to the higher density of crosslinking polymeric matrix through glutaraldehyde. Kim and Sun (2015) have also reported a linear regression between viscosity and dry shear bond strength of enzyme-modified soy protein adhesives with an R^2 value of 0.7678 and adjusted R^2 value of 0.6880 (at the 95% confidence level).

Modification of Rheology and Viscosity of Protein Products

Physical modification such as heating of dispersion during preparation enhances adhesive properties of soy proteins (Tanford, 1968). However, protein components have different thermal stability: temperature of denaturation near 74°C for β-conglycinin, and near 87°C for glycinin (Puppo et al., 2004). Heating leads to structural alterations such as dissociation of quaternary structure, protein subunit denaturation and more (Wang and Johnson, 2001). These structural modifications of protein alter rheological characteristics, i.e., viscosity and elasticity (Mirmoghtadaie et al., 2016).

Vnučec et al. (2015) studied the effect of thermal modification on viscosity of soy protein solutions under vacuum condition. They dispersed 10 g of thermally modified and unmodified SPI powders in 100 mL of distilled water, and stirred at different temperatures for 2 h at 150 rpm. The pH was adjusted with NaOH solution. They showed that vacuum thermal modification at temperatures higher than 150°C degraded the protein structure and SPI powder lost the adhesive properties. Furthermore, the thermal treatment changed the dependence of the viscosity with the pH. The adjustment of the pH to 10 produced an increase of the viscosity of all samples. Nevertheless, the difference between unmodified and thermal modified SPI remained. This difference was especially significant for thermal modification at 100°C. In addition, increasing dispersion preparation temperatures also increased the viscosity of pH adjusted adhesives. It was explained that the increase of viscosity could be due to increased hydration capacity and swelling of the protein (Xu et al., 2012).

Renkema et al. (2002) studied the influence of pH and heat denaturation on gel formation and gel properties of soy protein isolate (SPI). Heat denaturation is often a prerequisite for gel formation of globular proteins, and denaturation temperatures depend strongly on pH and ionic strength (Damodaran, 1988). Gel properties were determined by rheological measurements and related to the effect of heating on protein aggregation/ precipitation. Renkema et al. (2002) claimed that G' was a measure of stiffness of gels, i.e., the resistance to deformation. Furthermore, they also stated that heat denaturation was a prerequisite for gel formation by soy proteins. The stiffness of soy protein gels varied as a function of pH and ionic strength. Higher values for G' were obtained at pH values lower than 6. Low G' values correlated with a high amount of dissolved protein in heated 1% dispersions. It is expected that the protein that remained dissolved after heating, was not incorporated in the network.

In order to produce heat-induced protein gels a bio-polymer needs to unfold and expose protein active sites. This allows the formation of non-covalent interaction which produces protein molecule aggregation (Ziegler and Foegeding, 1990). Zhou et al. (2015) investigated the influence of pH on heat-induced formation of cottonseed protein gels. The authors claimed that the stiffness of cottonseed protein gels varied as a function of pH, and higher values of G' were obtained at pH < 7. The gel was formed due to the aggregation of protein molecules which was caused by a reduction in the repulsive forces (Zhou et al., 2015). Speroni et al. (2009) studied the gelation of SPI after high-pressure (HP) and thermal treatment. Gelation was characterized by a thermal cycle (G' as a function of time) that included a heating ramp, a temperature plateau and a cooling stage. The main increase in G' occurred during the cooling stage, and highest values of storage modulus were achieved for samples without HP treatment. The authors stated that gel matrix was sustained by hydrogen bonds and HP treatments reduced the ability to establish hydrogen bonds upon thermal cycle (Speroni et al., 2009). Rheological spectra of sequential HP-thermal treated samples showed that a gel-like behavior was observed for samples with and without HP treatment. However G' values decrease in HP-treated in comparison with untreated one. The authors suggested that HP treatment previous to a thermal one induced the formation of weaker gels, and HP interfered with the ability of soybean protein to establish inter-molecular interactions.

Qi and Sun (2011) investigated the compatibility of a modified soy protein (MSP) with commercial synthetic latex adhesives (SLAs) at different blending ratios. Apparent viscosity of MSP/SLAs blends was reduced significantly at 20–60% MSP, which improved flowability and spread rate (Qi and Sun, 2011). These authors also studied rheological behavior during curing reaction at 50°C. The gelling process promoted the formation of a three-dimensional network improving adhesion properties (Halasz et al., 2000). Their results showed that the modified MSP reacted with UF based resin to form a gel within 3–48 h depending on soy protein concentration. However, G' values for pure SLAs and MSP remained almost the same after 30 minutes of oscillation test. They stated that the addition of MSP to UF resin provided an acidic environment for the cross-linking reaction to occur. Their results indicated that MSP provided functional groups for the chemical reactions and acted as an acidic catalyst. Zhu et al. (2016) measured the rheological behaviors of camelina protein isolate (CPI) modified by sodium bisulfite and guanidine–HCl (Gdm.Cl). Both types of modified proteins exhibited lower viscosity than the native CPI. For Gdm. Cl products, the lowest viscosity was 130.5 cP for 5% Gdm.Cl modification (dry protein basis), and slowly increased to 151.6 cP for 250% Gdm.Cl modification. Unlike the continuous decreasing viscosity of bisulfite-modified proteins, protein aggregation caused the increase of the viscosities of Gdm.Cl-modified proteins at Gdm.Cl concentrations of 50, 100, 250%. The authors also observed that Gdm.Cl-modified proteins were less viscous than bisulfite-modified proteins, indicating that the protein's hydrogen bond was more influential than disulfide bond on shearing behaviors in this hydro-protein system. Gao et al. (2013) prepared wood adhesives based on soy meal modified with polyethylene glycol diacrylate (PEGDA) as a crosslinker and viscosity reducer. The viscosity of modified soy meal adhesives effectively decreased by 35% compared with the addition of PEGDA, and the wet shear strength of their bonded plywood increased; the wet shear strength of plywood bonded with 4% PEGDA-modified soy meal adhesive increased 114% compared to unmodified soy meal.

Kumar et al. (2004) also studied the effect of enzymatic treatments on the viscosity of soy based adhesives. The authors stated that enzyme modified adhesives showed a decrease in the Brookfield viscosity with storage time, which could be due to alkaline hydrolysis (Kumar et al., 2004). Furthermore, the entire enzyme modified adhesives showed an almost similar value of viscosity, suggesting that specificity of enzymes for hydrolysis did not affect the viscosity behavior of the samples. Moreover, enzymatic treatment also affected the mechanical properties of protein adhesives. For example, with chymotrypsin, an extensive hydrolysis leading to a low molecular weight protein resulted in a significant decrease in viscosity, which may be responsible for decreased adhesion. As the viscosity decreased, adhesive could easily penetrate into wood and very small amount was available on the surface for adhesion.

Effect of pH on Rheology and Viscosity of Soy Protein Adhesives

Bacigalupe et al. (2015) studied the bonding performance of soy protein concentrate (SPC) suspensions at different pH values (Table 2). Bacigalupe et al. (2015) grouped the pH effect on the dry bonding performance into three categories. The first weakest group included pH 8, 10 and 11 with bonding strength values around 2.45 MPa. The second category includes pH 9 and 13 with moderate bonding strength. The strongest performing group is pH 12 with the highest adhesive strength of 3.63 MPa. Wang et al. (2009) studied the effects of pH from 1.6 to 9.6 on the dry, wet and soaked

Table 2. Effects of pH on bonding performance and rheological parameters of soy protein concentrate (SPC) slurries (11.5% of solid content). Compiled per Bacigalupe et al. (2015).

Sample	pH	Bonding strength (MPa)	Flow consistency index (Pa s)	Flow behavior index	Flow point (Pa)
SPC-8	8.0	2.52	23.2	0.31	18.2
SPC-9	9.0	2.98	51.2	0.22	27.0
SPC10	10.0	2.58	105.7	0.17	134.5
SPC-11	11.0	2.41	216.2	0.08	194.5
SPC-12	12.0	3.63	51.9	0.25	63.3
SPC-13	13.0	3.24	28.9	0.32	13.0

strength of SPI. He et al. (2016) reported the effects of pH from 4.5 to 11.0 on the dry, wet and soaked strength of cottonseed-based adhesives. Both studies found the adhesives and viscosity changed with the slurries' pH conditions. However, their data were not comparable with Bacigalupe et al. (2015) due to different pH ranges tested.

Thus, Bacigalupe et al. (2015) carried out rheological studies with different pH conditions to analyze these suspensions' micro structures to explore the relationship between the rheological properties and bonding performance. They analyzed the effect of the pH by fitting the viscosity curves to the Ostwald I method (i.e., power-law). With it, two parameters, the flow consistency index (Pa s) and the flow behavior index (dimensionless) were obtained (Table 2). Based on the two parameters, these slurries should be classified as pseudoplastic fluids (flow consistency index > 0 and flow behavior index between 0 and 1). This means that the viscosity of all SPC adhesive slurries decreased with increasing shear load.

Frequency sweep analyses can be used to evaluate the degree of intermolecular interaction of protein solutions (Hu et al., 2013; Zhu et al., 2016). Figure 2 presents the amplitude sweep test for these SPC adhesives as this experiment can be used to analyze the alteration of the structural conformation of the protein, unrolling it and allowing weak intermolecular interactions (hydrogen bonds, disulfide bonds, etc.) introduced by pH changes (Leiva et al., 2007). Samples with pH 8, 9 and 13 presented the limit of the Linear Viscoelastic (LVE) region at low strain values (lower than 1%); samples with pH 10 and 12 exhibited this limit at 3% and sample with

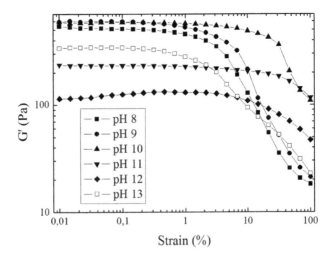

Figure 2. Amplitude sweep test and G' values as a dependence of pH for SPC adhesives at different pH values. Refer to Table 1 for sample information. Reprinted from Bacigalupe et al. (2015).

pH 11 displayed this limit at 20% of strain. This behavior could be explained through the modification of the protein induced by high pH values. The LVE of pH 8 and 9 decayed at a relatively low strain, indicating that the protein is partially unrolled. The LVE of samples with pH 11 decreased at higher strain values which indicated the appearance of non-covalent interactions producing a highly elastic structure. LVE decayed drastically for pH 13 due to the high alkali concentration that started to hydrolyze the protein, losing the elastic structure.

Figure 3 shows tan δ values as a dependence of strain for all samples. SPC adhesives with pH 8 and 9 partially maintained the secondary structure, therefore the backbone of the protein chain was relatively folded. At pH 13 it reached the flow point (tan δ = 1) at low strain values due to the appearance of shortened chains. The increase of G″ represents the deformation energy spent by the sample during the process of changing the material's structure. SPC adhesives at pH 10 and 11 reached the flow point at high strain values due to the unfolding of the protein and the interactions between the functional groups of the side chains, generating a gel-like structure. Table 2 presents the flow point values as a dependence of the pH. Alkali modified adhesives at pH 12 showed a desirable rheological behavior; consequently the adhesive tended to flow under application conditions. Nevertheless it recovered and regained its structure at rest, adhering to the surface and holding the bond between the two pieces of wood.

Bacigalupe et al. (2015) related these rheological data with bonding performance of these SPC adhesives. SPC samples at pH 8 and 9 were

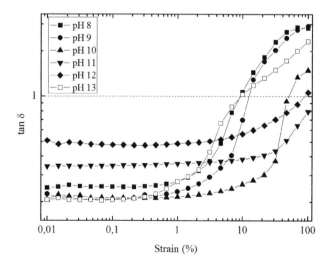

Figure 3. Tan δ values for SPC adhesives in the ranging strain from 0.1 to 100%. Refer to Table 1 for sample information. Reprinted from Bacigalupe et al. (2015).

at a low denaturation degree, which means that most of the functional groups are unavailable due to the native conformation of the protein. Protein tends to fold and impedes the interactions between adhesive and substrate (Ciannamea et al., 2010). The sample at pH 11 presented a very high viscosity and flow point values, which difficult the application onto the wood surface (leading to the lowest bond strength). The two samples at pH 12 and 13 displayed the best bonding properties because the protein was fully unfolded and permitted the functional groups of the protein side chain to establish interaction with the substrate by van der Waals forces, hydrogen bonds, disulfide bonds and more. However, for pH 13, the low viscosity showed that the degree of hydrolysis of the proteins produced the shortening of the backbone, and therefore lead to excessive and undesirable water absorption and decreased bonding properties.

Wang et al. (2009) reported that water resistance of SPI adhesive was improved by preparation of the adhesive slurry near SPI's isoelectric point (pI). However, at this pH, SPI slurry exhibited the lowest solubility with aggregation and precipitation occurred. Thus, Wang et al. (2009) used xanthan gum, pectin, dextrin D2006, and dextrin D2256 as suspension agents for the SPI slurry prepared at pH 4.6. They reported that the slurry with xanthan gum had the highest viscosity, resembled rheological properties of SPI at neutral pH, and gave the suspension good flowability. At pH 4.6, the addition of 0.25% xanthan gum solution was able to prevent soy protein from precipitation and also resulted in the most homogenous mixture with SPI. However, it should be made aware that there was about 12% reduction in wet strengths of SPI with xanthan addition. This reduction was attributed to interfering interaction between xanthan and proteins with weakened interactions between protein-protein and protein-wood.

Viscosity of Cottonseed-Based Adhesives

Like soy-based products, the interest in cottonseed-based wood adhesives has also been revived recently (Cheng et al., 2013; He et al., 2014a,b,c; He and Cheng, 2017). In this section, the relevant viscosity data of the cottonseed-based adhesive slurries are briefly discussed. Indeed, the viscosity of cottonseed protein and meal dispersions was studied long time ago with the purpose of improving the "working life" of these adhesives for plywood bonding (Cheng and Arthur, 1949; Hogan and Arthur, 1951). Due to the nature of on-site preparation of wood adhesive dispersions for veneer bonding, their stability for 24 h or so would be acceptable "working life" of these adhesive dispersions. Cheng and Arthur (1949) tested the effect of aging (up to 28 h at 25°C) on the viscosity of 2–20% of cottonseed protein dispersions with varying NaOH concentrations (12%, 14% and 15%). Trichloroacetate (2.5%) was also included to prevent gel formation. Cheng and Arthur (1949) found that the viscosity of cottonseed protein dispersions decreased on ageing and alkali concentrations, but increased with protein solid contents. Cheng and Arthur (1949) further tested the effect of sugars on the viscosity and stability of the protein dispersion. The addition (10%) of either dextrose or sucrose increased the initial viscosity of the dispersion. However, dextrose decreased the relative rate of decrease in viscosity with age, but sucrose did not. Further experiments indicated that improvement of the viscosity and stability was related to reducing sugars. As the ability of the reducing sugars to stabilize the protein dispersion was always accompanied by a large increase in the initial viscosity upon addition of the sugar, Cheng and Arthur (1949) assumed a reaction probably occurred between the two types of molecules to form larger and stable aggregates. Cheng and Arthur (1949) proposed to prepare a cottonseed protein dispersion with a "practical life" by the addition of 12.5 to 15% dextrose (on the weight of protein) as the viscosity of such dispersion should not decrease below 50 p during a 4 h period. Hogan and Arthur (1951) reported similar observations with cottonseed meal dispersions. Whereas they tested cottonseed meal preparations with different nitrogen solubility varied from 20% to 80% of total nitrogen, Hogan and Arthur (1951) found there was no substantial difference in viscosities of the meal adhesive dispersions. In other words, the viscosity of the meal dispersions was independent of the nitrogen solubility of the protein contained in the meal, which was due to the different processes of the oil removal to produce the meal.

Recently, it has been found that water washing could improve the adhesive performance of cottonseed meal (He et al., 2014a). The washed cottonseed meal showed adhesive strength and water resistance comparable to, or even better than cottonseed protein isolate (He et al., 2014b,c). He et al. (2016) investigated the effects of pH and storage time on the adhesive strengths and rheological properties of cottonseed meal-based products to increase the basic knowledge of cottonseed meal-based adhesives and optimize the two parameters for their practical application in wood bonding. The adhesive strength of washed cottonseed meal at pH 6.0 were basically constant, showing small fluctuation from 4.5 and 5.1 MPa, respectively, over the storage time (Table 3). Whereas the adhesive strength of the WCSM preparation at pH 7.5 decreased at day 2, and then recovered with the values near the initial levels for day 4 and/or 8, the slurry of washed cottonseed meal at pH 9.0 showed a decreasing trend in adhesive strength with increasing storage time. In contrast, the slurries of cottonseed protein isolate prepared at all three pH conditions showed a trend of decreasing adhesive strengths over the storage time. This data indicated that the adhesive slurries of washed meal at pH 6.0 were the most stable in all the samples tested.

Table 3. Effect of storage time on dry adhesive strength (MPa) and apparent viscosity (mPa.s at shear rate of 10 s^{-1}) of the adhesive slurries of washed cottonseed meal (WCSM) and cottonseed protein isolate (CSPI) prepared at pH 6.0, 7.5 and 9.0. Compiled per He et al. (2016).

pH	Storage time (day)							
	0		2		4		8	
	Strength	Viscosity	Strength	Viscosity	Strength	Viscosity	Strength	Viscosity
	WCSM							
6.0	4.53	21.2	4.34	66.0	4.65	88.2	4.23	195.9
7.5	3.82	48.2	2.57	126.6	3.01	165.5	3.42	251.4
9.0	4.26	133.9	3.68	379.0	3.37	379.9	5.38	436.2
	CSPI							
6.0	4.78	6.2	3.70	5.8	4.04	5.6	3.44	7.3
7.5	4.11	36.8	2.52	33.6	2.21	26.4	2.96	19.2
9.0	3.59	16.3	1.95	13.8	2.34	46.8	2.58	18.7

The slurry of washed cottonseed meal showed higher viscosity than protein slurry under the same conditions (Table 3). The viscosities of both meal and protein slurries increased at higher pH.

The highest viscosity was observed at pH 9.0 for the meal slurry, but at pH 7.5 for the protein slurry. The viscosity of WCSM slurries prepared at all three pH conditions increased with storage time. He et al. (2016) assumed that the observation of the increase in viscosity of meal slurries with storage is typical for the thixotropic behavior (increasing viscosity during time of shearing at constant shear rate) of hydrocolloidal starch and protein (Irani et al., 2016; Zhang et al., 2016). They observed that storage time affected the viscosity of protein slurries less than that of meal slurries. They assumed that the carbohydrate components in washed cottonseed meal (He et al., 2015) should have contributed to different rheological behaviors from cottonseed protein as Wang and Guo (2014) reported that sucrose made whey protein a viscous, flowable liquid rather than an unflowable slurry and paste. Per the observations in this work, He et al. (2016) concluded that the inexpensive washed cottonseed meal preparation was more feasible and showed a range of acceptable operational parameters than the relatively expensive cottonseed protein isolate when used for wood bonding purposes. The adhesive slurry of washed cottonseed meal could be prepared at its optimal pH of 6.0, but it was also usable in the pH range of 4.5 to 9.0.

Conclusions

The mechanical properties of protein-based adhesives depend strongly on the conditions of preparation and application thereof. The analysis of the rheological properties provides information about the internal structure of the adhesive and its relationship with the bonding performance.

Most research on protein-based adhesives focuses on the viscosity and its variation by different physical (temperature, enzymes) and chemical (pH, crosslinking) treatments. From the practical point of view, the rheological properties are related to the ease of spread and flow over a surface. An adhesive with a high flow point value will not flow easily on the surface, resulting in undesirable bond strength, while adhesives with very low viscosity lead to excessive and undesirable water absorption and decreasing bonding properties. It is expected that the literature review of this chapter enriches discussions and encourages the search for a relationship between the rheological properties and good bonding performance of protein-based adhesives.

Acknowledgements

Author AB and MME wish to thank the Unit of Rural Change (UCAR PIA 14034) of the Ministry of Agriculture, Livestock and Fisheries (MAGyP), of Argentina, for their financial support.

Keywords: Bioadhesive, Protein adhesive, Rheology, Viscosity, Rheological properties

References

Asghari, A.K., I. Norton, T. Mills, P. Sadd and F. Spyropoulos. 2016. Interfacial and foaming characterisation of mixed protein-starch particle systems for food-foam applications. Food Hydrocolloids 53: 311–319.

Bacigalupe, A., D.B. Garcia and O. Ferré. 2013. Rheological behavior of an environmentally friendly dry blood powder based adhesive for the wood industry. Annual Trans. Nordic. Rheol. Soc. 21: 45–48.

Bacigalupe, A., A.K. Poliszuk, P. Eisenberg and M.M. Escobar. 2015. Rheological behavior and bonding performance of an alkaline soy protein suspension. Int. J. Adhes. Adhes. 62: 1–6.

Cheng, F.W. and J.C. Arthur Jr. 1949. Viscosity of cottonseed protein dispersions. J. Am. Oil Chem. Soc. 26: 147–150.

Cheng, H.N., M.K. Dowd and Z. He. 2013. Investigation of modified cottonseed protein adhesives for wood composites. Ind. Crop. Prod. 46: 399–403.

Ciannamea, E.M., P.M. Stefani and R.A. Ruseckaite. 2010. Medium-density particleboards from modified rice husks and soybean protein concentrate-based adhesives. Bioresour. Technol. 101: 818–825.

Damodaran, S. 1988. Refolding of thermally unfolded soy proteins during the cooling regime of the gelation process: Effect on gelation. J. Agric. Food Chem. 36: 262–269.

Frihart, C.R. and H. Satori. 2013. Soy flour dispersibility and performance as wood adhesive. J. Adhes. Sci. Tech. 27: 2043–2052.

Gao, Q., Z. Qin, C. Li, S. Zhang and J. Li. 2013. Preparation of wood adhesives based on soybean meal modified with PEGDA as a crosslinker and viscosity reducer. BioResources 8: 5380–5391.

Gardner, D.J., M. Blumentritt, L. Wang and N. Yildirim. 2014. Adhesion theories in wood adhesive bonding. Rev. Adhes. Adhes. 2: 127–172.

Halasz, L., O. Vorster, J. Pizzi and J. Alphen. 2000. A rheological study of the gelling of UF polycondensates. J. Appl. Polym. Sci. 75: 1296–1302.

He, Z., H.N. Cheng, D.C. Chapital and M.K. Dowd. 2014a. Sequential fractionation of cottonseed meal to improve its wood adhesive properties. J. Am. Oil Chem. Soc. 91: 151–158.

He, Z., D.C. Chapital, H.N. Cheng and M.K. Dowd. 2014b. Comparison of adhesive properties of water- and phosphate buffer-washed cottonseed meals with cottonseed protein isolate on maple and poplar veneers. Int. J. Adhes. Adhes. 50: 102–106.

He, Z., D.C. Chapital, H.N. Cheng, K.T. Klasson, M.O. Olanya and J. Uknalis. 2014c. Application of tung oil to improve adhesion strength and water resistance of cottonseed meal and protein adhesives on maple veneer. Ind. Crop. Prod. 61: 398–402.

He, Z., H. Zhang and D.C. Olk. 2015. Chemical composition of defatted cottonseed and soy meal products. PLoS One 10(6): e0129933. DOI:10.1371/journal.pone.0129933.

He, Z. and H.N. Cheng. 2017. Preparation and utilization of water washed cottonseed meal as wood adhesives. pp. 156–178. *In*: Z. He (ed.). Bio-based Wood Adhesives-Preparation, Characterization, and Testing. Science Publishers, CRC Press/Taylor & Francis Group. This volume.

He, Z., D.C. Chapital and H.N. Cheng. 2016. Effects of pH and storage time on the adhesive and rheological properties of cottonseed meal-based products. J. Appl. Polymer Sci. 13: 43637. doi: 10.1002/APP.43637.

Hogan, J.T. and J.C. Arthur Jr. 1951. Viscosity of cottonseed meal dispersions. J. Am. Oil Chem. Soc. 28: 436–438.

Hu, H., J. Wu, E.C.Y. Li-Chan, L. Zhu, F. Zhang, X. Xu, G. Fan, L. Wang, X. Huang and S. Pan. 2013. Effects of ultrasound on structural and physical properties of soy protein isolate (SPI) dispersions. Food Hydrocolloids 30: 647–655.

Irani, M., S.M.A. Razavi, E.-S.M. Abdel-Aal and M. Taghizadeh. 2016. Influence of variety, concentration and temperature on the steady shear flow behavior and thixotropy of canary seed (Phalaris canariensis) starch gels. Starch: doi:10.1002/star.201500348.

Kalapathy, U., N. Hettiarachchy, D. Myers and K. Rhee. 1996. Alkali-modified soy proteins: effect of salts and disulfide bond cleavage on adhesion and viscosity. J. Am. Oil Chem. Soc. 73: 1063–1066.

Kim, M.J. and X.S. Sun. 2015. Correlation between physical properties and shear adhesion strength of enzymatically modified soy protein-based adhesives. J. Am. Oil Chem. Soc. 92: 1689–1700.

Kumar, R., V. Choudhary, S. Mishra, I.K. Varma and B. Mattiason. 2002. Adhesives and plastics based on soy protein products. Ind. Crops. Prod. 16: 155–72.

Kumar, R., V. Choudhary, S. Mishra and I.K. Varma. 2004. Enzymatically-modified soy protein part 2: adhesion behaviour. J. Adhes. Sci. Technol. 18: 261–273.

Leiva, P., E. Ciannamea and R.A. Ruseckaite. 2007. Medium-density particleboards from rice husks and soybean protein concentrate. J. Appl. Polym. Sci. 106: 1301–1306.

Lin, H. and S. Gunasekaran. 2010. Cow blood adhesive: Characterization of physicochemical and adhesion properties. Int. J. Adhes. Adhes. 30: 139–144.

Luo, J., J. Luo, C. Yuan, W. Zhang, J. Li, Q. Gao and H. Chen. 2015. An eco-friendly wood adhesive from soy protein and lignin: performance properties. RSC Advances 5: 100849–100855.

Mirmoghtadaie, L., S. Shojaee Aliabadi and S.M. Hosseini. 2016. Recent approaches in physical modification of protein functionality. Food Chem. 199: 619–627. DOI: 10.1016/j.foodchem.2015.12.067.

Moubarik, A., B. Charrier, A. Allal, F. Charrier and A. Pizzi. 2010. Development and optimization of a new formaldehyde-free cornstarch and tannin wood adhesive. Eur. J. Wood Prod. 68: 167–177.

Moubarik, A. 2015. Rheology study of sugar cane bagasse lignin-added phenol–formaldehyde adhesives. J. Adhes. 91: 347–355. DOI: 10.1080/00218464.2014.903803.

Nordqvist, P., F. Khabbaz and E. Malmström. 2010. Comparing bond strength and water resistance of alkali-modified soy protein isolate and wheat gluten adhesives. Int. J. Adhes. Adhes. 30: 72–79.

Nordqvist, P., M. Lawther, E. Malmström and F. Khabbaz. 2012. Ind. Crop. Prod. 38: 139–145.

Nordqvist, P., N. Nordgren, F. Khabbaz and E. Malmström. 2013. Plant proteins as wood adhesives: Bonding performance at the macro- and nanoscale. Ind. Crop. Prod. 44: 246–252.

Puppo, M.C., N. Chapleau, F. Speroni, M. de Lamballerie-Anton, F. Michel, C. Añón and M. Anton. 2004. Physicochemical modifications of high-pressure-treated soybean protein isolates. J. Agr. Food Chem. 52: 1564–1571.

Qi, G. and X.S. Sun. 2011. Soy protein adhesive blends with synthetic latex on wood veneer. J. Am. Oil Chem. Soc. 88: 271–281. DOI: 10.1007/s11746-010-1666-y.

Qi, G., N. Li, D. Wang and X.S. Sun. 2013. Physicochemical properties of soy protein adhesives modified by 2-octen-1-ylsuccinic anhydride. Ind. Crop. Prod. 46: 165–72.

Renkema, J.M.S., H. Gruppen and T. van Vliet. 2002. Influence of pH and ionic strength on heat-induced formation and rheological properties of soy protein gels in relation to denaturation and their protein compositions. J. Agric. Food Chem. 50: 6064–6071. DOI: 10.1021/jf020061b.

Schultz, J. and M. Nardin. 1994. Theories and mechanisms of adhesion. In: A. Pizzi and K.L. Mittal (eds.). Handbook of Adhesive Technology, Marcel Dekker, New York.

Speroni, F., V. Beaumal, M. de Lamballerie, M. Anton, M.C. Añón and M.C. Puppo. 2009. Gelation of soybean proteins induced by sequential high-pressure and thermal treatments. Food Hydrocoll. 23: 1433–1442. DOI: 10.1016/j.foodhyd.2008.11.008.

Sun, X.S. 2005. Soy protein adhesives. Bio-Based Polym. Compos. 327–368.

Tanford, C. 1968. Protein denaturation. pp. 121–282. In: C.B. Anfinsen, M.L. Anson, John T. Edsall and Frederic M. Richards (eds.). Advances in Protein Chemistry.

Turunen, M., L. Alvila, T.T. Pakkanen and J. Rainio. 2003. Modification of phenol–formaldehyde resol resins by lignin, starch, and urea. J. Appl. Polym. Sci. 88: 582–588.

Ugovsek, A. and M. Sernek. 2013. Characterization of the curing of liquefied wood by rheometry, DEA and DSC. Wood Sci. Technol. 47: 1099–1111.

Vnučec, D., A. Goršek, A. Kutnar and M. Mikuljan. 2015. Thermal modification of soy proteins in the vacuum chamber and wood adhesion. Wood Sci. Technol. 49: 225–239. DOI: 10.1007/s00226-014-0685-5.

Wang, C. and L.A. Johnson. 2001. Functional properties of hydrothermally cooked soy protein products. J. Am. Oil Chem. Soc. 78: 189–195.

Wang, D., X.S. Sun, G. Yang and Y. Wang. 2009. Improved water resistance of soy protein adhesive at isoelectric point. Am. Soc. Agric. Biol. Eng. 52: 173–7.

Wang, G. and M. Guo. 2014. Property and storage stability of whey protein-sucrose based safe paper glue. J. Appl. Poly. Sci. 131: 39710.

Xu, J.T., H. Liu, J.H. Ren and S.T. Guo. 2012. Assessment and distinction of commercial soy protein isolate product functionalities using viscosity characteristic curves. Chinese Chem. Lett. 23: 1051–1054.

Zhang, Z., V. Arrighi, L. Campbell, J. Lonchamp and S.R. Euston. 2016. Properties of partially denatured whey protein products 2: Solution flow properties. Food Hydrocolloids 56: 218–226.

Zhou, J., H. Zhang, L. Gao, L. Wang and H.F. Qian. 2015. Influence of pH and ionic strength on heat-induced formation and rheological properties of cottonseed protein gels. Food Bioprod. Process. 96: 27–34. DOI: 10.1016/j.fbp.2015.06.004.

Zhu, X., D. Wang and X.S. Sun. 2016. Physico-chemical properties of camelina protein altered by sodium bisulfite and guanidine-HCl. Ind. Crop. Prod. 83: 453–461.

Ziegler, G.R. and E.A. Foegeding. 1990. The gelation of proteins. Adv. Food Nutr. Res. 34: 203–298.

14

Effects of Nano-materials on Different Properties of Wood-Composite Materials

Hamid R. Taghiyari,[1,*] *Jack Norton*[2] *and Mehdi Tajvidi*[3]

ABSTRACT

Wood-composite building products offer homogeneous structures that address the ever-growing needs of different industries for new materials with consistent, predictable, and unique qualities. However, they are susceptible to biological degradation; they are vulnerable to changes caused by water and water vapor; and fire is always a great concern. Nanomaterials can easily be added to wood-composite materials during formation to improve their qualities. Nano-metals (nanosilver, nanocopper, and nano-zincoxide) and nano-minerals (nano-wollastonite and nano-clays) have successfully been used to improve the biological resistance of wood-composite products against wood-deteriorating fungi. Addition of highly thermally conductive nano-metals and nano-wollastonite increased the thermal conductivity coefficient of composite mats, thereby reducing hot-press time, and improving physical and mechanical properties. A number of organo-nanoparticles were used to improve the water resistance of wood-composites. Nano-cellulose and nanofibrills provided composites with unique properties. Novel composites based on nanofibrillated plant

[1] Wood Science & Technology Department, Faculty of Civil Engineering, Shahid Rajaee Teacher Training University (SRTTU), Tehran, Iran.
[2] Retired, Horticulture & Forestry Science, Queensland Department of Agriculture, Forestry and Fisheries, Australia.
[3] School of Forest Resources, University of Maine, Orono, ME, USA.
* Corresponding author: htaghiyari@srttu.edu, htaghiyari@yahoo.com

cellulose and bacterial cellulose embedded in natural and synthetic polymeric matrices are also promising due to their compatibility with the matrix. Different nanomaterials have also been employed to improve the bonding strength as well as water resistance of bio-adhesives such as soy protein, to replace formaldehyde-based and petrochemically derived adhesives. In this chapter, some of the applications of different nano-materials that have successfully been used to improve qualities of wood composites are presented.

Introduction

Recorded history reveals that wood is a material with unique properties and has helped mankind and nations build communities and develop their civilizations. There are hundreds of products that come from wood, including solid wood construction products, pulp and paper, chemicals, and different wood-composites. The term wood-composite (or wood panels, panel boards, etc.) refers to any product, which can be manufactured from mechanically chopped, peeled, milled, and ground, or refined wood (such as veneers, strands, particles, fibers, etc.) that are then bonded by adhesives, usually through a process at high temperature and pressure (Youngquist et al., 1997; Kharazipour, 2004). We take these products for granted, but it is to be noted that excluding wood from our lives would greatly diminish our quality of life. The main advantage is that wood is a renewable resource, unlike other resources such as metals, crude oil, stone and minerals. Moreover, trees and forests provide suitable habitats for wildlife. They are considered a primary part of recreation and are used in urban design for municipal wellbeing and pollution-reduction purposes (Milton, 1995).

Forest resources in many parts of the world are limited, and plantations and natural regeneration of forests cannot keep pace with the demand for wood and wood-composites. The total area of the world's forests is under four billion hectares (Kües, 2007) and is decreasing at an alarming rate whilst the consumption of wood is increasing with the exponential growth of the world's population. Effective utilization and protection of these limited resources are therefore necessary. Proper utilization of different agro-forestry products to produce wood-composites should also be considered as a part of better utilization and management of the present forestlands (Nayeri et al., 2014). In this regard, wood can be modified to improve its durability against biological deterioration by fungi (Hill, 2006; Schmidt, 2006; 2007; Abdolzadeh et al., 2015) and insects (Hickin, 1975), its susceptibility to fire (Taghiyari, 2012a), and its dimensional stability in moist and humid conditions. Nanotechnology has been used in many areas of science in recent decades (Ayesh and Awwad, 2012; Drelish, 2013; Saber et al., 2013) and is arguably the most actively and rapidly developing research area at the beginning of the present century (Guz, 2012). It has been used to improve the quality of many materials, including solid wood and wood-composite panels. It should, however, be kept in mind that when the size of a particle or the size of grains in a solid is reduced to the nano-scale, unusual changes in its properties may occur (Li, 2012). For instance, silver is a conductive metal, but silver nano-particles may become non-conductive. Nano-sized silica particles are conductive; and supermagnetism can be obtained when reducing the size of ferromagnetic particles to nanometer scale (Li, 2012).

Because of the actual and potential ability of nano-materials to improve characteristics of many materials, studies have so far been carried out to overcome some of the natural shortcomings of wood and wood based panels. One of the basic characteristics of solid wood is that its strength properties are not the same in different directions (Doost-hoseini et al., 2014). Homogenized raw materials can be formed in a desired shape, size, and dimension (Kües, 2007). In-process treatment (IPT) provides a simple way of applying different nano-materials during the production of wood-composite panels to address the limitation in untreated material. Different nano-materials can be used to improve, eliminate, or at least ameliorate, the negative effect of the unwanted properties (Taghiyari, 2014a; 2015; Taghiyari and Schmidt, 2014).

Although wood and wood-composites benefit from many unique properties, making them the suitable materials for many applications, they also suffer water absorption and resultant thickness swelling, vulnerability to fire, being susceptible to wood-deteriorating fungi, wood-boring insects and termite attack. This review chapter presents some recent studies of nanotechnology and nano-materials to overcome these shortcomings and to improve different properties in agricultural and wood-based composite panels.

Improving the Biological Resistance of Wood-Composites

Under suitable conditions, wood can last for centuries and provide good service. However, it is a biological material and therefore subject to biological attack by organisms seeking food and energy (Schmidt, 2006). These biological agents include different kinds of wood-deteriorating fungi, wood-boring insects, termites, and marine borers. These organisms attack wood for a variety of reasons; some consume it as food, others use it for shelter, while others use it for both. The many complex carbohydrates or sugars and starch present in untreated wood and wood-composite materials are a primary source of food for a lot of wood-deteriorating agents. Given suitable circumstances and conditions, these agents break down and consume different components of cell-walls in wood (Milton, 1995). Wood-staining fungi merely discolor, or stain the wood; while wood-deteriorating, or wood-rotting fungi can change its chemical composition, eventually affecting wood's physical, mechanical, and chemical properties. To grow, fungi have basic requirements, including temperature (10 to 30°C), adequate moisture (20–60%), oxygen (water-saturated timbers and logs are considered safe from fungal attack), and wood as a source of food (cellulose, hemicellulose, or lignin) (Milton, 1995).

To protect wood from biological attack, many materials have been introduced at laboratory and industrial scales (Schmidt, 2006; 2007). In recent years, nano-materials have been successfully added to preservatives to protect wood against wood-rotting fungi (Karimi et al., 2013; Lykidis et al., 2013; Taghiyari et al., 2014a,b,c; Mantanis et al., 2014). Nanosilver (NS) and nanocopper (NC) have been reported to be effective in limiting the growth of *Trametes versicolor* in commercial particleboard panels (Taghiyari et al., 2014a). They added 200 ppm silver and copper nano-suspensions to particleboard mats at 100 and 150 mL kg^{-1} dry weight wood particles, and then compared the levels of protection with control panels. Aqueous NS and NC suspensions were mixed with urea-formaldehyde (UF) resin before spraying the resin on wood particles. The size range of nano-particles was 10–80 nm. Specimens for fungal tests were prepared according to the European standard EN 113. Specimens were incubated at 25°C and 65% relative humidity (RH) for 16 weeks. Results of the fungal culture and mass loss measurement demonstrated that both NS and NC reduced the growth of *T. versicolor*. Comparison between the NS- and NC-treated panels showed lower mass loss values in NC-panels, indicating higher fungicidal properties of copper in hindering the growth of fungus hyphae compared to silver (Palanti et al., 2012). Taghiyari et al. (2014a) found that copper nanoparticles not only protected the test material against *T. versicolor*, but also contributed to increased hardness. This was attributed to better heat-transfer from the hot-plates to the particle mat, facilitating better cure of the resin. They also studied the effect of exposure to *T. versicolor* on hardness of particleboard panels treated with different concentrations of NS and NC. They found that panels treated with 150 mL kg^{-1} dry weight wood particles of 200 ppm nanosilver suspension demonstrated the lowest decrease in hardness after fungal exposure in comparison to other panels (panels containing NS-100, NC-100, and NC-150). They concluded that although the high heat-transfer property of silver nanoparticles had caused de-polymerization of UF-resin in the surface layers of particleboard panels, resulting in a significant decrease in the hardness values of NS-treated panels, its fungicidal property ameliorated the negative effect caused by exposure to the *T. versicolor* fungus. NC-treated panels showed the same trend. A significant negative correlation (85%) was reported between mass loss and hardness values. Taghiyari et al. (2014a) concluded that nanocopper would be better for the particleboard manufacturing industry because it provides two improved properties at the same time: improved biological resistance against *T. versicolor*, and increased hardness of panels. In another study, zinc oxide, zinc borate, and copper oxide nanoparticles were tested against mold and decay fungi and subterranean termites (Mantanis et al., 2014). The authors reported that mold fungi were slightly inhibited by nanozinc borate, while the other nanometal preparations did not inhibit mold fungi. Mass loss from fungal attack by *Trametes versicolor* was also inhibited by the zinc-based compounds. They also indicated that all pine specimens treated with nanozinc borate strongly inhibited termite feeding, with mass losses varying from 5.2% to 5.4%. In contrast, the nano-copper-based treatments were much less effective against subterranean termites, *Coptotermes formosanus*, with mass losses more than 10%, varying at 10–16% (Mantanis et al., 2014).

Although the addition of nanosilver and nanocopper were reported to be significantly effective in reducing the growth of wood-deteriorating fungi and improving hardness in particleboard panels, the panel manufacturing industry in Iran did not show much interest to use these metal nano-particles in their production program for two main reasons: the high cost of preparing silver or copper at nano-scale and the associated cost of applying these expensive metals during the manufacturing process. Therefore, researchers investigated other materials that are cheaper than silver or copper, as well as being abundant in many countries. Wollastonite is a calcium inosilicate mineral ($CaSiO_3$) that may contain small amounts of iron, magnesium, and manganese substituting for calcium. Its color is usually white and its thermal conductivity coefficient is high (Taghiyari et al., 2013a). It is abundant in many countries, including the US, Brazil, China, and Iran. The mineral wollastonite was found to be effective in reducing the effects of certain pathogens, including fungi (Aitken, 2010; Karimi et al., 2013). It is also nontoxic to humans or wildlife. Whilst there was no clear evidence to indicate wollastonite presented a health hazard, long term exposure to wollastonite dust was investigated (Huuskonen et al., 1983a; Maxim and McConnell, 2005). It was reported that the long term health effects due to inhalation of wollastonite appeared to be negligible because no correlation of serum angiotensin-covering enzymes in wollastonite workers with slight pulmonary fibrosis was found (Huuskonen et al., 1983b). Wollastonite gel or powder can be easily applied in the wood-composite manufacturing industry as an IPT, either mixed with the resin or sprayed on wood fibers and particles, and is not considered to be a serious hazard to human health or wildlife (Karimi et al., 2013).

Nanowollastonite (NW) gel was used for the production of medium-density fiberboard (MDF) (Taghiyari et al., 2013a; 2014b,c). The size range of wollastonite nanofibers was 30–110 nm. Table 1 shows the composition of NW mixed with UF resin and sprayed on to the wood fibers in a drum-mixer before the hot press. A magnetic mixer stirred the mixture (UF resin

Table 1. Composition of the nanowollastonite gel used (Taghiyari et al., 2013a; 2014b,c).

Nanowollastonite compounds	Composition by mass (%)
CaO	39.77
SiO_2	46.96
Al_2O_3	3.95
Fe_2O_3	2.79
TiO_2	0.22
K_2O	0.04
MgO	1.39
Na_2O	0.16
SO_3	0.05
Water	4.67

and NW gel) for 30 minutes for each load of the drum-mixer. Consumption level of wollastonite gel was 10% based on the dry weight of wood fibers or chicken feathers. There are many chicken farms in Iran and therefore there is an abundant source of chicken feathers. Chicken feathers are considered to be a waste material, and therefore, it offers a large, cheap, renewable fiber source for use as an additive for medium-density fiberboard to satisfy the increasing needs for wood fiber substitutes (Taghiyari et al., 2014b,c; Taghiyari and Sarvari Samadi, 2015). Once produced, panels were kept in a conditioning chamber at 30°C, and 45±2% RH for three months before physical and mechanical tests were carried out.

The produced MDF panels were exposed to brown-rot fungus *Antrodia vaillantii* (DC.: Fr.) Ryv. isolate 240 from the Hamburg Wood Biology collection. Specimens were exposed to the fungus in Kolle flasks on 4.8% malt extract agar. The isolate was re-identified by rDNA-ITS sequencing (Schmidt et al., 2012). Incubation was carried out at 25°C and 65% RH for 16 weeks. Mass loss (ML) was determined according to EN 113 standard specifications. Five replicate specimens were prepared for each treatment. At the end of the incubation, surface mycelia were removed and the specimens were re-dried at 103°C for 24 h and weighed to determine mass loss caused by fungal attack. The investigation showed that mass loss as a result of exposure to the fungus in panels treated with wollastonite was significantly decreased. It was determined that NW could ameliorate some negative effects caused by mixing wood fibers with chicken feather fibers. The authors concluded that NW can be used to improve resistance to fungal attack in MDF and can facilitate using chicken feather fibers in MDF to provide a reliable and renewable resource for natural fibers for MDF-manufacturing industry. A similar effect of wollastonite nanofibers on the growth of *A. vaillantii* was reported in particleboard panels made from wood and chicken-feather fibers (Taghiyari et al., 2014b,c).

Wollastonite was also reported to be effective in providing biological resistance in solid wood (Karimi et al., 2013). In a study, poplar specimens (*Populus nigra*), cut and prepared according to DIN-52176 standard specifications, were impregnated with a wollastonite nano-suspension at the concentration of 6.3%. Results of exposure to the white-rot fungus *Trametes versicolor* showed that mass loss in the control specimens was 47.5% after 16 weeks of exposure to the decaying fungi, while the mean mass loss in the NW-impregnated specimens was only 3.6%. According to the Findlay scale, ASTM D-2017 (1981), wood with a mass loss of more than 30% is classified as perishable or without resistance. The NW-impregnated specimens, with less than 5% mass loss, were classified in the very resistant category. The authors concluded that impregnation with NW greatly improved the biological durability of poplar wood specimens against deterioration by *T. versicolor.* They added that the impregnation process did not have negative effects on the mechanical properties of poplar wood.

In wood-plastic-composite (WPC) products although wood particles are presumed to be completely encapsulated by the plastic in a WPC matrix, thereby protecting them from both moisture and fungal attack, observations of early WPC decks in Florida showed fungal fruiting bodies on the surface (Morrell et al., 2006). Subsequent studies clearly proved that the wood particles in many WPCs remained susceptible to fungal degradation (Laks et al., 2000; Mankowski and Morrell, 2000; Pendleton et al., 2002) mainly due to inadequate compatibility between the two phases.

In a recent study, nanoclay was added to wood-plastic composites and exposed to five different fungi (Bari et al., 2015). Nanoclay was used at 2, 4, and 6% based on the dry weight of wood flour. Specimens were exposed to three white-rot fungi *Physisporinus vitreus, Pleurotus ostreatus* and *Trametes versicolor* and two brown-rot species *Antrodia vaillantii* and *Coniophora puteana.* The authors reported that the maximum mass loss (3.2%) was caused by *T. versicolor* in the control treatment. The minimum mass loss (0.2%) was found in the 6%-nanoclay treatment exposed to *P. vitreus.* Mass loss amounts were statistically dependent on the species of fungi. The fact that *T. versicolor* and *P. vitreus* demonstrated the highest and lowest mass loss in all WPC-treatments indicated *T. versicolor* was the most aggressive, and *P. vitreus* was the mildest fungi of the five fungi studied. Mass loss values gradually decreased in all five fungi as a result of the increase in nanoclay-content. This clearly illustrated the positive effect of nanoclay on the biological resistance against different wood-deteriorating fungi in WPCs. They also reported a significant correlation (R^2 higher than

70%) between the average values of ML versus water absorption based on the four WPC-treatments and the mass losses. They implied that the penetration of fungus mycelium into the WPC-matrix followed a similar pattern to the penetration of water.

The effect of fungal exposure (*T. versicolor*) was studied on the permeability of nanosilver- and nanozincoxide-impregnated Paulownia wood specimens heat-treated at 100 and 150°C (Taghiyari et al., 2015). The authors reported a statistically significant decrease in air permeability of all treatments; the decrease was attributed to the growth and accumulation of hyphae along vessel lumens, perforation plates, and pit openings, eventually blocking the fluid transfer path.

Other nano-materials were also reported to be effective in improving antibacterial properties and inhibiting biodegradation in the wood polymer nanocomposite (WPNC) simul wood (*Bombex ceiba*, L.) (Devi and Maji, 2013). The researchers used TiO_2 nanopowder (< 100 nm particle size), SiO_2 nanopowder (< 5–15 nm), γ-trimeghoxy silyl propyl meth-acrylate (MSMA) and nanoclay (nanomer, surface modified by 15–35 wt% octadecylamine and 0.5–5 wt% amino propyl triethoxy silane). *Phanerochaete* sp. was collected from locally available rotten *Bombex* wood. This fungal genus was previously reported as the main wood-deteriorating fungus of this wood species (Kersten and Cullen, 2007). They concluded that fungal growth was maximum and minimum for untreated wood and styrene acrylonitrile copolymer treated wood specimens, respectively.

Thermal Conductivity in the Wood-Composite Mat

Wood has a very low thermal conductivity coefficient in comparison to metal and mineral materials (0.055–0.17 W/mK depending on the direction of the wood grain, compared to 429 W/mK in silver) (Yu et al., 2010). Thermal conductivity coefficient was measured with an apparatus designed and built based on Fourier's Law for heat conduction (Fig. 1). Circular specimens 30 mm in diameter and 16 mm in length were completely covered with silicone adhesive to insulate them from the surrounding atmosphere. Thermal conductivity was calculated using Equations 1 and 2.

$$Q = KA \frac{\Delta T}{L} \qquad \text{(Equation 1)}$$

$$K = \frac{Q \times L}{A \times \Delta T} \qquad \text{(Equation 2)}$$

Where:

K = Thermal conductivity (W/m.K)
Q = Heat transfer (W)
L = Specimen thickness (m)
A = Cross section area of specimens (m^2)

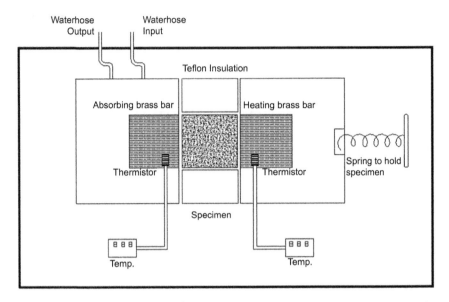

Figure 1. Schematic drawing of the thermal conductivity measurement apparatus for the wood-composite mat (Taghiyari et al., 2013b).

$\Delta T =$ Temperature difference (T_1-T_2) (°K)

In a study on MDF (Taghiyari et al., 2013b), the thermal conductivity coefficient of control MDF boards was 0.099 W/mK. Adding 10% of nano-wollastonite (NW) to the MDF-mat (based on the dry weight of wood fibers) increased thermal conductivity coefficient to 0.110 (W/mK); that is an increase of 11.5%. Addition of NW decreased standard deviation in thermal conductivity between the replicates of NW-treated MDF panels. This means that NW treatment resulted in an increased homogeneity in the composite matrix. The moisture content was equal in all treatments due to the thermo-hygromechanical behavior of wood (Figueroa et al., 2012).

Taghiyari et al. (2013b; 2014d) showed that addition of NW decreased the time for curing of the UF-resin in the core section of the MDF mat. Physical and mechanical properties were increased significantly (Taghiyari et al., 2013b; 2014d).

The heat-transfer property of metal (Saber et al., 2013) and mineral nanomaterials (Haghighi et al., 2013) improved some properties in solid wood and wood-composite materials. Taghiyari et al. (2013c) studied the effects of a 400 ppm aqueous suspension of silver nanoparticles on the heat-transfer rate from the hot-press platens to the core section of medium-density fiberboard (MDF). Nanosilver suspension was sprayed on the mat at three application levels of 100, 150, and 200 mL/kg based on the dry weight of wood fibers. SEM micro-graphs showed uniform spread of silver nanoparticles over wood fibers (Fig. 2). A digital thermometer with a temperature sensor probe was used to measure the temperature at the core section of the mat at 5-second intervals (Fig. 3). The probe of the thermometer was directly inserted 50 mm into the core of the mat (from the edge), in the horizontal direction. Temperature measurement was started immediately after the two hot platens reached the stop-bars. Measurement of temperature in the core section of the mat (immediately after the upper plate of the hot press reached the stop-bars) revealed statistically significant difference between the temperatures of the four treatments of control, NS100, NS150, and NS200 (Fig. 4). The authors reported that temperatures at the core section of NS150 and NS200 were higher than both NS100 and

Silver nanoparticles spread over the fiber

Figure 2. SEM micrograph of wood fibers showing silver nanoparticles (↓) scattered all over the fibers (Taghiyari et al., 2013b).

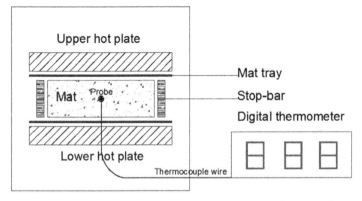

Figure 3. Temperature measurement in a wood-composite mat under press platens using a digital thermometer with its sensor probe inserted into the core section of the composite-board mat (Taghiyari et al., 2013b).

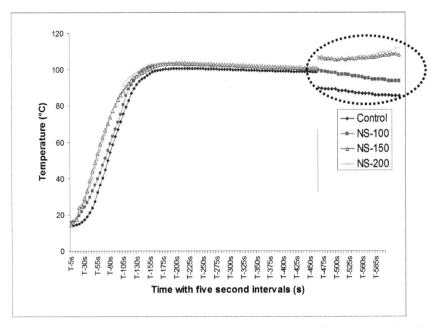

Figure 4. Temperature at the core section of the medium-density fiberboard mat after the third minute of hot-pressing with five-second intervals (NS = nanosilver content mL/kg) (Taghiyari et al., 2013b).

control treatments. The higher temperature resulted in a depolymerization of the resin bonds in the surface layers of panels; that is, in the final minutes when all moisture content was nearly evaporated from the surface layers, the heat resulted in the depolymerization and breaking down of resin bonds. The depolymerization decreased some of the physical and mechanical properties of the produced panels, and eventually increased the fluid flow in the composite-matrix. The authors suggested that further studies should be carried out on the application of metal nanoparticles or mineral nanofibers to the core section of composite mat to facilitate the heat transfer to this area. This would also prevent over-heating of the surface layers and the consequent resin break-down.

The authors concluded that addition of metal nanoparticles to increase the heat-transfer rate to the core of composite mats does not improve all physical and mechanical properties of the panel. The optimum application level for metal nanoparticles is dependent on many factors, including the hot press temperature, hot-press duration, thermal conductivity coefficient of metal nanoparticles, and the type and density of composite panels.

In a more recent study where the addition of chicken feathers to the furnish was examined as a way of extending the wood fiber resource, temperature at the core of the mat in MDF panels was significantly different between the MDF panels without wollastonite nanofibers (all three treatments of without feather, 5%-feather, and 10%-feather contents) and the three NW-treated panels (the three panels of NW-treated, NW-5%, and NW-10% feather content) (Fig. 5). All treatments showed nearly the same increase in temperature up to 90 seconds of heating; however, the three treatments with 10%-NW showed a clearly higher temperature at the core section after the same time period. This clearly showed the effects of higher thermal conductivity coefficient in NW-treated panels in the transfer of heat to the core section of the mat (Taghiyari et al., 2013b). Control treatments had a nearly flat rate from the 90th second, indicating that time did not significantly affect the core-temperature; however, NW-treated panels had a decreasing trend. This was due to the rapid evaporation of water from the mat and the consequent decrease in the overall moisture content, resulting in a lower heat-transfer rate; and secondly, the higher thermal conductivity of the NW-treated panels in which absorption of heat took place at a higher rate.

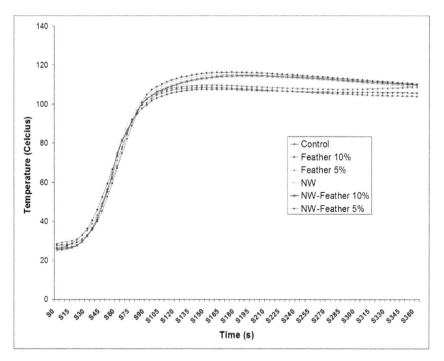

Figure 5. Temperature at the core section of the medium-density fiberboard (MDF) (NW: nanowollastonite; S: the time intervals).

Reduction of Hot Press Time

In most wood-composite manufacturing facilities, hot press time is considered to be the bottle-neck in the production process. The hot press time is dependant on many factors, including the thickness of the composite mat, press temperature, closing rate, and most importantly, moisture distribution throughout the mat (Taghiyari et al., 2011). Moisture of the mat cannot always be increased to facilitate rapid heat transfer as it in turn increases the hot press time. Increase in the moisture content also causes damage to panels. This is because a large volume of water vapor must leave the mat to the surrounding atmosphere in a short period of time. If the volume is too high, permeability in the composite matrix is not enough to allow the vapor and gases to escape. Accumulated vapor within the composite matrix eventually blows ruining the panels once the hot press platens open. Furthermore, for urea-formaldehyde (UF) resin, there is a limitation of moisture content (MC) level (Papadopoulos, 2006). A higher MC than the standard level for UF resin weakens the strength of resin. Finding new ways to increase the heat transfer rate to the core section of the composite mat has always been a challenge for the wood-composite manufacturing industry. Silver nano-particles decreased hot press time by 10.9% when 100 mL of nano-silver suspension was used for each kg of wood particles (dry weight basis) (Taghiyari et al., 2011). Nano-copper decreased hot press time by 5.7% when 100 mL nano-copper suspension was used. This is not unexpected considering the lower thermal conductivity coefficient of copper compared to silver (Bray, 1947; Menezes Nunes et al., 1991).

Permeability in Wood-Composite Panels

One of the characteristics of wood is its water absorption and dimensional change when exposed to rain, in soil contact, dipped in water or even placed in moist areas (Mantanis and Papadopoulos, 2010). This affects wood-composite products in different ways. First, the dimensions of wood-composites alter due to swelling. The swelling not only has undesirable side-effects on the design and dimensions of the wood-composite product that is used in the structure, but also the wood chips and fibers lose their integrity through micro-movements caused by swelling. Secondly, water breaks down some of the resin bonds that adhere wood chips or fibers together in the matrix, again decreasing the overall strength of the composite matrix. The way fluids flow through the porous structure of wood and wood-composites is dependent on the size, plurality, and arrangement of pores throughout the material (Shibata and Hirohashi, 2013; Taghiyari and Moradi Malek, 2014). These factors together determine impregnability of wood and wood-composites (Hermoso and Vega, 2016). Moreover, some pores are isolated and therefore, they are not active in the flow of fluids through the porous structure; while other pores are continuous, meaning they are inter-connected and form a continuous percolating pathway for the flow of fluid.

To investigate this, a gas permeability measurement apparatus was developed to measure the permeability values in porous media (including solid wood, wood-composite materials, carton and paper, and light-weight cement) (Taghiyari, 2012b; Taghiyari and Efhami, 2011) (Fig. 6). A falling-water volume-displacement method is used to calculate specific longitudinal gas permeability values based on the microstructure porosity of the specimen being evaluated (USPTO, 2009) (Siau, 1995; Taghiyari, 2013a,b). For the liquid permeability measurement, a special apparatus was designed and produced based on the Rilem test method II.4 specifications (Fig. 7). The procedure for liquid permeability was set by The International

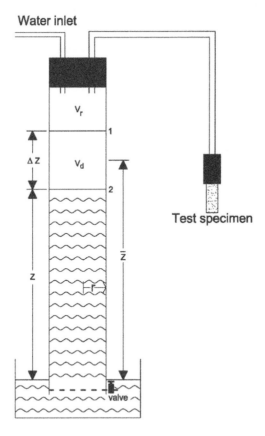

Figure 6. The schematic drawing of the gas permeability apparatus for porous media (USPTO No. US 8,079,249, B2) equipped with single-storey milli-second precision electronic time measurement device (Taghiyari et al., 2014e,f,g; 2016).

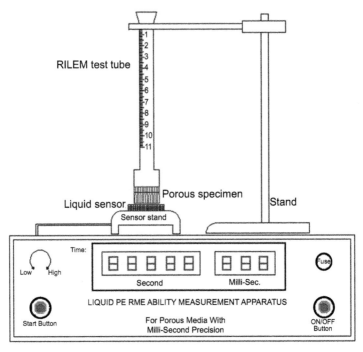

Figure 7. The schematic of the liquid permeability apparatus for porous media based on Rilem test method II.4 specifications.

Union of Laboratories and Experts in Construction Materials, Systems, and Structures. Penetration of liquid (in this case, distilled water) was conducted according to ASTM E-514 (Taghiyari et al., 2014e,f,g). For each specimen, two different times were recorded; the time when the first drop of water to fall from the lower end of the test specimens and the time when the column of water in the Rilem test tube dropped by 50 mm (6.6 cubic centimeters of water). Seven air permeability and two liquid permeability times were used to calculate correlation values.

In order to decrease water absorption in medium-density fiberboard (MDF), the water-repellent property of nano-silane (NS) was investigated (Taghiyari, 2013a; Taghiyari et al., 2013d). The NS liquid was the product of a reaction of organo silane, with an organic reactant. NS content was estimated based on the solid part in the suspension. For each treatment, the weight of NS solids was deducted from the fiber used; this way, the density of panels in different treatments with different fiber-content was kept constant. The final mixture of NS plus resin was evenly sprayed on the fibers (Fig. 8). The pH and viscosity of the resin were kept constant for all treatments. Density of all treatments was kept constant at $0.67 \text{ g}/\text{cm}^3$. The authors reported that nano-silane treatment significantly decreased liquid permeability in MDF although the amount of wood fibers was lower in NS-treated panels and micro-cavities formed in the composite-matrix (Fig. 9). This resulted in the NS 100-treatments being clustered with the control panels. In another study, NS decreased water absorption and thickness swelling in MDF (Taghiyari et al., 2013d). NS treatment affected the wood-composite in two ways. First, the water-repellant property of silane nano-particles acted as a physical barrier to penetration of water. Second, NS contributed to the process of sticking wood fibers together. However, silane-treated panels were susceptible to molds and therefore they are not recommended for humid climates (Fig. 10). Most of the fungi were identified as *Aspergillus niger* and *Penicillium* spp. (Taghiyari, 2014b; 2015).

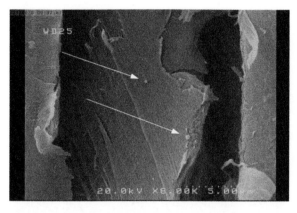

Figure 8. SEM image of wood fibers showing nanozycosil (↓) on the cell wall (× 6,000) (Taghiyari, 2014b).

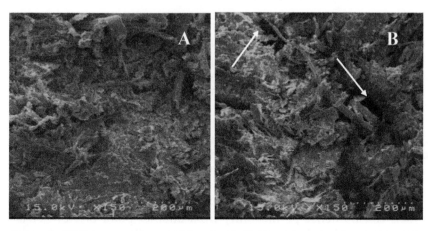

Figure 9. MDF texture (A) control specimen: fibers are integrated more intensely; (B) NZ-150: some void spaces (↓) are observed in the texture leading air to pass through much easier (Taghiyari, 2013a).

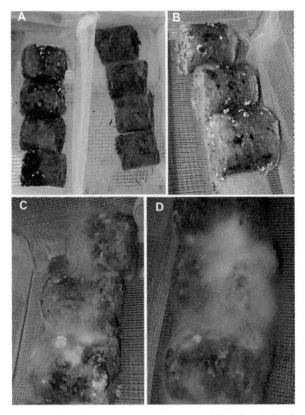

Figure 10. Photographs of molded MDF specimen in vapor chamber; A, B: after 14 weeks; C, D: progressed growth after 18 months (Taghiyari, 2014b).

Silver and copper nano-particles (NS and NC) decreased both gas and liquid permeability in particleboard produced at an industrial scale (Taghiyari, 2011; Taghiyari and Farajpour Bibalan, 2013). NS and NC suspensions were added to the mat at two levels of 100 and 150 milli-liters/kg dry weight wood particles. Permeability values were decreased in all nano-treated composite panels. The decrease was considered to be due to the high thermal conductivity coefficient of metal nanoparticles, resulting in better heat-transfer to the mat, leading to better cure of the resin. However, the optimum application levels of NS and NC were not the same. Significant differences in the thermal conductivity coefficients of silver and copper were reported to be the reason.

Accelerated heat-transfer in the NS- and NC-treated wood-composite panels can influence permeability from another point of view. Heat-treatment has significant effects on the permeability of different wood species (Taghiyari, 2013b). Structural modification and chemical changes of carbohydrates and lignin occur as wood is heated (Repellin and Guyonnet, 2005) and there is irreversible hydrogen bonding in the course of water movements within the pore system that also affects the fluid transfer process (Borrega and Kärenlampi, 2010). In this phenomenon, some of the polar sites that are usually available for water sorption will be irreversibly bonded within the hemicelluloses and amorphous regions of cellulose. These processes increase permeability as wood is heated from about 70°C up to 150°C. Higher temperatures increase high internal stresses that are released as cracks (Oltean et al., 2007). These micro-cracks facilitate the process of fluid transfer through the porous material causing the gradual increase in permeability. Nanoclay had no significant effect on permeability in plywood; however, moisture diffusion decreased significantly (Dashti et al., 2012). They used nanoclay at two application levels of 3 and 5% based on the weight of the wood flour, and press time was also studied at 4 and 5 minutes. It was concluded that due to the hydrophobic property of clay nanoparticles, increased level of nanoclay as filler resulted in reduction in thickness swelling and diffusion coefficient.

Nano-Cellulose in Wood-based Composites

The use of nano-cellulose is an interesting area in the optimization and improvement of many composite materials, including paper and boards, paper coatings, and organic electronics. In recent years, novel composites have been developed based on bacterial cellulose and nanofibrillated plant material in natural and synthetic polymeric matrices such as poly(lactic acid), chitosan, starch, and pullulan (Freire et al., 2012). These composites have great potential for a variety of applications including in the automobile and aviation industries, packaging, paper coating, flexible electronics, construction, as well as biomedical products and devices (Freire et al., 2012). The unique and outstanding properties of cellulose fibers, as a renewable source, are well documented (Fengel and Wegener, 1989). The properties include their high mechanical strength, chemical and biocompatibility characteristics, as well as non-toxicity to human and wildlife (Freire et al., 2012). These properties have so far been well addressed in the paper and textile industries (Klemm et al., 2005). More recently, the unique properties of cellulose have led to a focus on polymer-based nanocomposites (multiphase materials) consisting of a polymer matrix and a nanofiller (Freier et al., 2012). Some processes like esterification, etherification, urethane formation, and cross-linking or graft copolymerization can expand potential applications of cellulose (Yu and Chen, 2009). Moreover, the controlled heterogeneous modification of cellulose fibers can be considered as a particular characteristic. Here, the reaction is limited to the most accessible regions of the fibers while its bulk mechanical properties can still be preserved. In fact, this is considered one of the main and most practical strategies in utilizing cellulose as a reinforcing element in various composites (Bledzki and Gassan, 1999; Schurz, 1999; Belgacerm and Gandini, 2005; Samir et al., 2005; Freire and Gandini, 2006; Teeri et al., 2007; Dufresne, 2008). In this process, cellulose replaces some inorganic (mineral)-based fibers (Wang and Zhang, 2009). Automotive, construction, and packaging industries are some of the large industrial sectors that have utilized these composites in recent years (Freier et al., 2012).

Polymer-based nanocomposites (multiphase materials) consisting of a polymer matrix and a nanofiller, have attracted special attention (Freier et al., 2012). Compared to conventional polymer composites, they result in products with unique properties (Bordes et al., 2009) including improved mechanical, thermal, and barrier properties and transparency (Zimmermann et al., 2004; Hubbe et al., 2008; Nogi and Yano, 2008; Azeredo, 2009). Micro- and nanofibrillated cellulose (MFC and NFC, respectively), cellulose whiskers, and bacterial cellulose (BC) play an essential role in these nano-composites. The applications for these nano-composites are extremely wide (Nakagaito and Yano, 2004; Jung et al., 2008; Nogi et al., 2009). BC is also known as microbial cellulose (Freier et al., 2012) and is produced by different bacteria genera, such as *Gluconacetobacter*, *Sarcina*, and *Agrobacterium*, with *Gluconacetobacter xylinus* being the most common strain (Budhiono et al., 1999; Shoda and Sugano, 2005; Pecoraro et al., 2008). The yield of BC produced by *G. sacchari* has been reported to be very high (Trovatti et al., 2011). They can convert different substrates into cellulose in a few days. Some substrates such as glucose, glycerol, and other organic materials have so far been studied (Chawla et al., 2009; Carreira et al., 2011). Recently, advances in nano-composite research at the University of Averiro were achieved with NFC and BC as reinforcing elements (Freier et al., 2012). Based on the recent research on nanofibrillated plant cellulose and BC embedded in natural and synthetic polymeric matrices such as PLA (poly-lactic acid), CH (chitosan), starch, and pullulan, it can be concluded that the novel composites using nanofibrillated cellulose show promise as they present high compatibility within the matrix (Freier et al., 2012). They provide different industries with unique physical and thermal properties that may have applications in areas such as packaging, electronic devices, and biomedicine. These composites and their applications can still be expanded by combination with inorganic nanophases (Goncalves et al., 2008; 2009; Pinto et al., 2012; Vilela et al., 2010). Development of more novel cellulose-based nano-composites is still needed. One highly interesting application of cellulose nanofibrils has been recently developed at the University of Maine, U.S.A. where cellulose nanomaterials are used as the sole binder in the formulation of conventional particleboards completely replacing urea-formaldehyde resin. Cellubound, a novel laminate system made be laminating sheets of paper using cellulose nanofibrils as the sole binder is another application of cellulose nanomaterials to produced new composite panels (Tajvidi et al., 2016).

Improving Fire Properties in Wood-Composites by Nano-materials

Fire safety is an important concern in all types of construction, whether built with cement and iron or with wood. Wood and wood composites are especially important because they burn when exposed to heat and air. Thermal degradation of wood occurs in stages. The degradation process and the exact products of thermal degradation depend upon several factors, including the rate of heating as well as the temperature.

Inorganic salts showed promising results to render wood fire resistant (White and Dietenberger, 1999; Ayrilmis et al., 2007). A pressure process was developed to impregnate wood with mineral salts and deposit them in the wood cells, resulting in a building product known as Fire Retardant Treated (FRT) Lumber. However, it should be noted that fire-retardant chemicals reduce the strength on wood, cause corrosion of fasteners, increase hygroscopicity, and produce toxic gases and smoke (Winandy et al., 2002; Taghiyari, 2012a).

In the case of fire retardant treated wood composite materials, the effect of fire on the composite resin will eventually weaken its physical and mechanical properties. Although all these features have been considered and examined over the years, nano-based products have also been investigated for use as modern fire-retardants.

In a recent study, specimens were impregnated with silver nano-suspension (SNS) and the fire properties were measured with a low cost purpose built apparatus (Taghiyari, 2012a) (Fig. 11). The author compared the results of SNS-impregnated specimens with those of a control, borax and Celcune-impregnated specimens. Celcune is a newly developed fire-retardant in Iran based on a mixture of chemicals and minerals. SNS suspension has a couple of advantages. It is not acidic and consequently does not reduce the strength of wood or cause corrosion of fasteners. The silver ions may decrease the potential of wood to absorb water as it may bond with free hydroxyl sites in the microstructure of wood components (Rassam et al., 2010). The ability of silver nano treatment to transfer heat is of great importance in wood thermal degradation (Taghiyari, 2011) as it reduces accumulation of heat at one point of wood and consequently may delay thermal degradation, carbonization, and even pyrolysis. Taghiyari (2012a) reported that some fire-retarding properties were significantly improved in the SNS-impregnated specimens. He indicated that the high thermal conductivity coefficient of silver delayed accumulation of heat

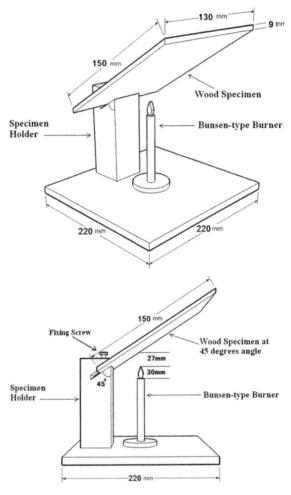

Figure 11. Fire-retardant testing apparatus (Iranian patent No. 67232) especially designed for solid wood and wood composites (Taghiyari, 2012a).

at the surrounding spot near to the flame and therefore, some properties were improved.

In another study, wollastonite nano-fibers (NW) were applied at 5, 10, 15, and 20 g/kg dry weight basis of wood fibers in MDF and compared with control specimens without NW (Taghiyari et al., 2013e). NS was applied superficially and internally. The authors reported that most fire-retarding properties were significantly improved with the increase in NW-content up to 15%. The properties included mass loss, ignition and glowing times, and fire endurance. Higher NW-content resulted in a decrease in both fire-retarding and physico-mechanical properties (Taghiyari et al., 2013e; Taghiyari and Nouri, 2015). This was partly attributed to lower fiber-content, and partly to the absorption of resin by NW nano-fibers. The work showed that surface application of NW was more effective in improving the fire-retarding properties in MDF than bulk application.

Other nano-materials, such as titanium and silicon dioxide (TiO_2 and SiO_2) nanoparticles, were also reported to have promising results in flame retardancy of wood polymer nanocomposites (WPNCs) (Devi and Maji, 2013). Based on their results, it can be concluded that there are potential characteristics in different nano-materials that can enhance fire-retarding properties in composite materials.

Nanomaterials in Biobased Wood Adhesives

Formaldehyde emission is considered a problem with the increasing concerns over environmental and health threats. Deployment of biodegradable and sustainable biomass as adhesives in the production of wood-based composites is therefore not only inevitable but also responsive to reducing impact of formaldehyde and petrochemical adhesives (Pizzi, 2006). Soy protein has been used as wood adhesives for centuries because of its low cost, fast curing time, and environmental friendliness (Kumar et al., 2008). Even though they are biodegradable, non-toxic, renewable, and commercially available, adhesives based on soy protein have limitations that prevent their industrial applications in producing wood-based composite panels. Low glue strength and poor water-resistance are the main weaknesses (Liu et al., 2010). Different materials and methods have been utilized to overcome the limitations and improve the performance of soy protein-based adhesives. In a study, bio-mimetic soy protein calcium carbonate ($CaCO_3$) hybrid wood glue was developed in an attempt to improve the adhesion strength (Liu et al., 2010). The authors reported that formation of compact rivets or interlocking links, and ion crosslinking of calcium, carbonate, and hydroxyl ions in the adhesive greatly improved the water-resistance and bonding strength of soy protein adhesives. Glue strength of soy protein hybrid adhesive was higher than 6 MPa even after three water-immersion cycles. They concluded that the green soy protein hybrid adhesive they developed, provided high glue strength and good water-resistance as a substitute for formaldehyde wood glues.

Thermal-caustic degraded soybean proteins with modifications by polyisocyanate and nano-scale montmorillonite (MMT) were used to develop a novel soybean protein-based wood adhesive with improved water resistance (Zhang et al., 2014). The authors indicated that the thermal-caustic degradation process effectively reduced the viscosity of the high-concentration soybean protein solution due to the sharp decrease in the molecular weight of the proteins. Crosslinking by polyisocyanate increased the molecular weight and produced soybean protein with a crosslinked structure. This improved the water resistance but shortened the work life of the polyisocyanate-modified soybean protein adhesive. They reported that addition of 3% MMT effectively prolonged the work life of the modified soybean protein adhesive and reduced the bond strength to some extent. Due to the fact that the exfoliated nano-scale MMT platelets not only reduced the number but also increased the steric hindrance of some active groups of soybean proteins via hydrogen and electrostatic bonds, a nano-scale blocking mechanism of MMT was presented to retard the effects on some properties of soybean protein adhesive.

Sadare et al. (2015) developed an environmentally friendly adhesive from soy protein isolate (SPI) by incorporating carbon nanotubes (CNTs) to replace formaldehyde and petrochemical adhesives. This study was based on the fact that the multifunctional properties of CNTs as nanofiller in the polymer matrix had the potential to improve bond strength and water resistance of the adhesive. The authors utilized sonication with mechanical (shear) stirring to obtain an adhesive with uniformly dispersed CNTs in the SPI matrix and no noticeable clusters. They indicated that thermogravimetric analysis of the sample showed the incorporated CNTs increased the thermal stability of the adhesive nanocomposite at higher loading fraction. The FTIR spectra of their work showed the surface functionalities of the SPI adhesive nanocomposite and the attachment of COO^- functional group on the surface of the adhesive which contributed to effective dispersion of the CNTs in the prepared nanocomposite adhesive. It was finally concluded that the preparation of a new adhesive can replace the formaldehyde and petrochemical adhesives in the market to be used in industry (Sadare et al., 2015).

In another research project, nanocrystaline cellulose (NCC) was used to improve the performance of polyvinyl acetate (PVA) as a wood adhesive (Kaboorani et al., 2012). NCC was added to PVA at different loadings, namely 1%, 2%, and 3% based on the weight of the adhesive. The blends were then used as a binder to evaluate block shear tests. Bonding strength of PVA was tested under dry and wet conditions, at 100°C. The mechanical properties of PVA film and its composites with NCC were measured by nanoindentation technique. Thermal stability and structure of nanocomposites were studied by thermogravimetric analysis and atomic force microscopy (AFM). The authors report that block shear tests demonstrated NCC can improve bonding strength of PVA in all conditions. Hardness, modulus of elasticity, and creep of PVA film were also improved slightly by the addition of NCC. Thermal stability of PVA was reported to be significantly improved by the addition of NCC. It was also reported that the quality of NCC dispersion in the PVA matrix can affect the variations in shear strength and other properties (Kaboorani et al., 2012).

Conclusions

Wood-composite panels have the advantage of offering a homogeneous structure that can be used in many industrial, construction, artistic, and public applications. However, their biological nature makes them susceptible to biological wood-deteriorating agents. There are other limitations that make application of wood-based composite materials difficult including water absorption and consequent swelling, and vulnerability to fire. The application of nano-materials readily reduces these limitations and improves the characteristics when nano-material suspensions are applied as in-process treatments. An overview of research carried out applying various nano-materials in wood-composite panels revealed numerous potential applications of nano-technology to the forest products industry. The use of metal (nanosilver, nanocopper, and nano zinc-oxide) and mineral nanomaterials (nano-wollastonite) with high thermal conductivity coefficients help in improving the thermal conductivity of composite mats and better cure of the resin especially in the core of the mat in hot-press. This led to a significant decrease in hot-press time, improvement in physical and mechanical properties, as well as decrease in gas and liquid permeability values. The water repellent properties of organo silane nanoparticles (nano-zycosil) can prevent penetration of water and vapor into the wood-composite matrix, resulting in an increased service life of the components used in furniture or in structures. Different anti-fungal or antibacterial nano-materials can significantly improve not only the biological resistance of wood-composite materials against different wood-deteriorating fungi, but they also improve physical and mechanical properties of the materials. Nano-cellulose and nanofibrills have also provided special composites with unique properties. Novel composites based on nanofibrillated plant cellulose and BC embedded in natural and synthetic polymeric matrices are also promising. The physical and thermal properties of these materials are unique, making them suitable for applications such as packaging, electronic devices, and biomedicine. Different nanomaterials have also been utilized to improve the bonding strength and water resistance of bio-adhesives such as soy protein to replace formaldehyde and petrochemical adhesives.

Over the past decade, some applications of nanotechnology and nano-materials have been studied from different perspectives in solid wood and wood-composite materials. While the application and use of nanotechnology has been extensively investigated, there are a large number of opportunities that are yet to be pursued to improve and extend wood-composite materials, eventually achieving better and higher quality results.

The effects of nano-materials on veneers and how they can better adhere to the composite panels is really important. Improvements in ice-blasting, as a new cleaning technology, can be achieved by nano-materials (Taghiyari and Lotfinejad-Sani, 2014). Different nano-materials can be used to strengthen the pull-off strength of paints and varnishes to the substrate solid wood and wood-composites.

Acknowledgement

The first author is grateful for the scientific support of Prof. Olaf Schmidt (University of Hamburg, Germany). We greatly appreciate the financial support of Shahid Rajaee Teacher Training University.

Keywords: Composite materials, Nanotechnology, Porous Structure, Renewable materials, Thermal conductivity, Wood-composite panels

References

Abdolzadeh, H., Gh. Ebrahimi, M. Layeghi and M. Ghassemieh. 2015. Analytical and experimental studies on stress capacity with modified wood members under combined stresses. Maderas Ciencia y Tecnología. doi.org/10.4067/S0718-221X2015005000025.

Aitken, E. 2010. Analyses of the Effect of Silicon on Fusarium Wilt on Banana. Horticulture Australia, Sydney, NSW, Australia.

Ayesh, A.I. and F. Awwad. 2012. Opportunity for DNA detection using nanoparticle-decorated graphene oxide. Journal of Nanomaterials & Molecular Nanotechnology 1: 1. dx.doi. org/10.4172/2324-8777.1000e101.

Ayrilmis, N., Z. Candan and R. White. 2007. Physical, mechanical, and fire properties of oriented strandboard with fire retardant treated veneers. Holz Roh Werkst. DOI 10.1007/s00107-007-0195-3, 65: 449–458.

Azeredo, HMC. 2009. Nanocomposties for packaging applications. Food Res. Int. 42: 1240–1253.

Bari, E., H.R. Taghiyari, O. Schmidt, A. Ghorbani and H. Aghababaei. 2015. Effects of nano-clay on biological resistance of wood-plastic composite against five wood-deteriorating fungi. Maderas. Ciencia y Tecnologia. 17(1): 2015–212.

Belgacerm, M.N. and A. Gandini. 2005. The surface modification of cellulose fibers for use as reinforcing elements in composite materials. Compos. Interface. 12: 41–75.

Bledzki, A. and J. Gassan. 1999. Composites reinforced with cellulose based fibers. Prog. Polym. Sci. 24: 221–274.

Bordes, P., E. Pollet and L. Averous. 2009. Nano-biocomposites: biodegradable polyester/nanoclay systems. Prog. Polym. Sci. 34: 125–155.

Borrega, M. and P.P. Karenlampi. 2010. Hygroscopicity of heat-treated norway spruce (*Picea abies*) wood. Eur. J. Wood Prod. 68(2): 233–235. DOI 10.1007/s00107-0090371-8.

Bray, J.L. 1947. Non-ferrous production metallurgy. 2nd Ed., New York: John Wiley & Sons, Chapter 26.

Budhiono, A., B. Rosidi, H. Taher and M. Iguchi. 1999. Kinetic aspects of bacterial cellulose formation in nata-de-coco culture system. Carbohydr. Polym. 40: 137–143.

Carreira, P., J.A.S. Mendex, E. Trovatti, L.S. Serafim, C.S.R. Freire, A.J.D. Silvestre and C.P. Neto. 2011. Utilization of residues from agro-forest industries in the production of high value bacterial cellulose. Bioresource Technol. 102: 7354–7360.

Chawla, P.R., I.B. Bajaj, S.A. Survase and R.S. Singhal. 2009. Microbial cellulose: Fermentative production and applications. Food Technol. Biotech. 47: 107–124.

Dashti, H., Sh. Salehpur, H.R. Taghiyari, F. Akbari Far and S. Heshmati. 2012. The effect of nanoclay on the mass transfer properties of plywood. Digest Journal of Nanomaterials and Biostructures 7(3): 853–860.

Devi, R.R. and T.K. Maji. 2013. Effect of nanofillers on flame retardancy, chemical resistance, antibacterial properties and biodegradation of wood-styrene acrylonitrile co-polymer composites. Wood Science and Technology 47: 1135–1152.

Doost-hoseini, K., H.R. Taghiyari and A. Elyasi. 2014. Correlation between sound absorption coefficients with physical and mechanical properties of insulation boards made from sugar cane bagasse. Journal of Composites Part B. 58: 10–15. DOI 10.1016/j. compositesb.2013.10.011.

Drelish, J. 2013. Nanoparticles in a liquid: New state of liquid? Journal of Nanomaterials & Molecular Nanotechnology 2: 1. doi: 10.4172/2324-8777.1000e105.

Dufresne, A. 2008. Cellulose-based composites and nanocomposites. pp. 401–418. *In*: M.N. Belgacem and A. Gandini (eds.). Monomers, Polymers and Composites From Renewable Resources. Elsevier, London.

Fengel, D. and G. Wegener. 1989. Wood–Chemistry, Ultrastructure, Reactions. Walter de Gruyter, Berlin/New York.

Figueroa, M., C. Bustos, P. Dechent, L. Reyes, A. Cloutier and M. Giuliano. 2012. Analysis of rheological and thermo-hygro-mechanical behaviour of stress-laminated timber bridge deck in variable environmental conditions. Maderas. Ciencia y Tecnologia. 14(3): 303–319.

Freire, C.S.R. and A. Gandini. 2006. Recent advances in the controlled heterogeneous modification of cellulose for the development of novel materials. Cell. Chem. Technol. 40: 691–698.

Freire, C.S.R., S.C.M. Fernandes, A.J.D. Silvestre and C.P. Neto. 2012. Novel cellulose-based composites based on nanofibrillated plant and bacterial cellulose: recent advances at the University of Avaeiro–a review. Holzforschung. DOI 10.1515/hf-2012-0127.

Goncalves, G., P.A.A.P. Marques, T. Trindade, C.P. Neto and A. Gandini. 2008. Superhydrophobic cellulose nanocomposites. Journal Colloid Interf. Sci. 324: 42–46.

Goncalves, G., P.A.A.P. Marques, C.P. Neto, T. Trindade, M. Peres and T. Monteiro. 2009. Growth, structural, and optical characterization of ZnO-coated cellulosic fibers. Cryst. Growth Des. 9: 386–390.

Guz, I.A. 2012. Continuum solid mechanics at nano-scale: How small can it go? Journal of Nanomaterials & Molecular Nanotechnology 1: 1. dx.doi.org/10.4172/2324-8777.1000e103.

Haghighi Poshtiri, A., H.R. Taghiyari and A.N. Karimi. 2013. The optimum level of nano-wollastonite consumption as fire-retardant in poplar wood (*Populus nigra*). International Journal of Nano Dimension (IJND) 4(2): 141–151.

Hermoso, E. and A. Vega. 2016. Effect of microwave treatment on the impregnability and mechanical properties of *Eucalyptus globulus* wood. Maderas-Cienc Tecnol. 18(1): DOI 10.4067/S0718-221X2016005000006.

Hickin, N.E. 1975. The insect factor in wood decay. 3rd Ed. The Rentokil Library. Associated Business Programmes London, 383pp.

Hill, C. 2006. Wood Modification Chemical, Thermal and Other Processes; John Wiley & Sons, Ltd., ISBN: 0-470-02172-1, 239pp.

Huuskonen, M.S., J. Jarvisalo, H. Koskinen, J. Nickels, J. Rasanen and S. Asp. 1983a. Preliminary results from a cohort of workers exposed to wollastonite in a Finnish limestone quarry. Scandinavian Journal of Work, Environment, and Health 9(2): 169–175.

Huuskonen, M.S., A. Tossavainen, H. Koskinen, A. Zitting, O. Korhonen, J. Nickels, K. Korhonen and V. Vaaranen. 1983b. Wollastonite exposure and lung fibrosis. Environmental Research 30: 291–304.

Jung, R., H.S. Kim, Y. Kim, S.M. Kwon, H.S. Lee and H.J. In. 2008. Electrically conductive transparent papers using multiwalled carbon nanotubes. Journal of Polymer Science Pol. Phys. 46: 1235–1242.

Kaboorani, A., B. Riedl, P. Blanchet, M. Fellin and O. Hosseinaei. 2012. Nanocrystalline cellulose (NCC): a renewable nano-material for polyvinyl acetate (PVA) adhesive. Eur. Polymer J. 48: 1829–1837.

Karimi, A., H.R. Taghiyari, A. Fattahi, S. Karimi, Gh. Ebrahimi and A. Tarmian. 2013. Effects of wollastonite nanofibers on biological durability of poplar wood (*Populus nigra*) against *Trametes versicolor*. BioResources 8(3): 4134–4141.

Kharazipour, A. 2004. Pflanzensubstrat, Verfahren zu seiner Herstellung und dessen Verwendung. German Patent, DE 102 004016 666.8.

Klemm, D., B. Heublein, H.P. Fink and A. Bohn. 2005. Cellulose: fascinating biopolymer and sustainable raw material. Angew. Chem. Int. Ed. 30: 3358–3393.

Kersten, P. and D. Cullen. 2007. Extracellular oxidative systems of the lignin-degrading Basidiomycete *Phanerochaete chrysosporium*. Fungal Genet Biology 44: 77–87.

Kumar, R., D. Liu and L. Zhang. 2008. Advances in proteinous biomaterials. J. Biobased Mater. Bio. 2: 1–24.

Kües, U. 2007. Wood Production, Wood Technology, and Biotechnological Impacts. Universitätsverlag Göttingen, 646pp.

Laks, P.E., D.L. Richter and G.L. Larkin. 2000. Biological deterioration of wood-base composite panels. Wood Design Focus 11(4): 7–14.

Li, D. 2012. Nanostructuring materials towards conventionally unachievable combination of desired properties. Journal of Nanomaterials & Molecular Nanotechnology. 1: 1. dx.doi. org/ 10.4172/2324-8777.1000e102.

Liu, D., H. Chen, P.R. Chang, Q. Wu, K. Li and L. Guan. 2010. Biomimetic soy protein nanocomposites with calcium carbonate crystalline arrays for use as wood adhesive. Bioresour. Technol. 101: 6235–6241.

Lykidis, C., G. Mantanis, S. Adamopoulos, K. Kalafata and I. Arabatzis. 2013. Effect of nanosized zinc oxide and zinc borate impregnation on brown-rot resistance of black pine (Pinus nigra L.) wood. Wood Material Science & Engineering 8(4): 242–244.

Mankowski, M. and J.J. Morrell. 2000. Patterns of fungal attach in wood-plastic composites following exposure in a soil block test. Wood and Fiber Science 32(3): 340–345.

Mantanis, G.I. and A.N. Papadopoulos. 2010. The sorption of water vapour of wood treated with a nanotechnology compound. Wood Science and Technology 44(3): 515–522.

Mantanis, G., E. Terzi, S.N. Kartal and A. Papadopoulos. 2014. Evaluation of mold, decay and termite resistance of pine wood treated with zinc- and copper-based nanocompounds. International Biodeterioration and Biodegradation 90: 140–144.

Maxim, L.D. and E.E. McConnell. 2005. A review of the toxicology and epidemiology of wollastonite. Inhalation Toxicology 17(9): 451–466.

Menezes Nunes, F., T. Arai, G.M. Baker, C.E. Bates, B.A. Becherer, T. Bell, E.L. Bird et al., 1991. Heat Treating, Vol. 4 ASTM Handbook.

Milton, F.T. 1995. The Preservation of Wood, A Self Study Manual for Wood Treaters. Minnesota Extension Service, University of Minnesota, College of Natural Resources, BU-6413-S, Revised 1995, 103pp.

Morrell, J.J., N.M. Stark, D.E. Pendleton and A.G. McDonald. 2006. Durability of wood-plastic composites. Wood Design Focus 16(3): 7–10.

Nayeri, M.D., P.Md. Tahir, H.R. Taghiyari, A.H. Alias, A. Karimi, L.C. Abdullah, E.S. Bakar and F. Namvar. 2014. Medium-density fiberboard made from kenaf bast and core: Effects of refining pressure and time on specific gas permeability. BioResources 9(4): 7198–7208.

Nogi, M. and H. Yano. 2008. Transparent nanocomposites based on cellulose produced by bacteria offer potential innovation in the electronics device industry. Advanced Materials 20: 1849–1852.

Nori, M., S. Iwamoto, A.N. Nakagalto and H. Yano. 2009. Optically transparent nanofiber paper. Advanced Materials 21: 1595–1598.

Oltean, L., A. Teischinger and C. Hansmann. 2007. Influence of temperature on cracking and mechanical properties of wood during wood drying–A review. BioResources 2(4): 789–811.

Palanti, S., E. Feci, G. Predieri and V. Francesca. 2012. Copper complexes grafted to amino-functionalized silica gel as wood preservatives against fungal decay: Mini-blocks and standard test. BioResources 7: 5611–5621.

Papadopoulos, A.N. 2006. Property comparisons and bonding efficiency of UF and PMDI bonded particleboards as affected by key process variables. Bioresources 1(12): 201–208.

Pecoraro, E., D. Manzani, Y. Messaddeq and S.J.L. Ribeiro. 2008. Bacterial cellulose from *Glucanacetobacter xylinus*: preparation, properties and applications. pp. 369–383. *In*: M.N. Belgacem and A. Gandini (eds.). Monomers, Polymers and Composites From Renewable Resources. Elsevier, London.

Pendleton, D.E., T.A. Hoffard, T. Adcock, B. Woodward and M.P. Wolcott. 2002. Durability of an extruded HDPE/wood composite. Forest Products Journal 52(6): 21–27.

Pinto, R.J.B., S.C.M. Fernandes, C.S.R. Freire, P. Sadocco, J. Causio, C.P. Neto and T. Trindade. 2012. Antibacterial activity of optically transparent nanocomposite films based on chitosan or its derivatives and silver nanoparticles. Carbohydr. Res. 348: 77–83.

Pizzi, A. 2006. Recent developments in eco-efficient bio-based adhesives for wood bonding: opportunities and issues. J. Adhes. Sci. Technol. 20: 2724–3729.

Rassam, Gh., H.R. Taghiyari, B. Jamnani and M.A. Khaje. 2010. Effect of Nano-Silver Treatment on Densified Wood Properties, Part One: Swelling, Recovery Set, Bending Strength, International Research Group on Wood Protection, IRG/ WP 10-40533, Section 4.

Repellin, V. and R. Guyonnet. 2005. Evaluation of heat treated wood swelling by differential scanning calorimetry in relation with chemical composition. Holzforschung. 59(1): 28–34.

Saber, R., Z. Shakoori, S. Sarkar, Gh. Tavoosidana, Sh. Kharrazi and P. Gill. 2013. Spectroscopic and microscopic analyses of rod-shaped gold nanoparticles interacting with single-stranded DNA oligonucleotides. IET Nanobiotechnol. 7: 42–49.

Sadare, O.O., M.O. Daramola and A.S. Afolabi. 2015. Preparation and Characterization of Nanocomposite Soy-Carbon Nanotubes (SPI/CNTs) Adhesive from Soy Protein Isolate. Proceedings of the World Congress on Engineering Vol. 2. WCE 2015, July 1–3: London, U.K.

Samir, M.S.S.A., F. Alloin and A. Dufresne. 2005. Review of recent research into cellulosic whiskers: their properties and their application in nanocomposite field. Biomacromolecules 6: 612–626.

Schmidt, O. 2006. Wood and Tree Fungi: Biology, Damage, Protection, and Use. Springer-Verlag Berlin Heidelberg, 334pp.

Schmidt, O. 2007. Indoor wood-decay basidiomycetes: damage, causal fungi, physiology, identification and characterization, prevention and control. Mycological Progress 6(4): 261–279. DOI 10.1007/s11557-007-0534-0.

Schmidt, O., O. Gaiser and D. Dujesiefken. 2012. Molecular identification of decay fungi in the wood of urban trees. European Journal of Forest Research. 131: 885–891.

Schurz, J. 1999. Trends in polymer science: a bright future for cellulose. Prog. Polym. Sci. 24: 481–483.

Shibata, H. and Y. Hirohashi. 2013. Effect of segment scale in a pore network of porous materials on drying periods. Drying Technology 31(7). DOI: 10.1080/07373937.2012.752742.

Shoda, M. and Y. Sugano. 2005. Recent advances in bacterial cellulose production. Biotechnol. Bioprocess Eng. 10: 1–8.

Siau, J.F. 1995. Wood: Influence of Moisture on Physical Properties; Blacksburg, VA, Department of Wood Science and Forest Products Virginian Polytechnic Institute and State University, pp. 1–63.

Taghiyari, H.R. 2011. Study on the effect of nano-silver impregnation on mechanical properties of heat-treated *Populus nigra*. Wood Sci. and Tech. 45: 399–404. DOI 10.1007/s00226-010-0343-5.

Taghiyari, H.R. and D. Efhami. 2011. Diameter increment response of *Populus nigra* var. *betulifolia* induced by alfalfa. Austrian Journal of Forest Science. 128(2): 113–127.

Taghiyari, H.R., H. Rangavar and O. Farajpour Bibalan. 2011. Nano-silver in particleboard. BioResources 6(4): 4067–4075.

Taghiyari, H.R. 2012a. Fire-retarding properties of nano-silver in solid woods. Springer: Wood Science and Technology 46(5): 939–952. DOI 10.1007/s00226-011-0455-6.

Taghiyari, H.R. 2012b. Correlation between gas and liquid permeability in some nano-silver-impregnated and untreated hardwoods. Journal of Tropical Forest Science JTFS 24(2): 249–255.

Taghiyari, H.R. 2013a. Nano-zycosil in MDF: gas and liquid permeability. Eur. J. Wood Prod. 71(3): 353–360. DOI 10.1007/s00107-013-0691-6.

Taghiyari, H.R. 2013b. Effects of heat-treatment on permeability of untreated and nanosilver-impregnated native hardwoods. Maderas Ciencia y Tecnología. 15(2): 183–194.

Taghiyari, H.R. and O. Farajpour Bibalan. 2013. Effect of copper nanoparticles on permeability, physical, and mechanical properties of particleboard. Eur. J. Wood Prod. 71(1): 69–77. DOI 10.1007/s00107-012-0644-5.

Taghiyari, H.R., K. Mobini, Y. Sarvari Samadi, Z. Doosti, F. Karimi, M. Asghari, A. Jahangiri and P. Nouri. 2013a. Effects of nano-wollastonite on thermal conductivity coefficient of medium-density fiberboard. J. Molec. Nanotechnol. 2: 1. http://dx.doi.org/10.4172/2324-8777.1000106.

Taghiyari, H.R., A. Moradiyan and A. Farazi. 2013b. Effect of nanosilver on the rate of heat transfer to the core of the medium density fiberboard mat. International Journal of Bio-Inorganic Hybrid Nanomaterials 2(1): 303–308.

Taghiyari, H.R., A. Karimi and P.Md. Taher. 2013c. Nano-wollastonite in particle board: Physical and mechanical properties. BioResources 8: 5721–5732.

Taghiyari, H.R., A. Karimi and P.Md. Taher. 2013d. Organo-silane compounds in medium density fiberboard: Physical and Mechanical Properties. Journal of Forestry Research 26(2): 495–500.

Taghiyari, H.R., H. Rangavar and P. Nouri. 2013e. Fire-retarding properties of nanowollastonite in MDF. Eur. J. Wood Prod. 71(5): 573–581.

Taghiyari, H.R. 2014a. Nanotechnology in wood and wood-composite materials. J. Nanomater. Mol. Nanotechnol. 3: 1. dx.doi.org/10.4172/2324-8777.1000e106.

Taghiyari, H.R. 2014b. Nano-silane in MDF: Effects of vapor chamber on fluid flow. IET Nanobiotechnology doi: 10.1049/iet-nbt.2013.0064.

Taghiyari, H.R. and O. Schmidt. 2014. Nanotechnology in wood-based composite panels. International Journal of Bio-Inorganic Hybrid Nanomaterials 3(2): 65–73.

Taghiyari, H.R. and B. Moradi Malek. 2014. Effect of heat treatment on longitudinal gas and liquid permeability of circular and square-shaped native hardwood specimens. Heat and Mass Transfer 50(4): DOI 10.1007/s00231-014-1319-z.

Taghiyari, H.R. and Y. Lotfinejad-Sani. 2014. Effects of ice blasting on some mechanical properties of composite boards. Philippine Journal of Science 143(2): 177–185.

Taghiyari, H.R., B. Moradi Malek, M. Ghorbani Kookandeh and O. Farajpour Bibalan. 2014a. Effects of silver and copper nanoparticles in particleboard to control *Trametes versicolor* fungus. International Biodeterioration & Biodegradation 94: 69–72.

Taghiyari, H.R., E. Bari, O. Schmidt, M.A. Tajick Ghanbary, A. Karimi and P.Md. Tahir. 2014b. Effects of nanowollastonite on biological resistance of particleboard made from wood chips and chicken-feather against *Antrodia vaillantii*. International Biodeterioration & Biodegradation 90: 93–98.

Taghiyari, H.R., E. Bari and O. Schmidt. 2014c. Effects of nanowollastonite on biological resistance of medium-density fiberboard against *Antrodia vaillantii*. Eur. J. Wood Prod. 72(3): 399–406.

Taghiyari, H.R., M. Ghorbanali and P.Md. Tahir. 2014d. Effects of the improvement in thermal conductivity coefficient by nano-wollastonite on physical and mechanical properties in medium-density fiberboard (MDF). BioResources 9(3): 4138–4149.

Taghiyari, H.R., S. Habibzadeh and S.M. Miri Tari. 2014e. Effects of wood drying schedules on fluid flow in Paulownia wood. Drying Technology 32(1): 89–95.

Taghiyari, H.R., R. Oladi, S.M. Miri Tari and S. Habibzade. 2014f. Effects of diffusion drying schedule on gas and liquid permeability in *Paulownia fortunei* wood. Bosque. 35(1): 101–110. DOI: 10.4067/S0717-92002014000100010.

Taghiyari, H.R., H. Zolfaghari, M.E. Sadeghi, A. Esmailpour and A. Jaffari. 2014g. Correlation between specific gas permeability and sound absorption coefficient in solid wood. Journal of Tropical Forest Science 26(1): 92–100.

Taghiyari, H.R. 2015. Effects of nano-materials on gas and liquid permeability in wood and wood composites. Journal of Nanomaterials & Molecular Nanotechnology 4: 1. doi. org/10.4172/2324-8777.1000e107.

Taghiyari, H.R. and Y. Sarvari Samadi. 2015. Effects of wollastonite nanofibers on fluid flow in medium-density fiberboard. Journal of Forestry Research 26(3): DOI 10.1007/s11676-015-0137-6.

Taghiyari, H.R. and P. Nouri. 2015. Effects of nano-wollastonite on physical and mechanical properties of medium-density fiberboard. Maderas Ciencia y Tecnología. 17: 833–842.

Taghiyari, H.R., A. Kalantari, M. Ghorbani, F. Bavaneghi and M. Akhtari. 2015. Effects of fungal exposure on air and liquid permeability of nanosilver- and nanozincoxide-impregnated *Paulownia* Wood. Internaional Biodeterioration & Biodegradation 105: 51–57.

Taghiyari, H.R., A. Esmailpour and H. Zolfaghari. 2016. Effects of nanosilver on sound absorption coefficients in solid wood species. IET Nanobiotechnology 10(3): 147–153. Doi: 10.1049/iet-nbt.2015.0019.

Tajvidi, M., D.J. Gardner and D.W. Bousfield. 2016. Cellulose Nanomaterials as Binders: Laminate and Particulate Systems. Journal of Renewable Materials. DOI: http://dx.doi.org/10.7569/JRM.2016.634103.

Teeri, T.T., H. Brumer III, G. Daniel and P. Gatenholm. 2007. Biomimetic engineering of cellulose-based materials. Trends Biotechnol. 25: 299–306.

Trovatti, E., L.S. Serafim, C.S.R. Freire, A.J.D. Silvestre and C.P. Neto. 2011. *Gluconacetobacter sacchari*: an efficient bacterial cellulose cell-factory. Carbohydr. Polym. 86: 1417–1420.

USPTO. 2009. Gas permeability measurement apparatus; patent number 8079249.

Vilela, C., C.S.R. Freire, P.A.A.P. Marques, T. Trindade, C.P. Neto and P. Fardim. 2010. Synthesis and characterization of new CaCO3/cellulose nanocomposites prepared by controlled hydrolysis of dimethylcarbonate. Carbohydr. Polym. 79: 1150–1156.

Wang, Y. and L. Zhang. 2009. Biodegradable Polymer Blends and Composites From Renewable Resources. Wiley, New Jersey.

White, R.H. and M.A. Dietenberger. 1999. Wood Handbook, Chapter 17: Fire Safety; Forest Products Laboratory, Gen. Tech. Rep. FPL–GTR–113, Madison, WI: U.S. Department of Agriculture, Forest Service, Forest Products Laboratory, 463p.

Winandy, J.E., P.K. Lebow and J.F. Murphy. 2002. Predicting current serviceability and residual service life of plywood roof sheathing using kinetics-based models, the 9th Durability of Building Materials and Components Conference, Brisbane, Australia, 7pp.

Youngquist, J.A., A.M. Krzysik, P. Chow and R. Meniman. 1997. Properties of composite panels. pp. 301–336. In: R.M. Rowell, R.A. Young and J.K. Rowell (eds.). Paper and Composites from Agro-based Resources. CRC/Lewis Publishers, Boca Raton, FL.

Yu, L. and L. Chen. 2009. Biodegradable Polymer Blends and Composites From Renewable Resources. Wiley & Sons, New Jersey.

Yu, W., H. Xie, L. Chen and Y. Li. 2010. Investigation on the thermal transport properties of ethylene glycol-based nanofluids containing copper nanoparticles. Powder Technol. 197: 218–221.

Zhang, Y., W. Zhu, Y. Lu, Z. Gao and J. Gu. 2014. Nano-scale blocking mechanism of MMT and its effects on the properties of polyisocyanate-modified soybean protein adhesive. Ind. Crop. Prod. 57: 35–42.

Zimmermann, T., E. Pohler and T. Geiger. 2004. Cellulose fibrils for polymer reinforcement. Adv. Eng. Mater. 6: 754–761.

15

Bio-based Wood Adhesives Research

Advances and Outlooks

Zhongqi He[1],* and *Hui Wan*[2]

ABSTRACT

In the past three decades, concerns related to the environment, human health risks, and interests in resource recycling and sustainability have propelled the resurgence of research on bio-based adhesives, especially those based on agricultural and forest products and byproducts. This concluding chapter showcases the progress on the developments and utilization of bio-based wood adhesives reported in the individual chapters, and also several latest reports not covered by these chapters. These advances have paved the paths for enhanced utilization of agricultural and forest products and byproducts for global sustainability and a greener environment. We also discuss several emerging topics that are worth future exploration by bio-based adhesive researchers.

Introduction

Human beings have apparently had a propensity for gluing things together since the dawn of recorded history (Lambuth, 2003). Using natural glues

---
[1] Southern Regional Research Center, USDA Agricultural Research Service, 1100 Robert E. Lee Blvd., New Orleans, LA 70124, USA.
[2] Department of Sustainable Bioproducts, Mississippi State University, 201 Locksley Way, Starkville, MS 39759, USA.
* Corresponding author: Zhongqi.He@ars.usda.gov

or adhesives, early Chinese and Egyptians (indeed virtually every other civilization) learned to implement papyrus laminating, decorative wood veneering, and furniture and musical instrument assembly. However, the ancient adhesive material choices were quite limited. The desire to use wood more efficiently was an impetus for adhesive development in the 19th and 20th centuries. Two major early developments were the use of casein for glulam production and soy adhesives for interior plywood. Despite some success with bio-based adhesives, they were replaced by synthetic adhesives starting in the 1930s due to economics, water resistance, and ease of use. Refinement of the synthetic adhesives led to a great expansion of their application in the bonded wood products industry by growing existing markets and allowing for the development of new bonded products. The 20th century also witnessed an increased understanding of wood adhesive chemistry and product performance knowledge (Frihart, 2015). In the past three decades, concerns related to the environment, human health risk, and interests in resource recycling and sustainability have propelled the resurgence of bio-based adhesive research. Adhesives based on agricultural and forest products and byproducts are of particular interest. In this concluding chapter, we have showcased the developments and utilization of bio-based wood adhesives reported in the individual chapters and also several latest reports not covered by these chapters. We have further discussed several emerging issues that are worth future exploration by bio-based adhesive researchers.

Advances in Bio-based Wood Adhesive Research

Over 50 percent of the dry weights of biomolecular cells are composed of proteins, and this type of biopolymer is the most abundant mass of macromolecules. They function as the principal organic building blocks in living organisms (Cozzone, 2001). Proteinaceous feedstocks for industrial applications can be plant-based (Frihart et al., 2013; He et al., 2016d) or animal-based (Mekonnen et al., 2014; Tousi et al., 2014; Zhao, 2013). In addition, the biomass of cultured microorganisms can be potentially used as an industrial source of protein (Dong et al., 2016; Roy et al., 2014). Hence, industrial proteins can be sourced from single cell-based, plant-based, and animal-based proteins. In Chapter 1, Adhikari et al. (2017) reviewed the general structures and fundamental physicochemical properties of proteins, then summarized the potential application of proteins as wood adhesives and current trends of protein-based adhesive formulations from various sources. Typical vegetable sources of seed proteins used in wood adhesives studies include, but are not limited to soy (Luo et al., 2015), wheat (Khosravi et al., 2015), peanut (Li et al., 2015), canola (or rapeseed) (Yang et al., 2014), and cottonseed (He et al., 2016c). Animal-based protein wood adhesives are mainly from four types of animal tissues: epithelial, connective, muscle, and nervous tissues (Tousi et al., 2014; Zhao, 2013). To develop better protein-based adhesives, the fundamental structure-property relationships related to adhesion should be further understood. Thus, in Chapter 2, Qi et al. (2017) reviewed soy protein subunit characterization and their adhesion properties used as wood adhesives. Glycinin was found to be the main contributor to adhesion strength of soy protein. The basic and β subunits with larger portions of hydrophobic amino acid were shown to have greater adhesion properties than their respective acidic and $\alpha'\alpha$ subunits. In addition, Zhang and Hua (2007) reported that the adhesive strength of unmodified and 1 M urea modified 11S proteins was greater than that of their counterparts of 7S protein, when tested with pine, walnut, and cherry wood strips. The adhesive strength of 3 M urea modified 11S protein was lower than 3 M urea modified 7S protein when tested with walnut and cherry plywood. Nordqvist et al. (2012) separated the wheat storage protein, gluten, into two subclasses: the aqueous ethanol (70%)-soluble gliadins and insoluble glutenins and they tested their performance as wood adhesives. They reported that the adhesive properties of glutenin fraction were similar to those of whole gluten, but the properties of the gliadin fraction were inferior to those of gluten, especially in regard to water resistance. However, the differences in the adhesive strength and the water resistance between different fractions of the same seed protein were not so great to the degree that the separation was costly. As a matter of fact, Nordqvist et al. (2012) demonstrated the adhesive performance of gliadin was equal to that of glutenin when over-penetration of the protein into the wood material was avoided. Thus, much more research should be conducted on the modification of protein products for better adhesive performance.

In Chapter 3, Shi et al. (2017) reviewed the current modification methods of soy-based adhesives, including soy flour, soybean meal, soy protein concentrate, and soy protein isolate. Chemical modifications included alkali denaturation, surfactant denaturation, organic solvent denaturation, enzymatic treatment, epoxy groups crosslinking, formaldehyde compounds crosslinking, acylation and silanation. Physical modifications are those by compression, high temperature treatment, ultrasound treatment and blending with hydrophobic additives. Recently, cellulosic nanofibers, inorganic nanoparticles and nano-clays were also applied for nanoparticle modification which is also covered by Chapter 14 (Taghiyari et al., 2017). Canola protein and canola oil exhibited adhesive potential for the wood industry. Chapter 4 (Li et al., 2017) focused on canola protein and oil-based wood adhesives. The canola meal-based adhesive was found suitable for making particle-board or veneer flooring for use in indoor environments since it showed relatively low water resistance compared to pure canola protein-based adhesive. The canola protein-based adhesive, however, had high protein content and was more hydrophobic, making it suitable for use in indoor and outdoor conditions. Currently, canola oil is underexploited compared to intensive studies conducted on protein-based adhesives, but it can be used to make adhesives by synthesizing canola oil into polyurethanes. These polyurethanes have been tested to bond porous or non-porous substrates, such as wood to wood, wood to metal, or metal to metal. Li et al. (2017) also provided the information and insight on the synthesis of other vegetable oil-based polymers and potential adhesive application in this chapter. Chapter 5 (Cheng and He, 2017) reviewed recent developments in the use of proteins and carbohydrates in wood adhesive formulations. This chapter focused on two types of blends in wood adhesive formulations (i.e., the blends of two proteins, and blends of protein with carbohydrates). Using their own data, Chapter 5 showed that cottonseed protein could be a viable alternative to soy protein, with stronger adhesive strength and hot water resistance as compared to soy protein. If the use of soy protein was preferred, a promising line of approach was to blend cottonseed protein into soy protein. The two proteins could be blended together in different ratios to give a range of adhesive properties and hot water resistance, thus providing flexibility in formulating different adhesives to meet different requirements. Cottonseed proteins were also added to a carbohydrate adhesive and improved both the bonding strength of the carbohydrate adhesive and the durability of the adhesion in a hot water test.

Water-washed cottonseed meal was made from defatted cottonseed meal by a simple washing procedure to remove water-soluble components and led to a more cost-effective product than cottonseed protein. Thus, Chapter 6 (He and Cheng, 2017) detailed the preparation and utilization of water washed cottonseed meal as a wood adhesive. Data presented in this case study showed that, compared to the defatted meal, the water resistance of water washed cottonseed meal was improved. Moreover in most cases, its water resistance was comparable, or even better than the cottonseed protein isolate. The washed cottonseed meal had stronger strength and was more flexible in operational conditions and parameter setup than protein isolate. Chapter 7 (Basta et al., 2017) is another case study which discussed the preparation of high performance agro-based composites using rice bran- and corn starch-modified urea formaldehyde adhesives. Comparative data presentation demonstrated that both modified urea formaldehyde adhesive systems provided agro-fiber composites that met or exceeded the physical/mechanical property requirements of high grade particleboards (H-3). Thus, incorporation of the byproduct of oil production (in this case rice bran) to commercial urea formaldehyde would be a potentially simple way to solve the drawback of increasing the viscosity of the synthetic adhesive for easily penetrating through an agro-fibers mat. The approach provides board products with good bonding behaviors and high performance and also minimizes harmful formaldehyde emission during board formation.

Chapters 8 (Brosse and Pizzi, 2017) and 9 (He and Umermura, 2017) covered the application of tannin and citric acid in wood bonding, respectively. Tannin-based adhesives for wood industry have been produced and commercialized for more than two decades. In addition to an updated overview of tannin research, Chapter 8 presented a new potential source of condensed tannins (the grape pomace) and the first results on the utilization of grape pomace as a source of tannins for wood adhesives. Citric acid functioned as a cross-linking catalyst, a cross-linking agent, or a dispersing agent to improve the adhesive strength and/or operational properties (such as viscosity) for wood bonding. Chapter 9 discussed the recent research on citric acid, with or without sucrose, as a major adhesive component in manufacturing fiber and particleboards with a variety of lignocellulose raw materials. Other renewable natural byproducts, such as chitosan (Patel, 2015), lignin (Ghaffar and Fan, 2014), gum (Norstrom et al., 2014) and xylan (Norstrom et al., 2015), can be used as environmentally friendly wood adhesive components. Podschun et al. (2016) recently tested phenolated lignins serving as reactive precursors in wood veneer and particleboard bonding. The phenolation of beech organosolv lignin was for the production of lignin–phenol–formaldehyde resins with a phenol substitution of up to 40% (w/w). The authors reported that particleboards bonded with activated lignin resins showed improved mechanical properties compared with panels prepared with raw lignin–phenol–formaldehyde resins and fulfilled the European requirements for particleboard classification of P5 (load-bearing in humid environments).

Wood and other agricultural lignocellulosic biomass materials are natural resources available in huge quantities. Biomass liquefaction is a unique thermochemical conversion process for biomass utilizations. Chapter 10 (Wan et al., 2017) focused on the liquefaction of wood and other lignocellulosic biomass materials and the utilization of these liquefied biomass-based polymers in wood bonding. The liquefied biomass products have been tested as an adhesive, or blended with synthetic polymers and chemicals to form four types (i.e., phenol formaldehyde, polyurethane, epoxy, amino) of adhesives or resins per the biomass sources and liquefaction solvents. These liquefied biomass-based adhesives have been applied in bonding plywood, or making fiber and particleboards or other composite materials to reduce the cost and formaldehyde emissions. In addition, wood pyrolysis bio-oil (Liu et al., 2014; Ozbay and Ayrilmis, 2015) was also blended with synthetic resins. Synthesized phenol formaldehyde adhesives with a phenol substitution of 50% by bio-oil have been obtained from fast pyrolysis of mixed hardwoods (maple, birch, and beech) (Himmelblau and Grozdits, 1999). Bio-oil obtained from the pyrolysis of pine wood was mixed with polymeric diphenylmethane diisocyanate (pMDI) to form an adhesive binder system for flakeboard (Mao and Shi, 2012). More information on preparation, properties, and bonding utilization of pyrolysis bio-oil can be found in Chapter 11 (Mao et al., 2017). Tree pitch was considered as weatherproof coatings and caulks but not as adhesives, due to their plastic-flow behavior (Lambuth, 2003). In contrast to this concept, as a case study, Chapter 12 (Vieira et al., 2017) discussed the application of an Amazonic tree rosin (white pitch) for use as wood adhesive. The white-pitch is a sweet-smelling rosin that comes from the heart of the tree trunk (Protium hepytaphyllum). Alpha and beta-amyrin are the major active components of the natural rosin. The authors expected that white pitch can serve as a new alternative as a wood adhesive, or at least, as an excellent material for future study.

Rheological characterization is useful to study the flowability and viscoelastic properties of adhesive materials. From the practical point of view, the rheological properties are related to the ease of spread and flow over a surface. Chapter 13 (Bacigalupe et al., 2017) reviewed and discussed the rheological behavior of protein-based adhesives and their relationship with their bonding properties. Most of the discussion was focused on blood-, soy- and cottonseed-based adhesives modified by solid content, heat treatments, varying pH, enzymatic treatment and blending with commercial latex. Nanomaterials can also be added to wood-composite materials during formation to improve adhesive qualities. Nano-metals (nanosilver, nanocopper, and nano-zincoxide) and nano-minerals (nano-wollastonite and nano-clays) have been used to improve the biological resistance of wood-composite products against wood-deteriorating fungi. Different nanomaterials have been employed to improve the bonding strength as well as water resistance of bio-adhesives such as soy protein, to replace petrochemical-based adhesives. Chapter 14 (Taghiyari et al., 2017) reviewed the effects of nano-materials on applied properties of wood-composite materials. Per recent studies of nanotechnology and nano-materials, information in this chapter is helpful in the future research effort on improvement of bonding and performing properties of agricultural and wood-based composite panels by nanotechnology.

Outlook of Future Research

Great progress has been made on developing bio-based wood adhesives from renewable natural resources over the last couple of decades. These advances have paved the paths for enhanced utilization of agricultural and forest products and byproducts for global sustainability and a greener environment. Nevertheless, many important challenges in the basic knowledge and practical application of bio-based adhesives remain. Here we would like to share our outlook on some emerging issues on future bio-based wood adhesive research.

1) Difference in Adhesive Performances Between Different Types of Protein Adhesives

Proteins are perhaps the most studied bio-based wood adhesives. In addition to soy protein, many other seed proteins have been tested for their adhesive performance. Proteins could serve as adhesives, but it is not very clear how the adhesive performance differs between different types of proteins since many studies have been focused on only one type of protein. Yang et al. (2006) compared three protein-based adhesives for wood composites and found the following adhesive performance: mixed porcine and bovine blood meal > low-fat soy soybean flour > low-fat peanut flour. Nordqvist et al. (2010) compared bond strength and water resistance of alkali-modified soy protein isolate and wheat gluten adhesives. They found a clear difference in bonding performance between the two types of adhesives. The adhesive properties of soy protein, particularly with regard to water resistance, are much better than those of the wheat gluten adhesive. Cheng et al. (2013; 2016a; 2016b) comparatively tested soy protein and cottonseed protein, with or without additives, for wood bonding. Their data always showed that cottonseed protein exhibited better wood adhesive properties than soy protein. For cottonseed protein/polysaccharide formulations, hot water adhesive resistance was retained when the blend contained about 50% of polysaccharides. Soy protein formulations and its polysaccharide blends generally exhibited somewhat lower hot water resistance. Testing an underutilized indigenous Southern African oilseed legume, Amonsou et al. (2013) reported that the marama protein adhesive had better adhesive properties than the soy protein in terms of strength and resistance to delamination in water. The adhesive strength of the marama protein was about 1.5 times higher than the soy over the protein concentrations while about 47% of marama protein-glued wood pieces were delaminated, compared to 90% for soy protein-bonded pieces, after 2 cycles of a 48 hour water soaking. Cottonseed and marama proteins could be promising alternatives to soy protein in bio-based adhesive applications where stronger adhesive performance is required. A systematical comparison of the bonding ability of more proteins from different sources may not only be helpful in optimal utilization of different proteins for appropriate bonding requirements, but also shed fresh light on the gluing mechanism of protein adhesives.

2) Water Washing to Improve the Adhesive Properties of Seed Meal Products

Whereas plant seed meals (or flours) showed poor bonding performance than purified protein isolates, the researchers did not give up efforts on developing meal-based adhesives due to their lower cost than protein isolates (Frihart and Lorenz, 2013; Shi et al., 2017). He et al. (2014a) have developed a simple water washing procedure to improve the adhesive strength and water resistance of cottonseed meal-based adhesive preparation. Their experimental data indicated that washed cottonseed meal behaved much better than unwashed cottonseed meal, as well as cottonseed protein isolate that was prepared more expensively under caustic alkaline and acidic conditions. Recently, washed cottonseed meal has been produced at the pilot scale (He et al., 2016d), and industrial utilization of the water-washed cottonseed meal is promising (Liu et al., 2016a). Perhaps due to the difference in chemical composition (He et al., 2015), the washing procedure did not greatly improve the adhesive performance of defatted soy meal (He and Chapital, 2015; He et al., 2016a). Unlike their cottonseed counterparts, water washed soy meals behaved more like un-washed meal than soy protein isolates. However, it is still worthwhile to test if the simple washing procedure can improve the adhesive properties of other types of seed meals to the extent of the corresponding protein isolates.

3) Utilization of Bioenerngy-related Byproducts

Global food security is a serious challenge within next four decades as predictions indicate that the world population will be between 8.0 and 10.4 billion people by 2050 (DeFauw et al., 2012). Soy and wheat comprise portions of humans' diets and the replacement of petroleum-based synthetic adhesives by soy and wheat protein isolates on large scales may undermine the global effort on food security (He et al., 2014b). To avoid the competition between utilization of such proteins for protein-based products or human food, Zhu et al. (2016) advocated that alternative bio-resources must be discovered. During the last few decades, bioenergy production has been radically promoted worldwide as a clean and eco-friendly source of energy (Abhilash and Edrisi, 2015). Thus, exploring the byproducts from bioenergy production for adhesive applications would be desirable. Sorghum is used for ethanol production. Distillers dried grains with solubles (DDGS) are a co-product of the distillation and dehydration process during ethanol production. Li et al. (2011) tested the adhesive performance of the proteins extracted from sorghum DDGS. They demonstrated that acetic acid-extracted sorghum protein from DDGS had advantages such as significantly higher water resistance and lower energy input, compared with soy protein based adhesives. Bandara et al. (2013) examined the adhesive properties of five protein extracts from triticale distillers grain, a co-product of triticale based bioethanol processing. Similar to the observation of Li et al. (2011), they also found the highest adhesive strength with acetic acid extracted protein. Biodiesel-related byproducts (such as canola, camelina, and microalgal proteins) are also promising in the wood bonding application (Roy et al., 2014; Wang et al., 2014; Zhu et al., 2016).

Switchgrass is a viable biomass source for conversion to bioenergy for a large region of the United States and can improve soil preservation, carry a higher energy value, and reduce carbon emissions. Wei et al. (2014) conducted the liquefaction and substitution of switchgrass based bio-oil into epoxy resins. The authors proposed that the bio-derived adhesive is one value-added product that can generate revenue for companies and provide a profitable platform for beginning bioenergy facilities making commodity based chemicals such as pyrolysis oil, liquefaction oil, or phenols. This strategy seems applicable to other agricultural biomass. For example, cotton crop production produces several types of biomass byproducts (e.g., stalk, bur, gin trash, defatted cottonseed meal) (He et al., 2016b; Liu et al., 2016b). Some of these cotton byproducts (i.e., cotton stalk, cotton boll shell, cotton gin trash, whole cottonseed, cottonseed hull, and defatted cottonseed meal cake) have been tested for biochar and bioenergy (biogas and bio-oil) production. Incorporation of adhesive-oriented bio-oil production in the cotton byproduct biomass pyrolysis should expand the utilization of cotton byproducts, thus increasing their commercial values.

4) Durability of Wood Products Bonded with Bio-based Wood Adhesives

The durability of the bonded wood products is a crucial consideration in practical bio-based adhesive application. D'Amico et al. (2012) examined the effects of aging on mechanical properties of wood to wood bonding with wheat flour adhesives. They found no significant reduction in bonding properties, but a trend to lower adhesive strength after 12 months of storage was noticeable. Per their analytic data, they assumed that degradation of starch, not protein, is mainly responsible for the moderate loss of adhesive power during the 12-month storage of the bonded wood specimens. Bertolini et al. (2013) evaluated the performance of particleboards made with residues of chromium-copper-boron oxides impregnated *Pinus* sp. and castor oil-based polyurethane resin before and after artificial accelerated aging. In general, aged samples presented light dark surfaces, probably due to lignin decomposition by photochemical degradation. However, the particleboard samples subjected to artificial aging with the cycles equivalent to 1 year showed superior performance in relation to the parameters of modulus of rupture and modulus of elasticity. The researchers attributed the better mechanical properties after aging to an increase in resin crystallinity degree. In the meantime, the researchers cautioned that, after longer exposure cycles, more degradation in resin might occur, leading to subsequent reduction of these properties. Recently, Zeng et al. (2016) studied the aging resistance properties of poplar plywood bonded by soy protein-based adhesive. They conducted accelerating aging and natural aging tests. The shear strength of plywood bonded with soy meal/bisphenol epoxy adhesive was 1.19 and 1.09 MPa for the surface and core layer, respectively. The two values of the shear strength for the surface and core layer were reduced to 0.88 MPa and 0.71 MPa after eight 25°C wet-dry cycles, to 0.96 MPa and 0.79 MPa after eight 63°C wet-dry cycles, and 0.53 MPa and 0.27 MPa after eight 95°C wet-dry cycles, respectively. On the other hand, the shear strength of soy meal-only adhesive gradually decreased to 0 (surface and core layer) after six and five 25°C wet-dry cycles, respectively. Interestingly, such dramatic reduction in the shear strength was not observed in the 20-month natural ageing when the bonded plywood samples were placed indoor without environmental conditioning. The data points (taken every four months) of shear strengths of both surface and core layer generally fluctuated without substantial changes during the 20 months of aging. The researchers proposed the accelerated aging with 25°C wet-dry cycles was the best method with simulation to the natural aging process. However, rigorous data is needed to support this conclusion. More general research should also be encouraged to provide the durability- or aging-relevant parameters of a specific bio-based adhesive to promote its industrial application.

5) Developing a Systematic Approach for Transferring Bio-based Adhesives to Commercially Available Products

While a substantial amount of research work has been done on the adhesive performances of seed proteins, there is lack of information on how variations such as grow season and location of a plant affected the quality of a protein-based wood adhesive. Such concerns need to be adequately addressed before these biomaterials are used as industrial feedstocks for wood adhesive applications. This concern further shows that the conversion of a new bioproduct concept into a new market-available product needs a systematical approach. Generally, we need to be equipped with the capacity to address the issues coming from bio-based wood adhesive manufacturing and bio-based wood adhesive application. We need to work with wood adhesive manufacturers and wood composite manufactures to address the variation in raw material qualities, cost, and quantity and satisfy the need of end users, with a sustainable manufacturing approach. While making sure the bio-based wood adhesives meet the requirements of the existing market, we need to put in more efforts to develop new processes, new products, and thus create new markets that will adopt the features of bio-based wood adhesives, using statistical process control. Agricultural product researchers and forest product researchers can work together to create a broad future for using agricultural residues for new bio-based wood adhesive development.

Keywords: Aging, Biomass, Bio-oil, Byproduct, Pyrolysis, Sustainability

References

Abhilash, P. and S.A. Edrisi. 2015. Socio-economic implications of bioenergy production: A book review. Front. Bioeng. Biotechnol. 3: article 174. doi: 10.3389/fbioe.2015.00174.

Adhikari, B.B., P. Appadu, M. Chae and D.C. Bressler. 2017. Protein-based wood adhesives: current trends of preparation and application. pp. 1–58. *In*: Z. He (ed.). Bio-based Wood Adhesives: Preparation, Characterization, and Testing. CRC Press/ Taylor & Francis Group, Boca Raton, Fl.

Amonsou, E.O., J.R. Taylor and A. Minnaar. 2013. Adhesive potential of marama bean protein. Int. J. Adhes. Adhes. 41: 171–176.

Bacigalupe, A., Z. He and M.M. Escobar. 2017. Effects of rheology and viscosity of bio-based adhesives on bonding performance. pp. 293–309. *In*: Z. He (ed.). Bio-based Wood Adhesives: Preparation, Characterization, and Testing. CRC Press/Taylor & Francis Group, Boca Raton, Fl.

Bandara, N., L. Chen and J. Wu. 2013. Adhesive properties of modified triticale distillers grain proteins. Int. J. Adhes. Adhes. 44: 122–129.

Basta, A.H., H. El-Saied and J.E. Winandy. 2017. Comparative evaluation of rice bran- and corn starch-modified urea formaldehyde adhesives on improvements of environmental performance of agro-based composites. pp. 179–196. *In*: Z. He (ed.). Bio-based Wood Adhesives: Preparation, Characterization, and Testing. CRC Press/Taylor & Francis Group, Boca Raton, Fl.

Bertolini, M.D.S., F.A.R. Lahr, M.F.d. Nascimento and J.A.M. Agnelli. 2013. Accelerated artificial aging of particleboards from residues of CCB treated Pinus sp. and castor oil resin. Mat. Res. 16: 293–303.

Brosse, N. and A. Pizzi. 2017. Tannins for wood adhesives, foams and composites. pp. 197–220. *In*: Z. He (ed.). Bio-based Wood Adhesives: Preparation, Characterization, and Testing. CRC Press/Taylor & Francis Group, Boca Raton, Fl.

Cheng, H.N. and Z. He. 2017. Wood adhesives containing proteins and carbohydrates. pp. 140–155. *In*: Z. He (ed.) Bio-based Wood Adhesives: Preparation, Characterization, and Testing. CRC Press/Taylor & Francis Group, Boca Raton, Fl.

Cheng, H.N., M.K. Dowd and Z. He. 2013. Investigation of modified cottonseed protein adhesives for wood composites. Ind. Crop. Prod. 46: 399–403.

Cheng, H.N., C.V. Ford and M.K. Dowd. 2016a. Use of additives to enhance the properties of cottonseed protein as wood adhesives. Int. J. Adhes. Adhes. 68: 156–160.

Cheng, H.N., C.V. Ford, M.K. Dowd and Z. He. 2016b. Soy and cottonseed protein blends as wood adhesives. Ind. Crop. Prod. 85: 324–330.

Cozzone, A.J. 2001. Proteins: Fundamental Chemical Properties. pp. 1–10. Encyclopedia of Life Sciences. John Wiley & Sons, Ltd., Chichester, UK.

D'Amico, S., M. Hrabalova, U. Muller and E. Berghofer. 2012. Influence of ageing on mechanical properties of wood to wood bonding with wheat flour glue. Eur. J. Wood Wood Prod. 70: 679–688.

DeFauw, S.L., Z. He, R.P. Larkin and S.A. Mansour. 2012. Sustainable potato production and global food security. pp. 3–19. *In*: Z. He et al. (eds.). Sustainable Potato Production: Global Case Studies. Springer, Amsterdam, Netherlands.

Dong, T., E.P. Knoshaug, R. Davis, L.M.L. Laurens, S. Van Wychen, P.T. Pienkos and N. Nagle. 2016. Combined algal processing: A novel integrated biorefinery process to produce algal biofuels and bioproducts. Algal Res. doi:10.1016/j.algal.2015.12.021.

Frihart, C.R. 2015. Introduction to special issue: Wood adhesives: Past, present, and future. Forest Products J. 65: 4–8.

Frihart, C.R. and L. Lorenz. 2013. Protein modifiers generally provide limited improvement in wood bond strength of soy flour adhesives. Forest Products J. 63: 138–142.

Frihart, C.R., C.G. Hunt and M.J. Birkeland. 2013. Soy Proteins as wood adhesives. pp. 277–291. *In*: W.V. Gutowski and H. Dodiuk (eds.). Recent Advances in Adhesion Science and Technology. CRC Press, Boca Raton, FL.

Ghaffar, S.H. and M. Fan. 2014. Lignin in straw and its applications as an adhesive. Int. J. Adhes. Adhes. 48: 92–101.

He, Z., H.N. Cheng, D.C. Chapital and M.K. Dowd. 2014a. Sequential fractionation of cottonseed meal to improve its wood adhesive properties. J. Am. Oil Chem. Soc. 91: 151–158.

He, Z., D.C. Chapital, H.N. Cheng, K.T. Klasson, M.O. Olanya and J. Uknalis. 2014b. Application of tung oil to improve adhesion strength and water resistance of cottonseed meal and protein adhesives on maple veneer. Ind. Crop. Prod. 61: 398–402.

He, Z. and D.C. Chapital. 2015. Preparation and testing of plant seed meal-based wood adhesives. J. Vis. Exp. 97: e52557. doi:10.3791/52557.

He, Z., H. Zhang and D.C. Olk. 2015. Chemical composition of defatted cottonseed and soy meal products. PLoS One. 10(6): e0129933. DOI:10.1371/journal.pone.0129933.

He, Z. and H.N. Cheng. 2017. Preparation and utilization of water washed cottonseed meal as wood adhesives. pp. 156–178. *In*: Z. He (ed.). Bio-based Wood Adhesives-Preparation, Characterization, and Testing. CRC Press/Taylor & Francis Group, Boca Raton, Fl.

He, Z. and K. Umermura. 2017. Utilization of citric acid in wood bonding. pp. 221–238. *In*: Z. He (ed.). Bio-based Wood Adhesives-Preparation, Characterization, and Testing. CRC Press/Taylor & Francis Group, Boca Raton, Fl.

He, Z., D.C. Chapital and H.N. Cheng. 2016a. Comparison of the adhesive performances of soy meal, water washed meal fractions, and protein isolates. Modern Appl. Sci. 10(5): 112–120.

He, Z., S.M. Uchimiya and M. Guo. 2016b. Production and characterization of biochar from agricultural by-products: Overview and use of cotton biomass residues. pp. 63–86. *In*: M. Guo et al. (eds.). Agricultural and Environmental Applications of Biochar: Advances and Barriers. Soil Science Society of America, Inc., Madison, WI.

He, Z., D.C. Chapital, H.N. Cheng and O.M. Olanya. 2016c. Adhesive properties of water-washed cottonseed meal on four types of wood. J. Adhes. Sci. Technol. 30: 2109–2119.

He, Z., K.T. Klasson, D. Wang, N. Li, H. Zhang, D. Zhang and T.C. Wedegaertner. 2016d. Pilot-scale production of washed cottonseed meal and co-products. Modern Appl. Sci. 10(2): 25–33.

Himmelblau, D.A. and G.A. Grozdits. 1999. Production and performance of wood composite adhesives with air-blown, fluidized-bed pyrolysis oil. pp. 541–547. *In*: R.P. Overend and E. Chornet (eds.). Proceedings of the 4th biomass conference of the Americas, 29 August–2 September, Oakland, CA.

Khosravi, S., P. Nordqvist, F. Khabbaz, C. Ohman, I. Bjurhager and M. Johansson. 2015. Wetting and film formation of wheat gluten dispersions applied to wood substrates as particle board adhesives. Eur. Polymer J. 67: 476–482.

Lambuth, A.L. 2003. Protein adhesives for wood. pp. 457–478. *In*: A. Pizzi and K.L. Mittal (eds.). Handbook of Adhesive Technology, 2nd Ed. Marcel Dekker, Inc., New York, N.Y.

Li, J., X. Li, J. Li and Q. Gao. 2015. Investigating the use of peanut meal: a potential new resource for wood adhesives. RSC Adv. 5: 80136–80141.

Li, N., G. Qi, X.S. Sun and D. Wang. 2017. Canola protein and oil-based wood adhesives. pp. 111–139. *In*: Z. He (ed.). Bio-based Wood Adhesives-Preparation, Characterization, and Testing. CRC Press/Taylor & Francis Group, Boca Raton, Fl.

Li, N., Y. Wang, M. Tilley, S.R. Bean, X. Wu, X.S. Sun and D. Wang. 2011. Adhesive performance of sorghum protein extracted from sorghum DDGS and flour. J. Polym. the Environ. 19: 755–765.

Liu, M., X. Liu, Z. He and H. Wan. 2016a. Urea modified cottonseed protein adhesive for wood composite products. Abstract. The 70th Forest Products Society (FPS) international Convention, June 27–29, 2016, Portland, OR.

Liu, Y., Z. He, M. Shankle and H. Tewolde. 2016b. Compositional features of cotton plant biomass fractions characterized by attenuated total reflection Fourier transform infrared spectroscopy. Ind. Crop. Prod. 79: 283–286.

Liu, Y., J. Gao, H. Guo, Y. Pan, C. Zhou, Q. Cheng and B.K. Via. 2014. Interfacial properties of loblolly pine bonded with epoxy/wood pyrolysis bio-oil blended system. BioResources 10: 638–646.

Luo, J., J. Luo, C. Yuan, W. Zhang, J. Li, Q. Gao and H. Chen. 2015. An eco-friendly wood adhesive from soy protein and lignin: performance properties. RSC Advances 5: 100849–100855.

Mao, A. and S.Q. Shi. 2012. Dynamic mechanical properties of polymeric diphenylmethane diisocyanate/bio-oil adhesive system. Forest Prod. J. 62: 201–206.

Mao, A., Z. He, M. Wan and Q. Li. 2017. Preparation, properties, and bonding utilization of pyrolysis bio-oil. pp. 260–279. *In*: Z. He (ed.). Bio-based Wood Adhesives: Preparation, Characterization, and Testing. CRC Press/Taylor & Francis Group, Boca Raton, FL.

Mekonnen, T.H., P.G. Mussone, P. Choi and D.C. Bressler. 2014. Adhesives from waste protein biomass for oriented strand board composites: development and performance. Macromol. Mater. Engineer. 229: 1003–1012.

Nordqvist, P., F. Khabbaz and E. Malmstrom. 2010. Comparing bond strength and water resistance of alkali-modified soy protein isolate and wheat gluten adhesives. Int. J. Adhes. Adhes. 30: 72–79.

Nordqvist, P., D. Thedjil, S. Khosravi, M. Lawther, E. Malmstrom and F. Khabbaz. 2012. Wheat gluten fractions as wood adhesives-glutenins versus gliadins. J. Appl. Polymer Sci. 123: 1530–1538.

Norstrom, E., L. Fogelstrom, P. Nordqvist, F. Khabbaz and E. Malmstrom. 2014. Gum dispersions as environmentally friendly wood adhesives. Ind. Crop. Prod. 52: 736–744.

Norstrom, E., L. Fogelstrom, P. Nordqvist, F. Khabbaz and E. Malmstrom. 2015. Xylan-A green binder for wood adhesives. Eur. Polymer J.

Ozbay, G. and N. Ayrilmis. 2015. Bonding performance of wood bonded with adhesive mixtures composed of phenol-formaldehyde and bio-oil. Ind. Crop. Prod. 66: 68–72.

Patel, A.K. 2015. Chitosan: Emergence as potent candidate for green adhesive market. Biochem. Engineer. J. 74–81.

Podschun, J., A. Stucker, R.I. Buchholz, M. Heitmann, A. Schreiber, B. Saake and R. Lehnen. 2016. Phenolated lignins as reactive precursors in wood veneer and particleboard adhesion. Ind. Eng. Chem. Res. 55: 5231–5237.

Qi, G., N. Li, X.S. Suna and D. Wang. 2017. Adhesion properties of soy protein subunits and protein adhesive modification. pp. 59–85. *In*: Z. He (ed.). Bio-based Wood Adhesives: Preparation, Characterization, and Testing. CRC Press/Taylor & Francis Group, Boca Raton, Fl.

Roy, J.J., L. Sun and L. Ji. 2014. Microalgal proteins: a new source of raw material for production of plywood adhesive. J. Appl. Phycol. 26: 1415–1422.

Shi, S.Q., C. Xia and L. Cai. 2017. Modification of soy-based adhesives to enhance the bonding performance. pp. 86–110. *In*: Z. He (ed.). Bio-based Wood Adhesives: Preparation, Characterization, and Testing. CRC Press/Taylor & Francis Group, Boca Raton, Fl.

Taghiyari, H.R., J. Norton and M. Tajvidi. 2017. Effects of nano-materials on different properties of wood-composite materials. pp. 310–339. *In*: Z. He (ed.). Bio-based Wood Adhesives: Preparation, Characterization, and Testing. CRC Press/Taylor & Francis Group, Boca Raton, Fl.

Tousi, E.T., R. Hashim, S. Bauk, M.S. Jaafar, A.M. Al-Jarrah, H. Kardani, A. Arra, A.M. Hamdan and K.S.A. Aldroobi. 2014. A study of the properties of animal-based wood glue. Adv. Mat. Res. 935: 133–137.

Vieira, R.K., A.K. Vieiraa and A.N. Netravali. 2017. Application of the rosin from White Pitch (Protium heptaphyllum) for use as wood adhesive. pp. 280–292. *In*: Z. He (ed.). Bio-based Wood Adhesives: Preparation, Characterization, and Testing. CRC Press/Taylor & Francis Group, Boca Raton, Fl.

Wan, H., Z. He, A. Mao and X. Liu. 2017. Synthesis of polymers from liquefied biomass and their utilization in wood bonding. pp. 239–259. *In*: Z. He (ed.). Bio-based Wood Adhesives: Preparation, Characterization, and Testing. CRC Press/Taylor & Francis Group, Boca Raton, Fl.

Wang, C., J. Wu, G.M. Bernard and R.E. Wasylishen. 2014. Preparation and characterization of canola protein isolate -poly(glycidyl methacrylate) conjugates: a bio-based adhesive. Ind. Crop. Prod. 57: 124–131.

Wei, N., B.K. Via, Y. Wang, T. McDonald and M.L. Auad. 2014. Liquefaction and substitution of switchgrass (Panicum virgatum) based bio-oil into epoxy resins. Ind. Crop. Prod. 57: 116–123.

Yang, I., M. Kuo, D.J. Myers and A. Pu. 2006. Comparison of protein-based adhesive resins for wood composites. J. Wood Sci. 52: 503–508.

Yang, I., G.-S. Han, S.H. Ahn, I.-G. Choi, Y.-H. Kim and S.C. Oh. 2014. Adhesive properties of medium-density fiberboards fabricated with rapeseed flour-based adhesive resins. J. Adhes. 90: 279–295.

Zeng, X., J. Luo, J. Hu, J. Li, Q. Gao and L. Li. 2016. Aging resistance properties of poplar plywood bonded by soy protein-based adhesive. BioResour. 11: 4332–4341.

Zhang, Z. and Y. Hua. 2007. Urea-modified soy globulin proteins (7S and 11S): Effect of wettability and secondary structure on adhesion. J. Am. Oil Chem. Soc. 84: 854–857.

Zhao, X. 2013. Research on proteinous adhesive made of protein isolated from crayfish shell waste. Adv. Mat. Res. 815: 299–304.

Zhu, X., D. Wang and X.S. Sun. 2016. Physico-chemical properties of camelina protein altered by sodium bisulfite and guanidine-HCl. Ind. Crop. Prod. 83: 453–461.

Index

Printed and bound by CPI Group (UK) Ltd, Croydon, CR0 4YY

01/11/2024

01782624-0012